数控车铣加工

中级（中英双语版）

赵 慧 杨国星 主 编

王 称 史清卫 杨忠悦 张桂英 副主编

清華大学出版社

北京

内 容 简 介

本书是“数控车铣加工”系列教材中级分册，内容涵盖数控加工工艺、编程基础和机床操作三部分。本书教学内容紧密对接教育部发布的“1+X”数控车铣加工职业技能等级标准，重点介绍了数控车床、数控铣床和加工中心的编程指令、夹具、刀具、工艺流程、设备操作等专业知识。并且根据技能型人才培养需求，科学设计了典型教学案例及配套教学资源，通过渐进式的任务学习及训练，使学生掌握数控编程及加工的基本方法、工艺常识和操作技能。本系列教材及配套资源可用于机械及相关专业的本科、专科、高职和专业培训院校的数控技术等课程的教材。

本书根据高职高专的教学特点，结合高职高专学生的实际学习能力和教学培养目标编写，可作为高职高专机械加工专业或其他相关专业的通用教材，也可作为成人院校的培训教材，还可供从事机械加工的工程技术人员参考。

图书在版编目(CIP)数据

数控车铣加工：中级：汉、英/赵慧，杨国星主编. —北京：清华大学出版社，2024.5
ISBN 978-7-302-66282-2

Ⅰ. ①数… Ⅱ. ①赵… ②杨… Ⅲ. ①数控机床－车床－加工工艺－职业技能－鉴定－教材－汉、英 ②数控机床－铣床－加工工艺－职业技能－鉴定－教材－汉、英 Ⅳ. ①TG519.1 ②TG547

中国国家版本馆 CIP 数据核字(2024)第 098085 号

责任编辑：王 芳 薛 阳
封面设计：刘 键
责任校对：刘惠林
责任印制：刘 菲

出版发行：清华大学出版社
网 址：https://www.tup.com.cn，https://www.wqxuetang.com
地 址：北京清华大学学研大厦 A 座 **邮 编**：100084
社 总 机：010-83470000 **邮 购**：010-62786544
投稿与读者服务：010-62776969，c-service@tup.tsinghua.edu.cn
质量反馈：010-62772015，zhiliang@tup.tsinghua.edu.cn
课件下载：https://www.tup.com.cn，010-83470236
印 装 者：三河市君旺印务有限公司
经 销：全国新华书店
开 本：185mm×260mm **印 张**：17.25 **字 数**：420 千字
版 次：2024 年 5 月第 1 版 **印 次**：2024 年 5 月第1 次印刷
印 数：1～1500
定 价：59.00 元

产品编号：098628-01

前　言

随着自动化、数字化、网络化、智能化技术的快速发展及广泛应用，制造业的人才需求发生了很大变化，其需求对象由某一个领域单一技术的技能人才转变为“通才＋专才”复合型技术的技能人才。为进一步落实中国共产党第十九次全国代表大会提出的深化“产教融合”的重大任务，国务院在发布的《国家职业教育改革实施方案》中明确提出在职业院校、应用型本科高校启动“学历证书＋若干职业技能等级证书”制度（即“1＋X”职业技能等级证书制度）试点工作，明确了开展深度“产教融合”“双元”育人的具体指导政策与要求。其中，“1＋X”职业技能等级证书制度是统筹考虑、全盘谋划职业教育发展、推动企业深度参与协同育人和深化复合型技术技能人才培养培训而做出的重大制度设计。

本书是“1＋X”职业技能等级证书（数控车铣加工）系列教材之一，是根据教育部数控技能型紧缺人才培养培训方案的指导思想，以及数控车铣加工职业技能等级证书的标准要求，结合当前数控技术的发展及教学规律编写而成的。本系列教材以数控车铣加工职业技能等级证书考核样题为基础，选用国内多种通用的CAD/CAM软件，从数控车和数控铣产品加工的典型任务入手，通过理解工程图样及工艺文件，编制零件的数控加工工艺和加工程序，特别是针对数控车铣综合加工工艺进行案例分析，使学习者能够掌握数控机床加工编程、完成定位及联动加工、检测，并控制产品的加工精度、对数控机床的精度进行检验及排除数控机床的一般故障等技能。

目前，已经出版的数控车铣机床高职教材存在教学思路老套、教学内容更新慢、配套资源形式单一，以及教材贴合度差等弊端，且缺少思政资源，不满足当前的立体化、多媒体化、电子化、思政元素深度化等教学要求。针对这些现状，本书以最新的车铣设备作为编写硬件，采用EPIP教学模式，突出实践技能训练，深度递进式写作方法，配套电子资源包括课程标准、PPT、微课、动画、教学录像等，是数控车铣教学的“交钥匙工程”项目，在图书市场上还未见到该类型出版物。本书出版后，在中职、高职学校及数控设备加工人员培训等市场上，会有很大的用户群体。

本书共分6个项目，项目一和项目二由赵慧编写，项目三由杨国星编写，项目四由王称编写，项目五由杨忠悦编写，项目六由史清卫编写，张桂英参与了本书的翻译工作。

由于编者水平有限，书中难免会有不足，恳请使用本书的师生和读者批评指正。

编　者

2023年11月

Preface

"Luban Workshop" is a well-known brand for cultural exchanges between China and foreign countries, which is first practiced in Tianjin under the strong support and guidance of the Ministry of Education of China. It is committed to cultivating technical and skilled talents who are familiar with Chinese technology, understand Chinese processing and recognize Chinese products for the cooperative countries. It is a landmark achievement of the National Modern Vocational Education Reform and Innovation Demonstration Zone and a major innovation in the international development of China's vocational education. Since the first "Luban Workshop" was established in Thailand in 2016, China has successively carried out cooperation in countries along "the Belt and Road" to build a new stage for Sino-foreign vocational education cooperation. In December 2018, Tianjin Light Industry Vocational Technical College, together with Tianjin Transportation Technical College, cooperated with Egypt's Ainshams University and Cairo Advanced Maintenance Technology School to jointly build "Luban Workshop", in which the CNC machining technology is one of the key majors at the secondary vocational education level of the "Luban Workshop" jointly built by China and Egypt. In order to cooperate with the theoretical and practical teaching of the "Luban Workshop" in Egypt, carry out exchanges and cooperation, improve the international influence of China's vocational education, innovate the international cooperation mode of vocational colleges, and export the excellent resources of China's vocational education, the research group has prepared this book.

The Engineering Practice Innovation Project (EPIP) teaching mode is the core content of "Luban Workshop". It integrates theoretical teaching and practical teaching, and forms and develops students' comprehensive professional ability and innovation ability in real work situations. This book is based on the EPIP teaching mode, taking the "Luban Workshop" CNC processing equipment as the carrier, combining with the current development of CNC turning and milling technology, taking the actual engineering project as the guide, and taking the practical application as the guide, while paying attention to the basic theory education, highlighting the practical skill training, and cultivating high-level skilled talents with excellent scientific research ability and problem-solving ability.

Stirling engine outputs power through a cycle of cooling, compression, heat absorption and expansion of the working medium in the cylinder, so it is also called a hot gas engine. It contains 26 typical parts and several standard parts. This book, as the intermediate engineering volume of the series of "CNC Turning and Milling", takes six typical parts of the stirling engine model as the carrier and six typical projects as the main line, focusing on the professional knowledge and operating skills required for the turning programming and processing of bushings, working pistons, heating chambers, and the milling programming and processing of flywheels, cylinder block mounting bases and eccentric cams.

This book adopts a deep and progressive writing method, and tries to make it profound but easily understood, highlighting the characteristics of higher vocational education. Supporting electronic resources include curriculum standards, PPT, micro-class, animation, teaching videos, etc. , which are convenient for teaching and suitable for engineering students in higher vocational colleges. This book is written in both Chinese and English, and is also suitable for the students of "Luban Workshop" in Egypt who major in CNC machining technology.

This book contains 6 projects. Project 1 and Project 2 are written by Zhao Hui. Project 3 is written by Yang Guoxing. Project 4 is written by Wang Chen. Project 5 is written by Yang Zhongyue. Project 6 is written by Shi Qingwei. Zhang Guiying participates in the translation of this book.

Due to the limited knowledge of the editors, some mistakes and errors in the book are unavoidable. Thus welcome all readers to give your kind feedback.

Editor

Nov. 2023

目　　录

Contents

项目引导

目前,数控加工技术在国内外的加工制造业中已经得到广泛应用。数控加工产业的高速发展对数控编程与操作技术人员也提出了更高的要求。以实际工程项目为导引、以实践应用为导向,可以更好地培养学生科学探究和解决问题的能力。斯特林发动机通过气缸内的工作介质经过冷却、压缩、吸热、膨胀为一个周期的循环来输出动力,因此,又称热气机。本书将斯特林发动机模型(见图 0-1)的相关零件作为载体,以 6 个典型项目(见图 0-2)为主线,重点学习数控车削和数控铣削的工艺制定、数控编程与操作的专业知识和操作技能。

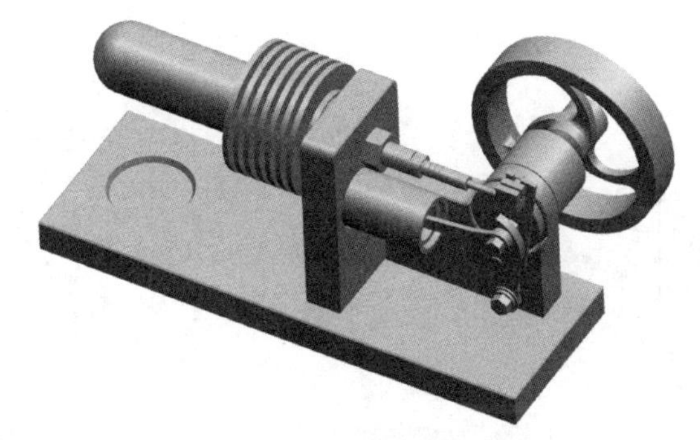

图 0-1 斯特林发动机模型

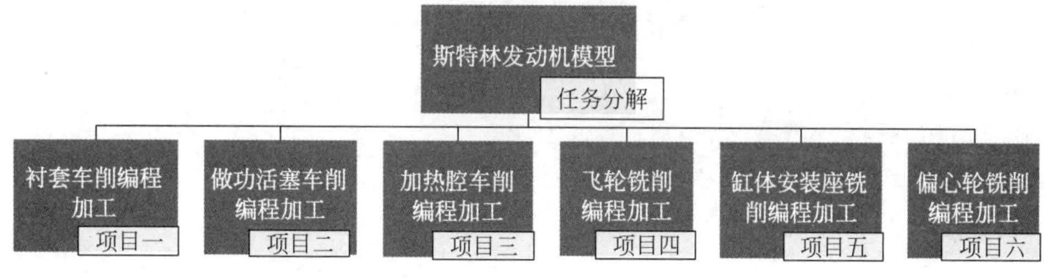

图 0-2 典型项目

斯特林发动机模型包括 26 个典型零件和若干标准件。针对其中 6 个典型零件数控加工所需的专业知识和操作技能,分别进行学习和训练,完成零件的加工,并最终完成机构装配。每个项目都包括完成零件加工所必需的专业知识和技能的学习,如设备、夹具、刀具、基本指令和机床操作等。根据数控加工的基本流程,进行有针对性的学习和训练,以达到熟练掌握数控车削和铣削的加工工艺设计、数控程序编写和机床操作的目的。

项目一　衬套车削编程加工训练

➢思维导图

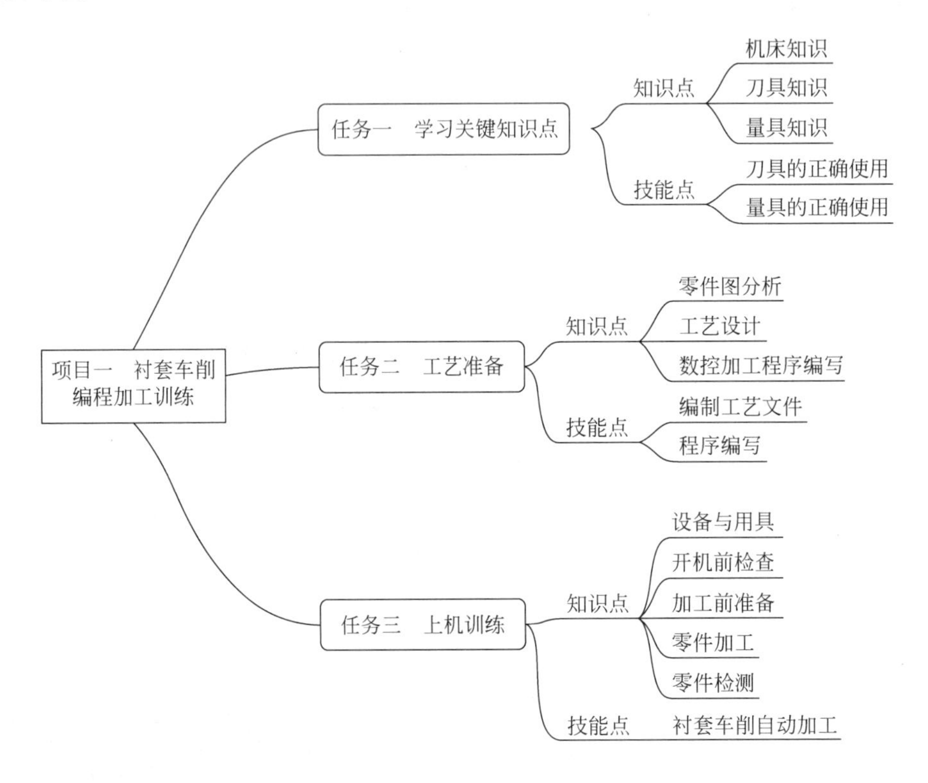

➢学习目标

知识目标

(1) 具备套类零件的识图能力。

(2) 理解衬套的用途和工艺特点。

(3) 理解纵向加工循环指令各参数的含义。

能力目标

(1) 掌握外圆车刀、镗孔刀、切断刀和外螺纹车刀的选择和使用方法。

(2) 能够独立确定加工工艺路线,并正确填写工艺文件。

(3) 能够正确操作数控车床,并根据加工情况调整加工参数。

(4) 能够根据零件结构特点和精度合理选用量具,并正确、规范地测量相关尺寸。

素养目标

(1) 培养学生的科学探究精神和态度。

(2) 培养学生的工程意识。

(3) 培养学生的团队合作能力。

➢任务引入

套类零件是指带有孔的零件，主要用于支撑和导向，在生产及生活中应用广泛。衬套零件为轴承支撑零件，对于各个孔之间的同轴度有着严格要求。

根据零件图(见图 1-1)要求，制定加工工艺、编写数控加工程序，并完成衬套零件的加工。该零件作为典型的套类零件，材料为 6061 铝合金，要求表面光整，无划伤。

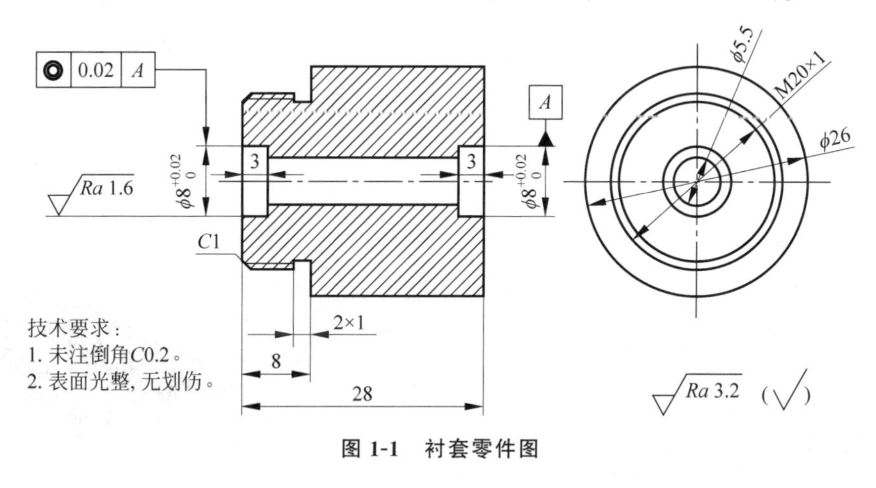

图 1-1　衬套零件图

任务一　学习关键知识点

1.1　机床知识

1.1.1　机床尾座

数控车床尾座按照控制方式分为自动尾座和手动尾座，其套筒中心线与主轴的中心线重合，主要用于安装顶尖和钻头。如图 1-2 所示，手动尾座主要由尾座体、套筒、手轮、尾座锁紧手柄和套筒锁紧手柄组成。

图 1-2　手动尾座

1.1.2　钻夹头

如图 1-3 所示，钻夹头主要用于装夹中心钻、直柄钻头和铰刀等刀具，配合数控车床尾座完成钻中心孔、钻孔和铰孔等工序的加工。

1.1.3　莫氏锥柄变径套

如图 1-4 所示，莫氏锥柄变径套根据尺寸规格分为1～5号，且具有自锁功能。其应用于尾座套筒与钻夹头、锥柄钻头和铰刀等刀具之间的安装，配合数控车床尾座完成钻孔和铰孔等工序的加工。莫氏锥柄变径套直接安装即可使用，但其拆卸过程需要借助楔铁完成。

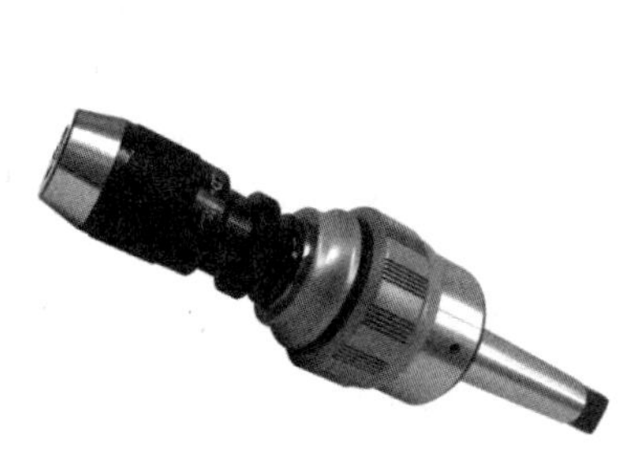

图1-3　钻夹头

图1-4　莫氏锥柄变径套

1.2　刀具知识

1.2.1　中心钻

如图1-5所示，中心钻的长径比较小，所以具有较好的刚度。其主要作用是引导麻花钻进行孔加工，同时也可减少加工中的误差。

图1-5　中心钻

1.2.2　钻头

如图1-6所示，麻花钻在孔加工中最为常见，其标准刀尖角度为118°。麻花钻用于钻孔加工，因其容屑槽成螺旋状形似麻花而得名。其常用材料为高速钢和硬质合金，根据刀柄形状分为直柄钻头和锥柄钻头。

图1-6　麻花钻

如图1-7所示，该刀具为镗孔刀，用于车削内孔。根据刀片安装位置分为左手车刀和右手车刀。在使用过程中，应该根据轮廓选择合适的刀片形状，并注意副切削刃是否会与工件发生碰撞。

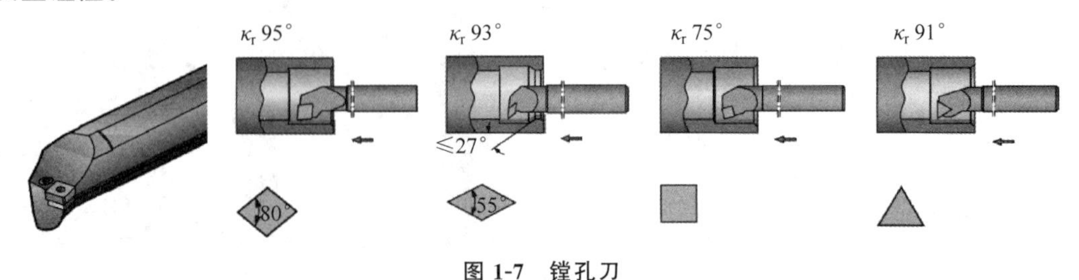

图1-7　镗孔刀

1.3　量具知识

如图 1-8 所示，内径百分表是采用比较测量法检测零件的内孔、深孔直径及其形状精度的量具。内径百分表由内测量杠杆式测量架和百分表组成，是将测头的直线位移变为指针角位移的计量器具。

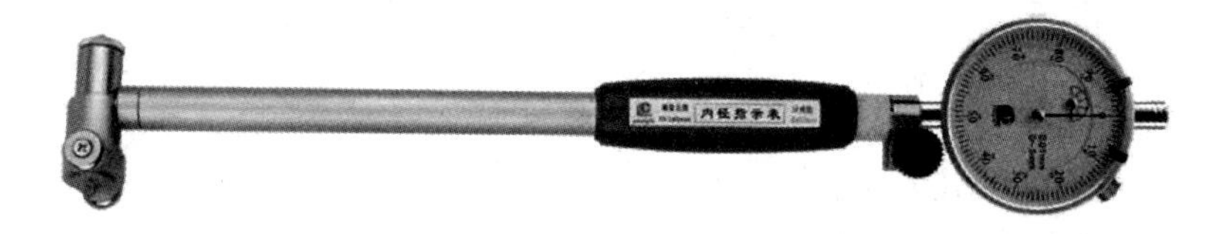

图 1-8　内径百分表

其使用方法如下。

(1) 根据被测尺寸的大小及公差，先选择一把合适的千分尺(普通千分尺分度值为 0.01mm，数显千分尺分度值为 0.001mm)。

(2) 把千分尺调整到被测值名义尺寸并锁紧。

(3) 右手握内径百分表，左手握千分尺。将内径百分表的测头放在千分尺内，使压表量在 0.3～0.5mm，且将表针进行校准并置零，注意要使内径百分表的测杆尽量垂直于千分尺。

(4) 用调整好的内径百分表测量零件孔径。

任务二　工艺准备

1.4　零件图分析

根据零件的使用要求，选择 6061 铝合金作为衬套零件的毛坯材料，毛坯下料尺寸定为 ϕ30×32。在加工时，以 ϕ30 毛坯外圆作为粗基准，先粗加工 ϕ5.5 通孔和右侧 ϕ8 阶梯孔，再粗加工 ϕ26 外圆，后精加工至尺寸要求。然后掉头，装夹 ϕ26 已精加工的外圆表面，使用内径百分表找正至同轴度要求，切削端面保证长度尺寸。最后加工左侧 ϕ8 阶梯孔、退刀槽和 M20 螺纹。

注意：在装夹毛坯时，应注意棒料伸出的长度，以免刀具与卡盘发生碰撞。

1.5　工艺设计

根据零件图分析，确定工艺过程，如表 1-1 所示。

表 1-1 工艺过程卡片

机械加工工艺过程卡片		产品型号	STL-00	零部件序号	STL-01		第 1 页
		产品名称	斯特林发动机模型	零部件名称	衬套		共 1 页
材料牌号	6061	毛坯规格	ϕ30×32	毛坯质量	kg	数量	1
工序号	工序名	工序内容		工段	工艺装备	工时/min	
						准结	单件
5	备料	按 ϕ30×32 尺寸备料		外购	锯床		
10	车加工	以 ϕ30 毛坯外圆作为粗基准,加工端面		车	车床	10	5
15	车加工	保持装夹位置不变,加工 ϕ5.5 通孔		车	车床	10	5
20	车加工	保持装夹位置不变,加工右侧 ϕ8 阶梯孔		车	车床内测千分尺	15	10
25	车加工	保持装夹位置不变,加工 ϕ26 外圆表面		车	车床游标卡尺	10	8
30	车加工	掉头,平端面,保证总长		车	车床游标卡尺	10	8
35	车加工	加工左侧 ϕ8 阶梯孔		车	车床内测千分尺	15	10
40	车加工	加工左侧螺纹大径及轴肩		车	车床千分尺	15	10
45	车加工	加工退刀槽		车	车床游标卡尺	10	6
50	车加工	加工 M20 螺纹		车	车床螺纹环规	10	8
55	清理	清理工件,锐角倒钝		钳			5
60	检验	检验工件尺寸		检			5

本训练任务针对第 10 工序车削加工,进行工序设计,制订工序卡片,如表 1-2 所示。

表 1-2 车削加工工序卡片

机械加工工序卡片	产品型号	STL-00	零部件序号	STL-01	第 1 页
	产品名称	斯特林发动机模型	零部件名称	衬套	共 1 页

项目	内容
工序号	01
工序名	车加工
材料	6061
设备	数控车床
设备型号	CK6150e
夹具	三爪自定心卡盘
量具	游标卡尺
	千分尺
	内测千分尺或内径百分表
准结工时	100min
单件工时	80min

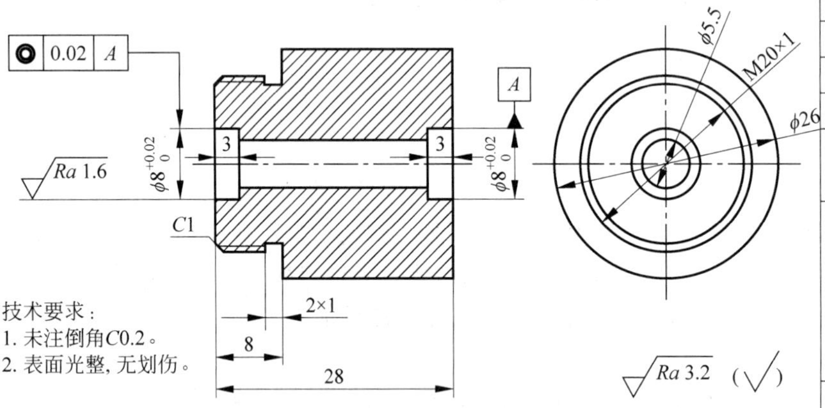

技术要求:
1. 未注倒角C0.2。
2. 表面光整,无划伤。

工步	工步内容	刀具	S/(r/min)	F/(mm/r)	a_p/mm	工步工时/min	
						机动	辅助
1	加工右侧端面	端面车刀	1200	0.18	1	5	5
2	钻中心孔	A3 中心钻	1600			5	5
3	钻孔	ϕ5.5 钻头	1300			5	5
4	粗加工 ϕ8 阶梯孔	镗孔车刀	900	0.18	1	2	5

续表

工步	工步内容	刀具	S/(r/min)	F/(mm/r)	a_p/mm	工步工时/min	
						机动	辅助
5	精加工 φ8 阶梯孔	镗孔车刀	1200	0.1	0.2	3	5
6	粗加工 φ26 外圆表面	外圆车刀	1100	0.2	1.5	2	5
7	精加工 φ26 外圆表面	外圆车刀	1300	0.12	0.2	5	5
8	掉头找正,并加工左侧端面	端面车刀	1200	0.18	1	5	15
9	粗加工左侧 φ8 阶梯孔	镗孔车刀	900	0.18	1	2	5
10	精加工左侧 φ8 阶梯孔	镗孔车刀	1200	0.1	0.2	3	5
11	粗加工左侧螺纹大径及轴肩	外圆车刀	1100	0.2	1.5	2	5
12	精加工左侧螺纹大径及轴肩	外圆车刀	1300	0.12	0.2	5	5
13	加工退刀槽	切断车刀(刀宽2mm)	800	0.06		3	3
14	加工 M20 螺纹	外螺纹车刀	600	1		5	5
15	拆卸、清理工件						5

1.6 数控加工程序编写

根据工序加工工艺,编写加工程序,如表 1-3 所示。

表 1-3　衬套数控加工程序

序　号	程序语句	注　解
	O0001;	**右侧数控加工程序**
N1	T0202;	调用镗孔车刀
	G97 G99 S900 M03;	设定恒转速控制指令,进给量单位为 mm/r,主轴转速为 900r/min,正转
	G0 X5 Z2 M8;	快速定位到循环起点($X5$,$Z2$)处,打开冷却液。注意,该点轴向位置应位于工件右侧,径向位置不大于底孔直径
	G71 U1 R0.5;	调用纵向粗加工循环指令
	G71 P10 Q20U-0.4 W0.05 F0.18;	径向精加工余量为 0.4mm(单边 0.2mm),轴向精加工余量为 0.05mm,进给量为 0.18mm/r,注意,在加工内孔时的径向精加工余量 U,以及在机床系统中的磨损补偿均为负值
N10	G0 G41 X8;	调用刀尖圆弧半径左补偿指令
	G1 Z-3;	
	X5.4;	
N20	G1 G40 X5;	取消刀尖圆弧半径左补偿指令
	G0 X100 Z150;	快速退刀至点($X100$,$Z150$)处
	M5;	停转主轴
	M9;	关闭冷却液
	M01;	选择性暂停,用于观察粗加工的完成情况
N2	T0202;	调用镗孔车刀
	G97 G99 S1200 M03;	设定恒转速控制指令,进给量单位为 mm/r,主轴转速为 1200r/min,正转

续表

序　　号	程序语句	注　　解
	G0 X5 Z2 M8；	快速定位到循环起点（$X5$，$Z2$）处，打开冷却液。注意，该点轴向位置应位于工件右侧，径向位置不大于底孔直径
	G70 P10 Q20 F0.1；	调用精加工循环指令
	G0 X100 Z150；	快速退刀至点（$X100$，$Z150$）处
	M5；	停转主轴
	M9；	关闭冷却液
	M01；	选择性暂停，用于观察精加工的完成情况
N3	T0101；	调用外圆车刀
	G97 G99 S1100 M03；	设定恒转速控制指令，进给量单位为 mm/r，主轴转速为 1100r/min，正转
	G0 X32 Z2 M8；	快速定位到循环起点（$X32$，$Z2$）处，打开冷却液
	G71 U1.5 R0.5；	调用纵向粗加工循环指令
	G71 P30 Q40 U0.4 W0.05 F0.2；	径向精加工余量为 0.4mm（单边 0.2mm），轴向精加工余量为 0.05mm，进给量为 0.2mm/r
N30	G0 G42 X7；	调用刀尖圆弧半径右补偿指令
	G1 Z0；	
	X26 C0.2；	插入 $C0.2$ 倒角
	Z-21；	
N40	G1 G40 X32；	
	G0 X100 Z150；	快速退刀至点（$X100$，$Z150$）处
	M5；	停转主轴
	M9；	关闭冷却液
	M01；	选择性暂停，用于观察粗加工的完成情况
N4	T101；	调用外圆车刀
	G97 G99 S1300 M03；	设定恒转速控制指令，进给量单位为 mm/r，主轴转速为 1300r/min，正转
	G0 X32 Z2 M8；	快速定位到循环起点（$X5$，$Z2$）处，打开冷却液。注意，该点轴向位置应位于工件右侧，径向位置不大于底孔直径
	G70 P30 Q40 F0.12；	调用精加工循环指令
	G0 X100 Z150；	快速退刀至点（$X100$，$Z150$）处
	M5；	停转主轴
	M9；	关闭冷却液
	M30；	程序结束
	O0002；	**左侧数控加工程序**
N1	T0202；	调用镗孔车刀
	G97 G99 S900 M03；	设定恒转速控制指令，进给量单位为 mm/r，主轴转速为 900r/min，正转
	G0 X5 Z2 M8；	快速定位到循环起点（$X5$，$Z2$）处，打开冷却液。注意，该点轴向位置应位于工件右侧，径向位置不大于底孔直径
	G71 U1 R0.5；	调用纵向粗加工循环指令

续表

序 号	程序语句	注 解
	G71 P10 Q20 U-0.4 W0.05 F0.18；	径向精加工余量为 0.4mm(单边 0.2mm)，轴向精加工余量为 0.05mm，进给量为 0.18mm/r，注意，在加工内孔时的径向精加工余量 U，以及在机床系统中的磨损补偿均为负值
N10	G0 G41 X8；	调用刀尖圆弧半径左补偿指令
	G1 Z-3；	
	X5.4；	
N20	G1 G40 X5；	取消刀尖圆弧半径左补偿指令
	G0 X100 Z150；	快速退刀至点(X100，Z150)处
	M5；	停转主轴
	M9；	关闭冷却液
	M01；	选择性暂停，用于观察粗加工的完成情况
N2	T0202；	调用镗孔车刀
	G97 G99 S1200 M03；	设定恒转速控制指令，进给量单位为 mm/r，主轴转速为 1200r/min，正转
	G0 X5 Z2 M8；	快速定位到循环起点(X5，Z2)处，打开冷却液。注意，该点轴向位置应位于工件右侧，径向位置不大于底孔直径
	G70 P10 Q20 F0.1；	调用精加工循环指令
	G0 X100 Z150；	快速退刀至点(X100，Z150)处
	M5；	停转主轴
	M9；	关闭冷却液
	M01；	选择性暂停，用于观察精加工的完成情况
N3	T0101；	调用外圆车刀
	G97 G99 S1100 M03；	设定恒转速控制指令，进给量单位为 mm/r，主轴转速为 1100r/min，正转
	G0 X32 Z2 M8；	快速定位到循环起点(X32，Z2)处，打开冷却液
	G71 U1.5 R0.5；	调用纵向粗加工循环指令
	G71 P30 Q40 U0.4 W0.05 F0.2；	径向精加工余量为 0.4mm(单边 0.2mm)，轴向精加工余量为 0.05mm，进给量为 0.2mm/r
N30	G0 G42 X7；	调用刀尖圆弧半径右补偿指令
	G1 Z0；	
	X19.9 C1；	插入 C1 倒角
	Z-8；	
	X27；	
N40	G1 G40 X30；	取消刀尖圆弧半径右补偿指令
	G0 X100 Z150；	快速退刀至点(X100，Z150)处
	M5；	停转主轴
	M9；	关闭冷却液
	M01；	选择性暂停，用于观察粗加工的完成情况
N4	T0101；	调用外圆车刀
	G97 G99 S1300 M03；	设定恒转速控制指令，进给量单位为 mm/r，主轴转速为 1300r/min，正转
	G0 X32 Z2 M8；	快速定位到循环起点(X32，Z2)处，打开冷却液

续表

序　　号	程序语句	注　　解
	G70 P30 Q40 F0.12；	调用精加工循环指令
	G0 X100 Z150；	快速退刀至点（X100，Z150）处
	M5；	停转主轴
	M9；	关闭冷却液
	M01；	选择性暂停，用于观察精加工的完成情况
N5	T0303；	调用切槽刀（刀宽 2mm）
	G97 G99 S800 M03；	设定恒转速控制指令，进给量单位为 mm/r，主轴转速为 800r/min，正转
	G0 X28 Z2 M8；	
	G0 Z-8；	
	G1 X18 F0.06；	
	G1 X28 F0.3；	
	G0 X100 Z150；	快速退刀至点（X100，Z150）处
	M5；	停转主轴
	M9；	关闭冷却液
	M01；	选择性暂停，用于观察粗加工的完成情况
N6	T0404；	调用外螺纹车刀
	G97 G99 S600 M03；	设定恒转速控制指令，进给量单位为 mm/r，主轴转速为 600r/min，正转
	G0 X22 Z2 M8；	定位至螺纹加工循环起点（X22，Z2）处
	G92 X19.3 Z-6.2 F1；	
	X18.9；	
	X18.7；	
	X18.7；	
	G0 X100 Z150；	快速退刀至点（X100，Z150）处
	M5；	停转主轴
	M9；	关闭冷却液
	M30；	程序结束

任务三　上机训练

1.7　设备与用具

设备：CK6150e 数控车床。

刀具：中心钻、钻头、外圆车刀、镗孔刀、切槽刀和螺纹车刀（刀宽 2mm）。

夹具：三爪自定心卡盘。

工具：卡盘扳手、刀架扳手等。

量具：0～150mm 游标卡尺、内测千分尺或内径百分表、M20×1 螺纹环规。

毛坯：ϕ30×32。

辅助用品：垫刀片、毛刷等。

1.8 开机前检查

可参考表1-4对机床状态进行点检。

表1-4　机床开机准备卡片

检查项目		检查结果	异常描述
机械部分	主轴部分		
	进给部分		
	刀架		
	三爪自定心卡盘		
电器部分	主电源		
	冷却风扇		
数控系统	电气元件		
	控制部分		
	驱动部分		
辅助部分	冷却系统		
	压缩空气		
	润滑系统		

1.9 加工前准备

在加工前，应先将本任务所需刀具准备齐全，并安装正确。根据工艺要求设定工件原点，录入数控加工程序，并进行图形校验。

1.10 零件加工

在图形校验过程验证无问题后，即可进行零件加工。在零件加工前，应详细了解机床的安全操作要求，穿戴好劳动保护服装和用具。在进行零件加工时，应熟悉数控车床各操作按键的功能和位置，了解紧急状况的处置方法。在加工过程中，尤其是在即将切削之前，应对照显示屏“剩余移动量”栏显示的剩余移动量，观察刀具与工件之间的实际距离。若实际距离与剩余移动量相差过大，则应果断停机检查，以免发生撞机事故。若有异常，则应及时停止机床运动。

1.11 零件检测

在零件加工完成后，应当认真清理工件，并按照质量管理的相关要求，对加工完成的零件进行相关检验，保证生产质量。机械加工零件“三级”检验卡片如表1-5所示。

表 1-5　机械加工零件“三级”检验卡片

零部件图号		零部件名称		工　序　号	
材料		送检日期		工序名称	
检验项目	自检结果	互检结果	专业检验	备注	
检验结论	□合格　□不合格　□返修　□让步接收 检验签章： 年　月　日				
不符合项描述					

项 目 总 结

轴套作为数控车床的典型加工零件，广泛应用于各种设备中。根据设备情况和精度的要求，其加工工艺也存在一些差别。编程人员及操作人员需要结合加工条件，合理制定加工工艺，以提高零件的加工精度和生产效率。

课 后 习 题

一、选择题

1. 加工表面多而复杂的零件，工序划分常采用（　　）。

　A. 按所用刀具划分　　B. 按加工部位划分

　C. 按安装次数划分　　D. 按粗精加工划分

2. 游标尺有 50 格刻度线，与主尺 49 格刻度线宽度相同，则此游标卡尺的最小读数是（　　）。

　A. 0.1mm　　B. 2cm　　C. 0.02mm　　D. 0.4mm

3. 内径百分表测量孔径时，应直到在轴向找出（　　）为止，以得出准确的测量结果。

　A. 最小值　　B. 平均值　　C. 最大值　　D. 极限值

4. 机床通电后，应首先检查（　　）是否正常。

　A. 加工路线　　B. 各开关按钮和按键

　C. 电压、油压、加工路线　　D. 工件精度

5. 对于位置精度要求较高的工件，不宜采用（　　）。

　A. 专用夹具　　B. 通用夹具

　C. 组合夹具　　D. 三爪自定心卡盘

6. （　　）指令是间断纵向切削循环指令。

A. G74　　B. G71　　C. G72　　D. G73

二、判断题

1. 刀具在某一坐标轴方向上远离工件的方向为该坐标轴的负方向。（　）
2. 一个程序段中应包含所有的功能字。（　）
3. 在数控车削加工中,进给速度大小直接影响表面粗糙度值和车削效率。（　）
4. 按照机床运动的控制轨迹分类,数控车床属于点位控制。（　）
5. 套类零件通常只起支撑作用。（　）

三、填空题

1. 数控机床上的坐标系采用________坐标系。
2. 数控编程一般分为________和________。
3. 外圆表面主要加工方法是________、________。
4. 数控车床的加工动作主要分为________的运动和________的运动两部分。
5. 数控编程的主要内容有________、确定加工工艺过程、________、编写零件加工程序单、________、________、首件试切。
6. ________是机械加工中经常碰到的一类零件,其应用范围很广。

四、简答题

1. 数控车削编程的特点是什么?
2. 钻头有哪些种类?
3. 莫氏变径套有哪些用途?
4. 简述内径百分表的使用方法。

五、综合编程题

根据零件图(见图 1-9)确定加工工艺,编写程序,并自动加工。

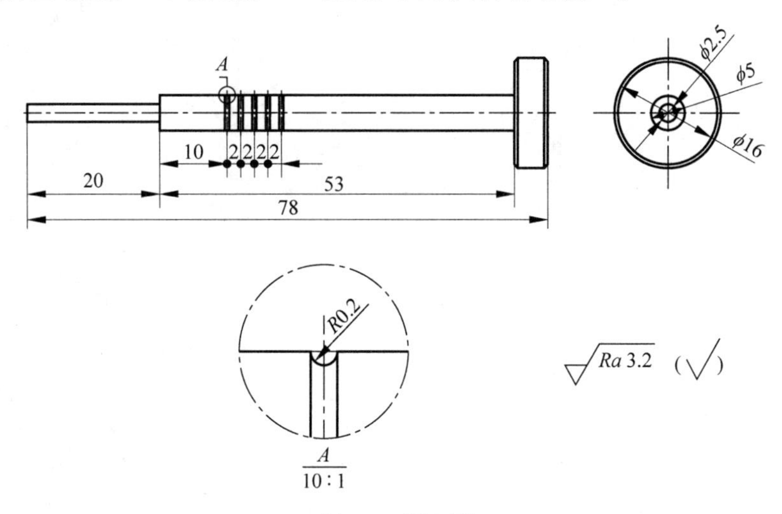

图 1-9　题五图

自我学习检测评分表如表 1-6 所示。

表 1-6　自我学习检测评分表

项　目	目标要求	分值	评分细则	得分	备注
学习关键知识点	(1) 了解机床的基本结构 (2) 掌握中心钻和钻头两种刀具的基础知识 (3) 掌握量具内径百分表的特点及使用方法	20	理解与掌握		
工艺准备	(1) 能够正确识读零件图 (2) 能够根据零件图分析、确定工艺过程 (3) 能够根据工序加工工艺，编写正确的加工程序	30	理解与掌握		
上机训练	(1) 会正确选择相应的设备与用具 (2) 能够正确操作数控车床，并根据加工情况调整加工参数	50	(1) 理解与掌握 (2) 操作流程		

思政小课堂

项目二　做功活塞车削编程加工训练

➢思维导图

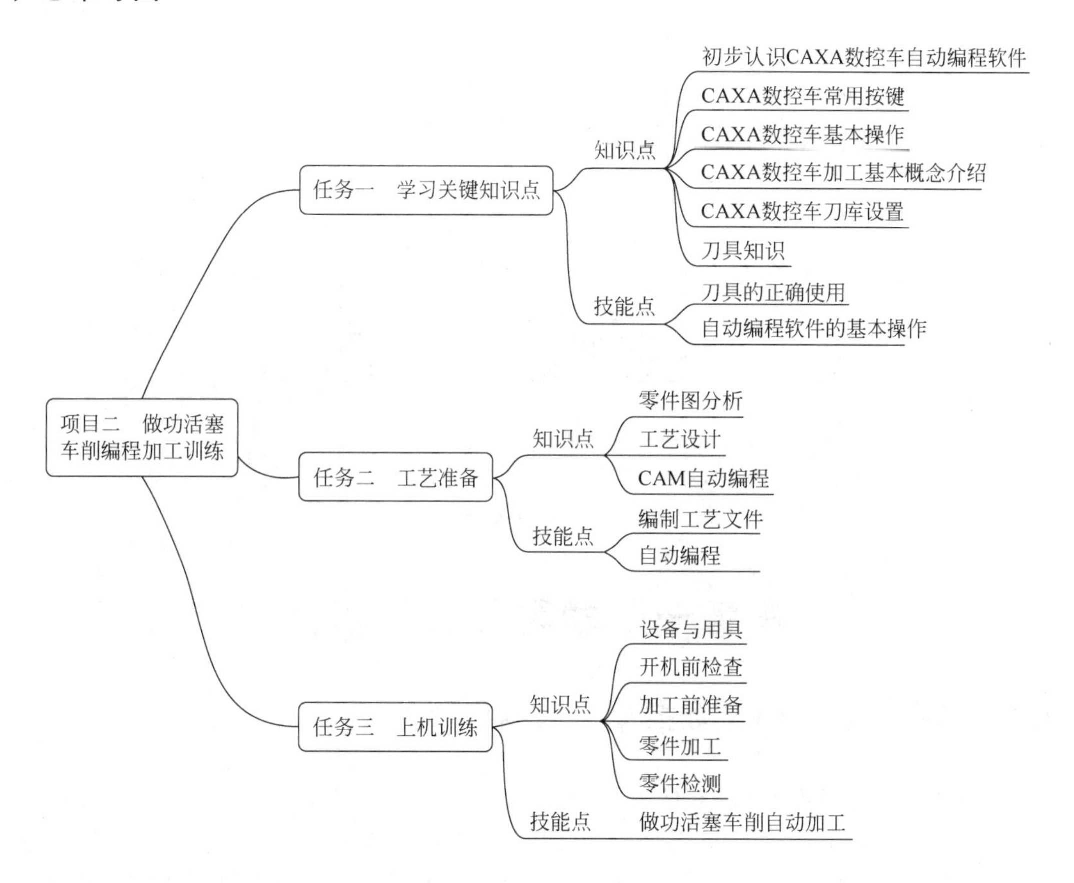

➢学习目标

知识目标

(1) 了解圆弧槽的加工特点。

(2) 了解 CAXA 数控车 CAM 软件的应用特点。

(3) 理解自动编程的基本原理。

能力目标

(1) 能够独立确定加工工艺路线,并正确填写工艺文件。

（2）能够合理选择 CAXA 数控车加工方法，合理设置加工参数，生成数控加工程序。

（3）能够根据零件结构特点和精度合理选用量具，并正确、规范地测量相关尺寸。

素养目标

（1）培养学生的科学探究精神和态度。

（2）培养学生的工程意识。

（3）培养学生的团队合作能力。

➢任务引入

活塞与缸体配合，经连杆带动完成进气、压缩、排气 3 个辅助行程作用，并将直线运动转换为圆周运动，从而实现动力输出。活塞作为发动机的主要零件，在发动机领域中应用广泛。

根据零件图（见图 2-1）要求，制定加工工艺、编写数控加工程序，并完成做功活塞零件的加工。该零件毛坯材料为 65 黄铜，要求表面光整，无划伤。

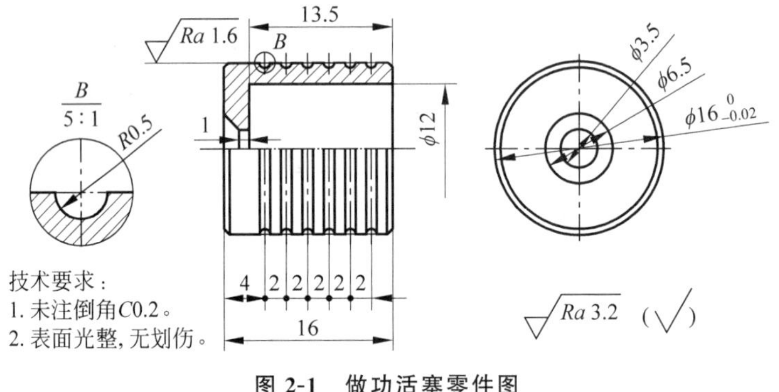

图 2-1 做功活塞零件图

任务一 学习关键知识点

2.1 初步认识 CAXA 数控车自动编程软件

CAXA 数控车具有 CAD 软件的强大绘图功能和完善的外部数据接口，可以绘制任意复杂的图形，可通过 DXF、IGES 等数据接口与其他系统交换数据。CAXA 数控车具备功能强大、使用简单的轨迹生成手段，以及通用的后置处理模块。该软件提供了功能强大、使用简单的轨迹生成手段，可按加工要求生成各种复杂图形的加工轨迹。通用的后置处理模块使 CAXA 数控车可以满足各种机床的代码格式，可输出 G 代码，并可对生成的代码进行校验及加工仿真。CAXA 数控车为二维绘图及数控车加工相结合提供了一个很好的解决方案。将 CAXA 数控车同 CAXA 专业设计软件与 CAXA 专业制造软件结合起来，将会全面地满足任何 CAD/CAM 软件的需求。数控机床是按照事先编制好的加工程序，自动对被加工零件进行加工。编程人员把零件的加工工艺路线、工艺参数、刀具的运动轨迹、位移量、切削参数及辅助功能，按照数控机床规定的指令代码及程序格式编写成加工程序单，再把该加工程序单中的内容记录在控制介质上，然后输入数控机床的数控装置中，从而控制数控机床实现零件的加工。

图 2-2 所示为 CAXA 数控车 2020 软件窗口，分为菜单栏、工具栏、标题栏、管理树栏、状态栏和绘图区。

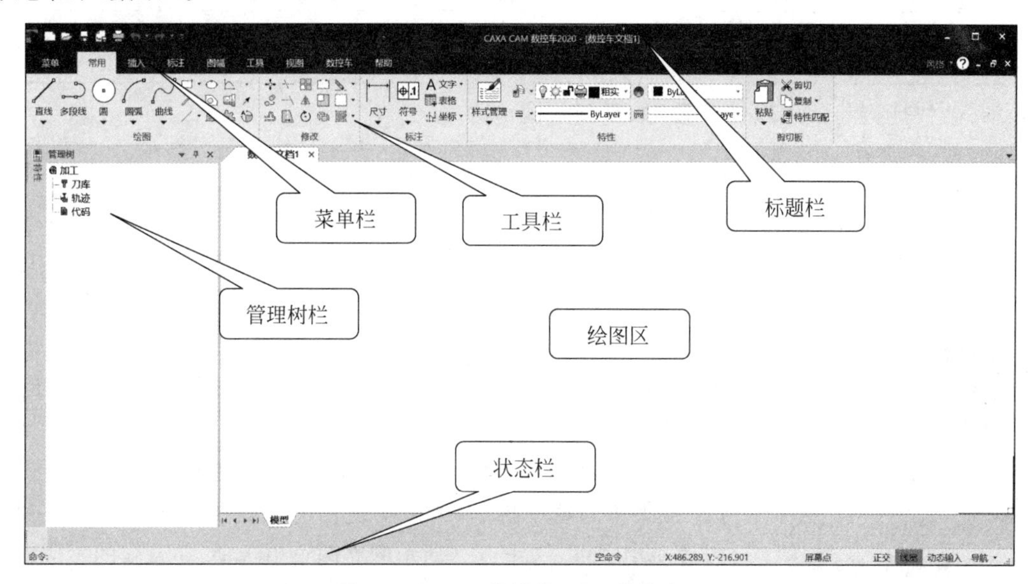

图 2-2　CAXA 数控车 2020 软件窗口

2.2　CAXA 数控车常用按键

2.2.1　基本按键和快捷键操作

鼠标左键可以用来激活菜单，以确定位置点、拾取元素等。例如，在运行画直线功能时，要先单击“直线”按钮，激活画直线功能。这时，在命令提示区出现下一步操作的提示“输入起点”，在绘图区内单击，输入一个位置点，再根据提示输入第二个位置点，就生成了一条直线。

鼠标右键用来确认拾取、结束操作终止命令。例如，在删除集合元素时，在拾取完毕要删除的元素后，右击就可以结束拾取，被拾取到的元素就被删除了；又如，在生成样条曲线的功能中，在顺序输入一系列点完毕后，右击就可以结束输入点的操作，样条曲线就生成了。

在 CAXA 数控车中，当系统要求输入点时，Enter 键和数值键可以激活一个坐标输入条，在输入条中可以输入坐标值。如果坐标值以@开始，则表示一个相对于前一个输入点的相对坐标。在某些情况下，也可以输入字符串。

按空格键可以弹出点工具菜单。当系统要求输入点时，按空格键可以弹出相应的菜单。

CAXA 数控车为用户提供了快捷键操作。对于一个熟练的 CAXA 数控车用户，快捷键将极大地提高工作效率。此外，用户还可以自定义想要的快捷键。在 CAXA 数控车中，设置了如表 2-1 所示的几种常用快捷键。

表 2-1 CAXA 数控车常用快捷键及定义方式

快捷键	定义	快捷键	定义	快捷键	定义
F1	帮助文件	F3	显示全部	F5	坐标系切换
F6	捕捉方式切换	F7	三维视图导航开关	F8	正交模式开关
F9	窗口切换	Delete	删除	Ctrl+P	打印

其余快捷键与常用软件一致,如复制快捷键为 Ctrl+C 等。

2.2.2 立即菜单

立即菜单是 CAXA 数控车提供的独特的交互方式。立即菜单的交互方式大幅度地改善了交互过程。传统的交互方式是完全顺序的逐级问答式交互,用户需按系统设定的交互方式路线逐项输入。而当采用立即菜单的交互方式时,系统提供立即菜单,在交互过程中,如果用户有需要,则可随时修改立即菜单中提供的默认选项,这样一来,就打破了完全顺序的逐级问答式交互过程。

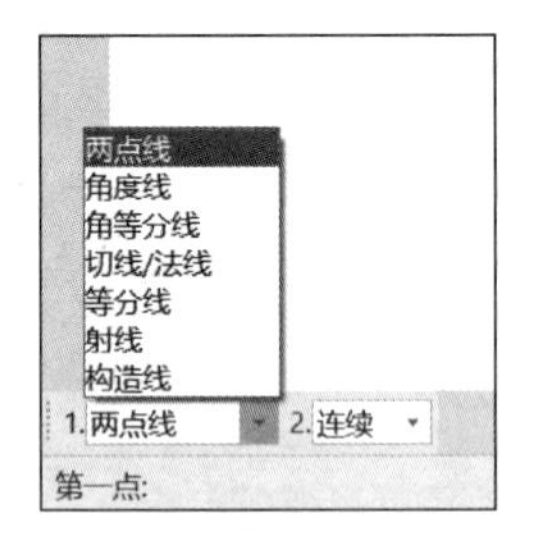

图 2-3 "直线"立即菜单示例

立即菜单的另一个主要功能是,对功能进行选项控制。得益于立即菜单的这种机制,可以实现功能的紧密组织。例如,在"直线"菜单中提供了如图 2-3 所示的立即菜单选项,在直线的生成方式中有两点线、角度线、角等分线、切线/法线、等分线、射线和构造线。

2.2.3 点的输入

在交互过程中,常常会遇到输入精确定位点的情况,这时系统提供了点工具菜单。可以利用点工具菜单来精确定位一个点。点工具菜单表现形式如图 2-4 所示。

按键盘的空格键可以激活点工具菜单。例如,在生成直线时,在系统提示"输入起点:"后,按空格键就会弹出点工具菜单,然后根据所需要的方式选择一种点定位方式。用户也可以使用命令字母,来切换到所需要的点状态。命令字母就是点工具菜单中每种点后面的字母。例如,在生成直线时,需要定位一个圆的圆心,那么在系统提示"输入起点:"后,按 C 键就可以将点状态切换到圆心点状态。下面是部分点状态的具体含义。

图 2-4 点工具菜单

(1) 屏幕点(S):鼠标在屏幕上单击当前平面上的点。

(2) 端点(E):曲线的起、终点,取离拾取点较近者。

(3) 中点(M):曲线的弧长平分点。

(4) 交点(I):曲线与曲线的交叉点,取离拾取点较近者。

(5) 圆心(C):圆或弧的中心。

(6) 象限点(Q):给定点的坐标象限点。

(7) 垂足点(P):用于作垂线。

(8) 切点(T):用于作切线和切圆弧。

(9) 最近点(N):曲线上离拾取点距离最近的点。

各类点均可以输入为象限点,可用直角坐标系、极坐标系和球坐标系之一来表示。在输入象限坐标时,系统提供立即菜单切换和输入数值。在默认点状态下系统根据鼠标光标位置自动判断端点、中点、交点和屏幕点。在进入系统时，系统的点状态为默认点。用户可在“系统参数设定”对话框里，选择是否对工具点状态进行锁定（用户可根据需要和习惯选择相应的选项，具体情况请参看以后的介绍）。在工具点状态锁定时，工具点状态一经指定即不改变，直到重新指定为止。但增量点例外，在使用完后即恢复到非相对点状态；在选择不锁定工具点状态时，工具点使用一次之后即恢复到默认点状态。

2.3 CAXA 数控车基本操作

2.3.1 对象概念

在电子图板中，绘制在绘图区内的各种曲线、文字、块等绘图元素实体，称为图元对象，简称对象。一个能够单独拾取的实体就是一个对象。在电子图板中，如块一类的对象还可以包含若干子对象。在电子图板绘图的过程中，除了编辑环境参数以外，还有生成对象和编辑对象的过程。

2.3.2 拾取对象

在电子图板中，如果想对已经生成的对象进行操作，则必须对对象进行拾取。拾取对象的方法可以分为点选、框选和全选。被选中的对象会被加亮显示，加亮显示的具体效果可以在系统选项中设置。

1. 点选

点选是指将光标移动到对象内的线条或实体上单击，该实体会直接处于被选中状态。

2. 框选

框选是指在绘图区内单击两个对角点形成选择框拾取对象。框选不仅可以选择单个对象，还可以一次选择多个对象。框选可分为正选和反选两种形式。

正选是指在选择过程中，第一角点在左侧、第二角点在右侧（即第一角点的横坐标小于第二角点），选择框色调为蓝色、框线为实线。在正选时，只有当对象上的所有点都在选择框内时，对象才会被选中。

反选是指在选择过程中，第一角点在右侧、第二角点在左侧（即第一角点的横坐标大于第二角点），选择框色调为绿色、框线为虚线。在反选时，若对象上有 1 个点在选择框内，则该对象就会被选中。

3. 全选

全选可以将绘图区内能够选中的对象一次全部拾取。全选也可采用快捷键 Ctrl＋A。应注意的是，拾取过滤设置等也会对全选功能选中的实体造成影响。

此外，在已经选择了对象的状态下，仍然可以利用上述方法直接在已有选择的基础上，添加拾取。

2.3.3 取消选择

在使用常规命令结束操作后，被选择的对象也会自动取消选择状态。如果想手工取消当前的全部选择，则可以按 Esc 键，也可以使用绘图区右键菜单中的“全部不选”命令取消全部选择。如果希望取消当前选择对象中的某一个或某几个的选择状态，则可以按 Shift 键选择需要剔除的对象。

2.3.4 命令操作

在电子图板中，无论进行什么样的操作都必须调用命令。调用命令的方法主要有单击主菜单或图标、键盘命令和快捷键命令 3 种。单击主菜单或图标是指在主菜单、工具栏或绘图区等位置，找到该命令的选项或图标，单击调用。

2.3.5 命令状态

电子图板的命令状态可以分为 3 种，即“空命令状态”“拾取实体状态”和“执行命令状态”。在“空命令状态”下，可以通过直接输入算术式求得计算结果，如下所示。

(1) 输入 8－2 命令，并按 Enter 键，输入区显示 8－2＝6。

(2) 输入 4 * 3 命令，并按 Enter 键，输入区显示 4 * 3＝12。

(3) 输入 2∧3 命令，并按 Enter 键，输入区显示 2∧3＝8。

2.4 CAXA 数控车自动编程软件基本操作

CAXA 数控车提供一体化刀库管理。刀库管理功能包括轮廓车刀、切槽车刀、螺纹车刀、钻头等多种刀具类型的管理，便于用户从刀库中获取刀具信息，并对刀库进行维护。

如图 2-5 所示，CAXA 数控车 2020 中直接通过“创建刀具”对话框，或者右击“菜单”，在弹出的下拉菜单中单击“创建刀具”命令就可以很方便地创建一把刀具。而且，将这把刀具所用到的切削用量和几何参数进行关联，在加工过程中调用这把刀具时，就可以同时调用这把刀具的切削要素，这样可以简化刀具在切削用量上重新设定参数的时间。

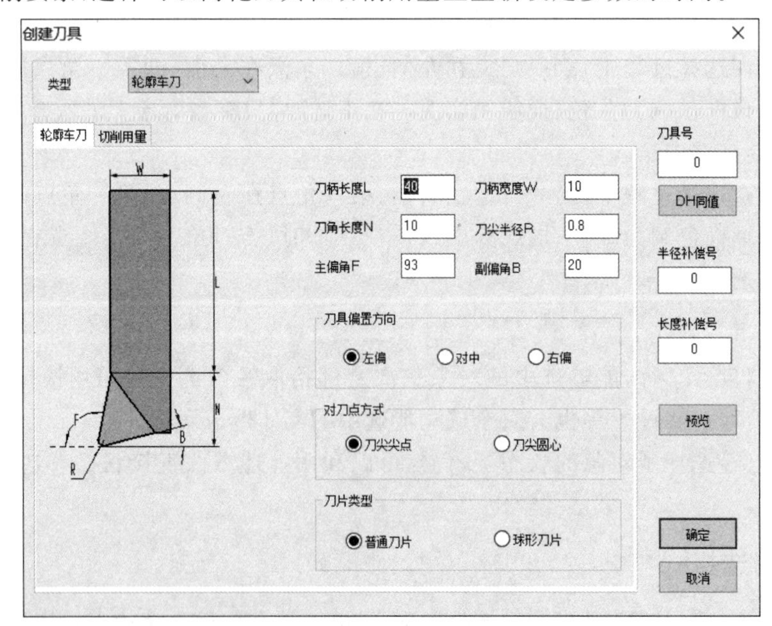

图 2-5 CAXA 数控车 2020“创建刀具”对话框

2.5　CAXA 数控车基本操作介绍

CAXA 数控车 2020 加工窗口布局如图 2-6 所示。

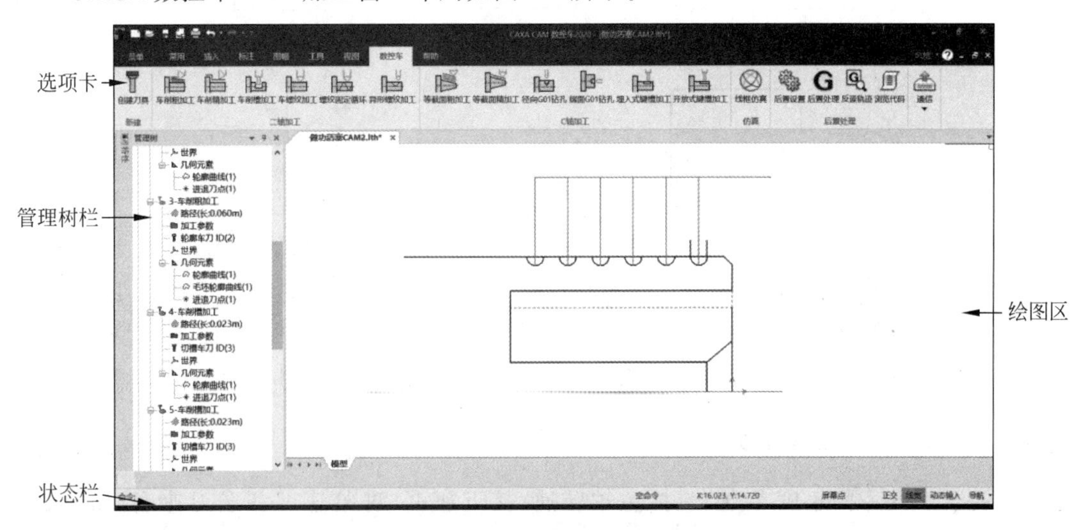

图 2-6　CAXA 数控车 2020 加工窗口布局

（1）选项卡：所有的功能命令可以在选项卡区域进行查找。

（2）管理树栏：所有的刀具、数控车加工轨迹、G 代码信息都会被记录并显示在管理树栏上。

（3）状态栏：选项卡功能的运行选项及操作命令提示。

（4）绘图区：支持多浏览，可在不同图纸间随意切换编辑。

2.6　CAXA 数控车加工基本概念介绍

用 CAXA 数控车实现加工的过程，首先，对图纸进行分析，确定需要数控加工的部分；然后，利用图形软件对需要加工的部分造型；接着，根据加工条件，选择合适的参数生成加工轨迹（包括粗加工轨迹、半精加工轨迹、精加工轨迹）；轨迹仿真检验；最后，配置好机床，生成 G 代码传输给机床加工。

2.6.1　两轴加工

在 CAXA 数控车中，机床坐标系的 *Z* 轴即绝对坐标系的 *X* 轴，平面图形均投影到绝对坐标系的 *XOY* 面。

2.6.2　轮廓

轮廓是一系列首尾相接的曲线集合，如图 2-7 所示。

2.6.3　毛坯轮廓

针对粗加工，需要指定被加工体的毛坯。毛坯轮廓是一系列首尾相接的曲线集合，如图 2-8 所示。

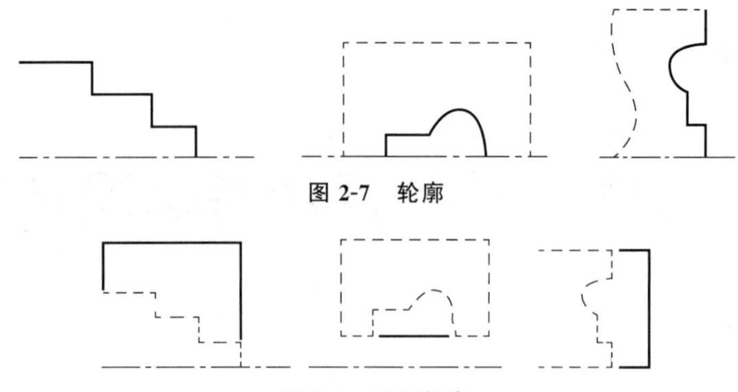

图 2-7　轮廓

图 2-8　毛坯轮廓

在进行数控编程，交互指定待加工的图形时，常常需要用户指定毛坯的轮廓，用来界定被加工的表面或被加工的毛坯本身。如果毛坯轮廓是用来界定被加工表面的，则要求指定的轮廓是闭合的；如果加工的是毛坯轮廓本身，则毛坯轮廓可以不闭合。

2.6.4　数控车床速度参数

数控车床的一些速度参数，包括主轴转速、接近速度、进给速度和退刀速度，如图 2-9 所示。

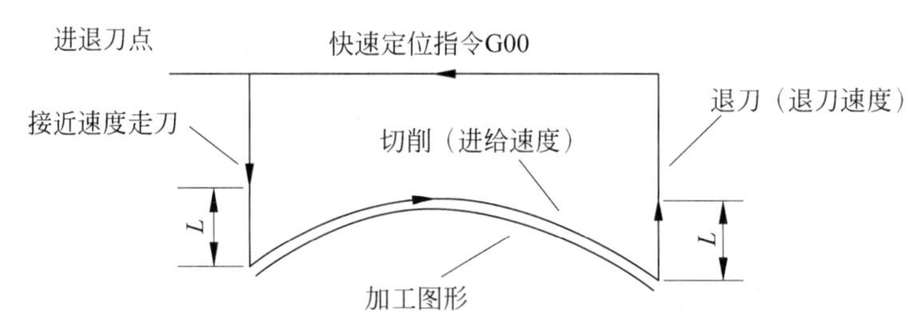

图 2-9　数控车床速度参数

（1）主轴转速为切削时机床主轴转动的角速度。

（2）进给速度为正常切削时刀具行进的线速度。

（3）接近速度为从进刀点到切入工件前刀具行进的线速度，又称进刀速度。

（4）退刀速度为刀具离开工件回到退刀位置时刀具行进的线速度。

这些速度参数的给定一般依赖于用户的经验，从原则上讲，它们与机床本身、工件的材料、刀具的材料、工件的加工精度和表面粗糙度要求等内容相关。

速度参数与加工的效率密切相关。

2.6.5　刀具轨迹和刀位点

刀具轨迹是系统按给定工艺要求生成的，对给定加工图形进行切削时，刀具行进的路线，如图 2-10 所示。系统以图形方式显示刀具轨迹。刀具轨迹由一系列有序的刀位点和连接这些刀位点的直线（直线插补）或圆弧（圆弧插补）组成。

本系统的刀具轨迹是按刀尖位置来显示的。

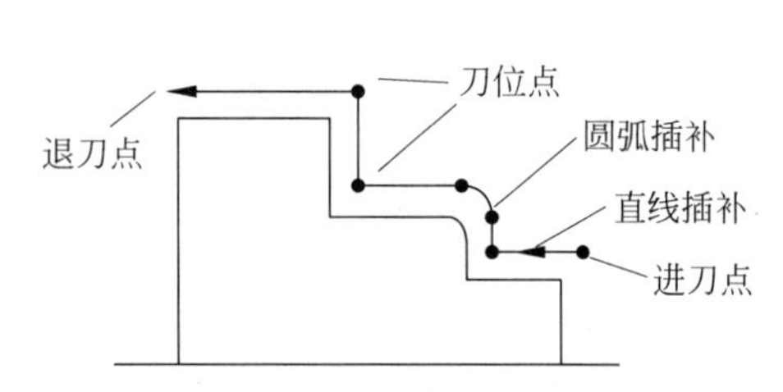

图 2-10　刀具轨迹和刀位点

2.6.6　加工余量

车加工是一个去余量的过程，即从毛坯开始逐步除去多余的材料，以得到需要的零件。这个过程往往由粗加工和精加工构成，在必要时还需要进行半精加工，即需要经过多道工序的加工。在前一道工序中，往往需要给下一道工序留下一定的余量。

实际加工模型是指定加工模型按给定的加工余量进行等距加工的结果，如图 2-11 所示。

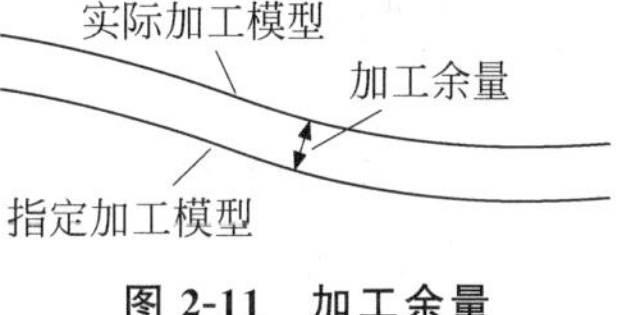

图 2-11　加工余量

2.6.7　加工误差

加工误差是刀具轨迹和实际加工模型的偏差。用户可通过控制加工误差来控制加工的精度。

用户给出的加工误差是刀具轨迹同加工模型之间的最大允许偏差，系统保证刀具轨迹与实际加工模型之间的偏离不大于加工误差。

用户应根据实际工艺要求给定加工误差，如进行粗加工，则加工误差可以较大，否则加工效率会受到不必要的影响；如进行精加工，则需根据表面要求等要求给定加工误差。

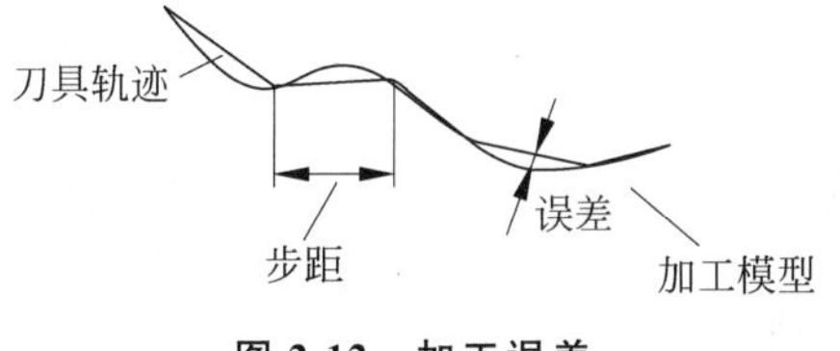

图 2-12　加工误差

在两轴加工中，对于直线和圆弧的加工不存在加工误差。加工误差是指对样条线进行加工时，用折线段逼近样条时的误差，如图 2-12 所示。

2.6.8　加工干涉

切削被加工表面时，如刀具切到了不应该切的部分，则称为出现干涉现象，或者称为过切。

在 CAXA 数控车中，干涉分为以下两种情况。

(1) 在被加工表面中存在刀具切削不到的部分时，存在的过切现象。

(2) 在切削时，刀具与未加工表面存在的过切现象。

2.7　CAXA 数控车刀库设置

在使用 CAXA 数控车软件进行加工前，需要对刀具、数控系统和机床进行设置，它们将直接影响到加工轨迹和生成的 G 代码。本章将对这些设置内容进行详细介绍。

刀库管理功能可定义、确定刀具的相关数据，以便于用户从刀库中获取刀具信息，并对刀库进行维护。刀库管理功能包括轮廓车刀、切槽车刀、螺纹车刀、钻头 4 种刀具类型的管理。

操作方法如下。

(1) 在菜单栏中单击“数控车”标签，再单击“创建刀具”按钮，弹出“创建刀具”对话框，用户可按自己的需要添加新的刀具。新创建的刀具列表会显示在绘图区左侧的“管理树”→“刀库”节点下。

(2) 双击“刀库”节点下的“刀具”节点，可以弹出“编辑刀具”对话框，来改变刀具参数。

（3）在“刀库”节点右击后，在弹出的下拉菜单中单击“导出刀具”命令，可以将所有刀具的信息保存到一个文件中。

（4）在“刀库”节点右击后，在弹出的下拉菜单中单击“导入刀具”命令，可以将保存到文件中的刀具信息全部读入文档中，并添加到“刀库”节点下。

（5）需要指出的是，刀库中的各种刀具只是对同一类刀具的抽象描述，并非是符合国标或其他标准的详细刀库。因此，只列出了对轨迹生成有影响的部分参数，其他与具体加工工艺相关的刀具参数并未列出。例如，将各种外轮廓、内轮廓、端面粗。精车刀均归为轮廓车刀，对轨迹生成没有影响。

2.7.1 轮廓车刀

“轮廓车刀”选项卡如图 2-13 所示，需要配置的参数具体如下。

（1）“刀具号”：刀具的系列号，用于后置处理的自动换刀指令。刀具号唯一，且对应机床的刀库。

（2）“半径补偿号”：刀具半径补偿值的序列号，其值对应于机床的数据库。

（3）“刀柄长度”：刀具可夹持段的长度。

（4）“刀柄宽度”：刀具可夹持段的宽度。

（5）“刀角长度”：刀具可切削段的长度。

（6）“刀尖半径”：刀尖部分用于切削的圆弧半径。

（7）“主偏角”：刀具前刃与工件旋转轴之间的夹角。

（8）“副偏角”：刀具后刃与工件旋转轴之间的夹角。

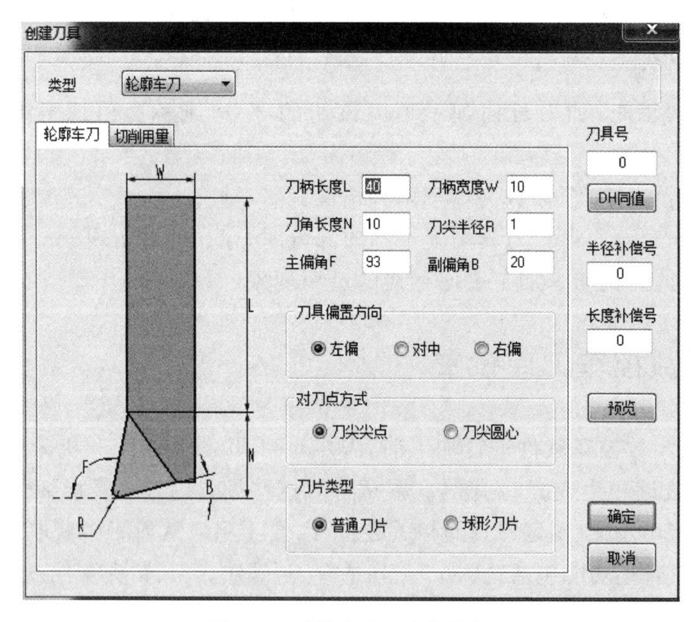

图 2-13 “轮廓车刀”选项卡

2.7.2 切槽车刀

“切槽车刀”选项卡如图 2-14 所示，需要配置的参数具体如下。

(1)“刀具号”：刀具的系列号,用于后置处理的自动换刀指令。刀具号唯一,且对应机床的刀库。

(2)“半径补偿号”：刀具半径补偿值的序列号,其值对应于机床的数据库。

(3)“刀具长度”：刀具整体长度。

(4)“刀具宽度”：刀具整体宽度。

(5)“刀刃宽度”：刀具切削刃的宽度。

(6)“刃尖半径”：刀具的刀尖部位具有的曲率半径大小。

(7)“刀具引角”：确定刀具切削部位的几何形状。

(8)“刀柄宽度”：刀具可夹持段的宽度。

(9)“刀具位置”：刀具与刀柄之间的对应位置。

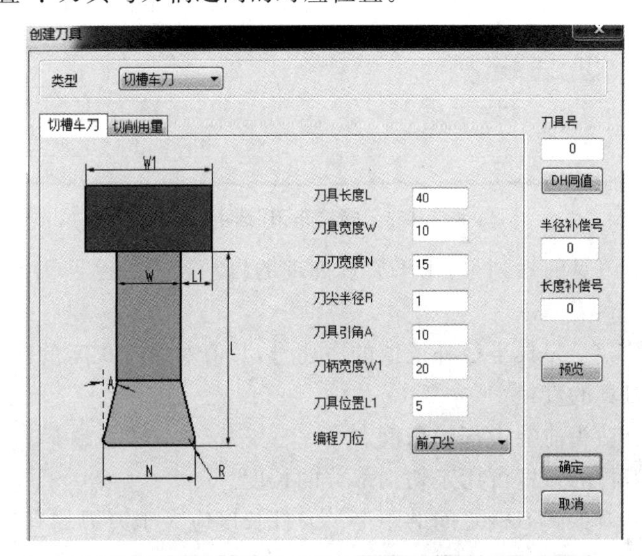

图 2-14　“切槽车刀”选项卡

2.7.3　螺纹车刀

“螺纹车刀”选项卡如图 2-15 所示,需要配置的参数具体如下。

(1)“刀具号”：刀具的系列号,用于后置处理的自动换刀指令。刀具号唯一,且对应机床的刀库。

(2)“半径补偿号”：刀具半径补偿值的序列号,其值对应于机床的数据库。

(3)“刀柄长度”：刀具可夹持段的长度。

(4)“刀柄宽度”：刀具可夹持段的宽度。

(5)“刀刃长度”：刀具主切削刃顶部的长度。

(6)“刀尖宽度”：螺纹牙底宽度。

(7)“刀具角度”：刀具切削段两侧边与垂直于切削方向的夹角,该角度决定了车削出的螺纹的牙型角。

2.7.4　钻头

“钻头”选项卡如图 2-16 所示,需要配置的参数具体如下。

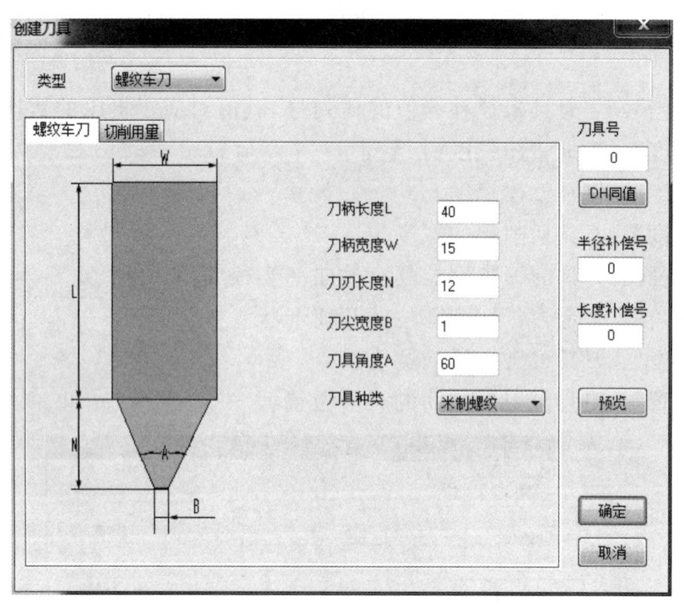

图 2-15 “螺纹车刀”选项卡

(1)“刀具号”：刀具的系列号，用于后置处理的自动换刀指令。刀具号唯一，且对应机床的刀库。

(2)“半径补偿号”：刀具半径补偿值的序列号，其值对应于机床的数据库。

(3)“直径”：刀具的直径。

(4)“刀尖角”：钻头前端尖部的角度。

(5)“刃长”：刀具的刀杆可用于切削部分的长度。

(6)“刀杆长”：刀尖到刀柄之间的距离。刀杆长度应大于刀刃的有效长度。

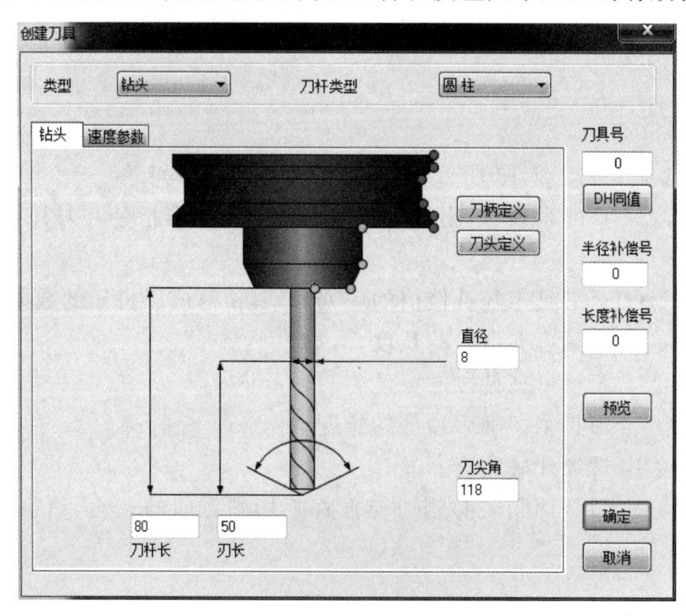

图 2-16 “钻头”选项卡

2.8　刀具知识

如图 2-17 所示，该刀具为圆弧切槽刀，属于成型刀具，用于车削加工圆弧槽。

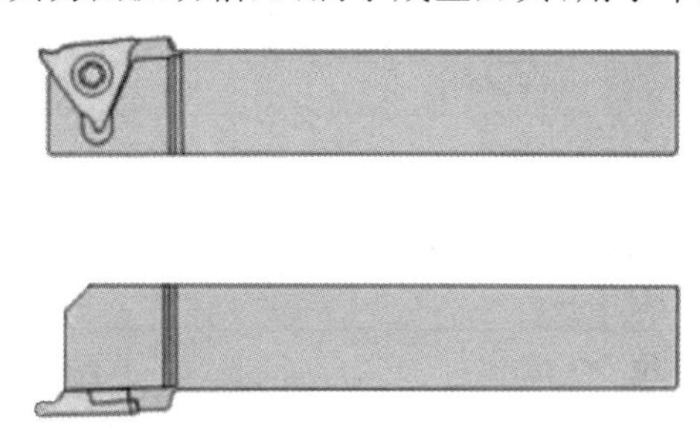

图 2-17　圆弧切槽刀

任务二　工艺准备

2.9　零件图分析

根据零件的使用要求，选择 65 黄铜作为做功活塞零件的毛坯材料，毛坯下料尺寸定为 $\phi20\times40$。在加工时，以 $\phi20$ 毛坯外圆作为粗基准，粗、精加工右侧外圆及内控部分至要求尺寸，然后切断，长度方向留加工余量，最后掉头装夹 $\phi16$ 外圆处（在装夹时，注意做好保护，以防表面夹伤），加工零件左端面至要求尺寸。

注意： *在车削 $\phi16$ 外圆时，车削长度要足够。另外，在装夹毛坯时，应注意棒料伸出的长度，以免刀具与卡盘发生碰撞。*

2.10　工艺设计

根据零件图分析，确定工艺过程，如表 2-2 所示。

表 2-2　工艺过程卡片

<table>
<tr><td colspan="2" rowspan="2">机械加工工艺过程卡片</td><td>产品型号</td><td>STL-00</td><td colspan="2">零部件序号</td><td colspan="2">STL-02</td><td>第 1 页</td></tr>
<tr><td>产品名称</td><td>斯特林发动机模型</td><td colspan="2">零部件名称</td><td colspan="2">做功活塞</td><td>共 1 页</td></tr>
<tr><td>材料牌号</td><td>H65</td><td>毛坯规格</td><td>φ20×50</td><td>毛坯质量</td><td>kg</td><td>数量</td><td colspan="2">1</td></tr>
<tr><td rowspan="2">工序号</td><td rowspan="2">工序名</td><td colspan="2" rowspan="2">工序内容</td><td rowspan="2">工段</td><td rowspan="2" colspan="2">工艺装备</td><td colspan="2">工时/min</td></tr>
<tr><td>准结</td><td>单件</td></tr>
<tr><td>5</td><td>备料</td><td colspan="2">按 φ20×50 尺寸备料</td><td>外购</td><td colspan="2">锯床</td><td></td><td></td></tr>
<tr><td>10</td><td>车加工</td><td colspan="2">以 φ20 毛坯外圆作为粗基准，加工 φ6.5 通孔</td><td>车</td><td colspan="2">车床游标卡尺</td><td>15</td><td>10</td></tr>
<tr><td>15</td><td>车加工</td><td colspan="2">保持装夹位置不变，加工右侧内腔底孔</td><td>车</td><td colspan="2">车床游标卡尺</td><td>10</td><td>5</td></tr>
<tr><td>20</td><td>车加工</td><td colspan="2">保持装夹位置不变，粗、精加工右侧 φ16 外圆部分</td><td>车</td><td colspan="2">车床千分尺</td><td>15</td><td>10</td></tr>
</table>

续表

工序号	工序名	工序内容	工段	工艺装备	工时/min	
					准结	单件
25	车加工	保持装夹位置不变，粗、精加工右侧 ϕ12 内孔部分	车	车床内径千分尺或内径百分表	20	15
30	车加工	加工外圆表面上的圆弧槽	车	车床	15	10
35	车加工	切断并保证总长	车	车床游标卡尺	10	5
40	车加工	掉头加工左侧锥孔	车	车床游标卡尺	15	10
45	清理	清理工件，锐角倒钝	钳			
50	检验	检验工件尺寸	检			

本训练任务针对第 10、第 15 和第 20 工序车削加工，进行工序设计，制订工序卡片，如表 2-3 所示。

表 2-3 车削加工工序卡片

机械加工工序卡片	产品型号	STL-00	零部件序号	STL-02	第 1 页
	产品名称	斯特林发动机模型	零部件名称	做功活塞	共 1 页

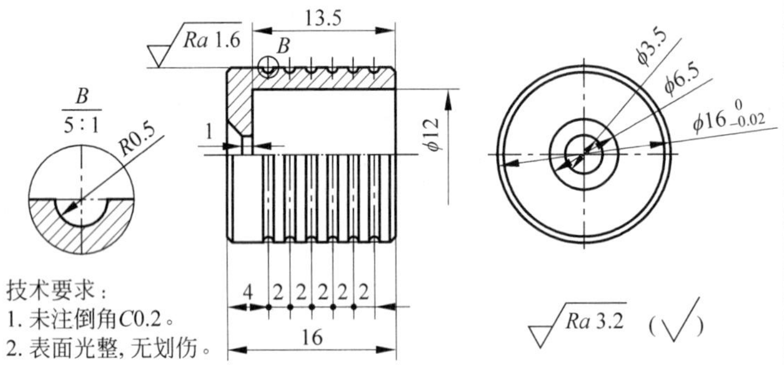

技术要求：
1. 未注倒角C0.2。
2. 表面光整，无划伤。

项目	内容
工序号	15
工序名	车加工
材料	6061
设备	数控车床
设备型号	CK6150e
夹具	三爪自定心卡盘
量具	游标卡尺 内测千分尺
准结工时	100min
单件工时	65min

工步	工步内容	刀具	S/(r/min)	F/(mm/r)	a_p/mm	工步工时/min	
						机动	辅助
1	工件安装						5
2	以 ϕ20 毛坯外圆作为粗基准，钻中心孔	A3 中心钻	2000			5	5
3	加工 ϕ6.5 通孔	ϕ6.5 钻头	1200			5	5
4	钻右侧内腔底孔	外圆粗车刀镗孔车刀	1000			5	5
5	粗加工右侧 ϕ16 外圆部分	外圆车刀	1000	0.2	1.5	5	5
6	精加工右侧 ϕ16 外圆部分	外圆车刀	1300	0.12	0.3	5	5
7	粗加工右侧 ϕ12 内孔部分	镗孔车刀	1200	0.15	1	10	5
8	精加工右侧 ϕ12 内孔部分	镗孔车刀	1400	0.1	0.2	5	5
9	加工外圆表面上的圆弧槽	圆弧切槽车刀	700	0.07		10	5
10	加工左侧倒角，并切断保证总长	切断车刀	1000	0.06		5	5
11	掉头粗加工左侧锥孔	倒角钻头	1100			10	5
12	清理工件，锐角倒钝						
13	检验工件尺寸						

2.11 CAM 自动编程

利用 CAXA 数控车建模，选择合适的加工方法，并生成数控加工程序。

2.11.1 外形轮廓编程

CAXA 数控车具备功能强大、使用简单的轨迹生成手段，以及通用的后置处理模块。

1. CAD 建模

如图 2-18 所示，首先绘制零件外形轮廓及毛坯轮廓，即定义加工区域。模型原点应该与编程原点保持一致，CAD 软件当前版本不支持自定义加工原点。

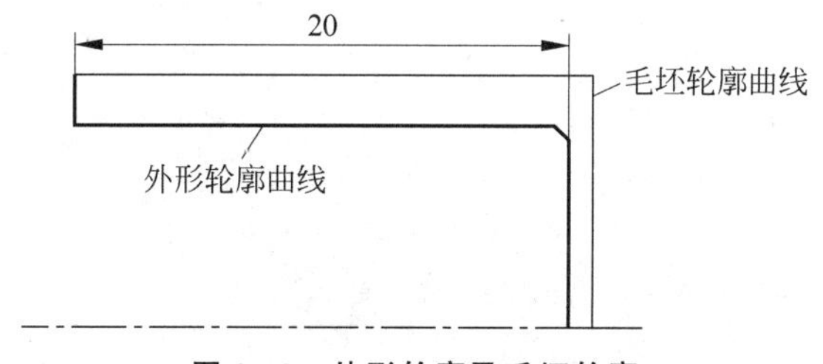

图 2-18　外形轮廓及毛坯轮廓

2. CAM 自动编程

如图 2-19 所示，在"二轴加工"选项组单击"车削粗加工"按钮。

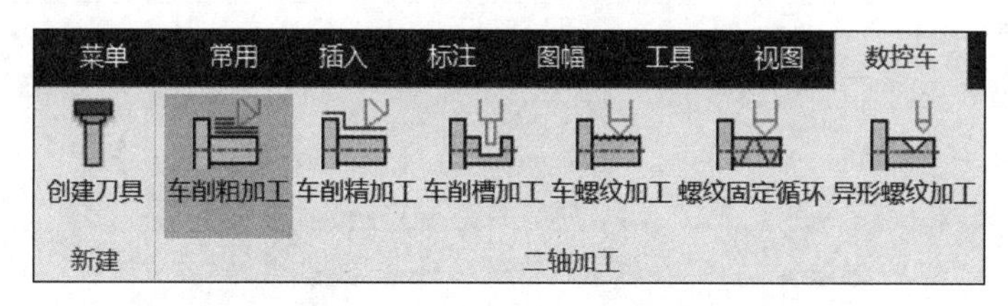

图 2-19　"二轴加工"选项组

如图 2-20 所示，设置加工参数。

如图 2-21 所示，设置进退刀方式参数。

如图 2-22 所示，设置刀具参数，单击"入库"按钮，以便后续加工直接调用。

如图 2-23 所示，"几何"选项卡用于选择"轮廓曲线"和"毛坯轮廓曲线"。在选择曲线时，应注意所选曲线并非实际工件轮廓及毛坯轮廓，选择曲线的不同将直接影响刀具轨迹的生成效果。

分别单击"轮廓曲线"按钮和"毛坯轮廓曲线"按钮，以"单个拾取"的方式选择"轮廓曲线"和"毛坯轮廓曲线"，在选择完成后右击结束选取。"进退刀点"的选择要注意刀具是否会与工件发生干涉，在外形轮廓加工时的进退刀点通常在毛坯的外侧。

生成的外形轮廓刀具轨迹如图 2-24 所示。

如图 2-25 所示，设置精加工参数，"几何"选项卡无须选择"毛坯轮廓曲线"，只需选择"轮廓曲线"即可，然后生成刀具轨迹，如图 2-26 所示。

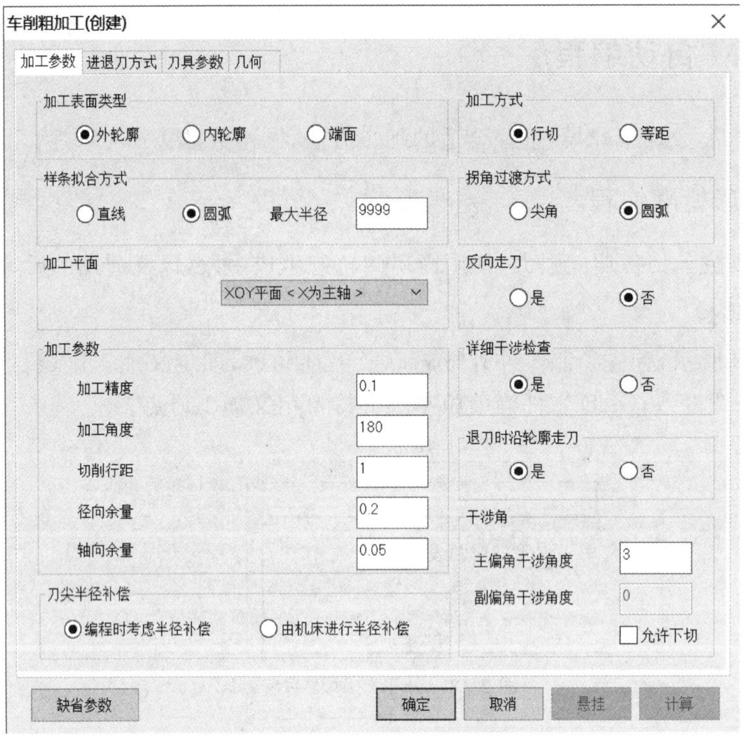

车削粗加工(创建)

加工参数 进退刀方式 刀具参数 几何

加工表面类型：外轮廓（选中） 内轮廓 端面

加工方式：行切（选中） 等距

样条拟合方式：直线 圆弧（选中） 最大半径 9999

拐角过渡方式：尖角 圆弧（选中）

加工平面：XOY平面 <X为主轴>

反向走刀：是 否（选中）

加工参数

参数	值
加工精度	0.1
加工角度	180
切削行距	1
径向余量	0.2
轴向余量	0.05

详细干涉检查：是（选中） 否

退刀时沿轮廓走刀：是（选中） 否

干涉角：主偏角干涉角度 3；副偏角干涉角度 0；允许下切

刀尖半径补偿：编程时考虑半径补偿（选中） 由机床进行半径补偿

缺省参数 确定 取消 悬挂 计算

图 2-20 “加工参数”选项卡

车削粗加工(创建)

加工参数 进退刀方式 刀具参数 几何

快速退刀距离 1

每行相对毛坯进刀方式：与加工表面成定角（长度 1，角度 45）；垂直（选中）；矢量（长度 1，角度 45）

每行相对加工表面进刀方式：与加工表面成定角（长度 1，角度 45）；垂直（选中）；矢量（长度 1，角度 45）

每行相对毛坯退刀方式：与加工表面成定角（选中）（长度 1，角度 45）；垂直；矢量（长度 1，角度 45）

每行相对加工表面退刀方式：与加工表面成定角（长度 1，角度 45）；垂直（选中）；矢量（长度 1，角度 45）

缺省参数 确定 取消 悬挂 计算

图 2-21 “进退刀方式”选项卡

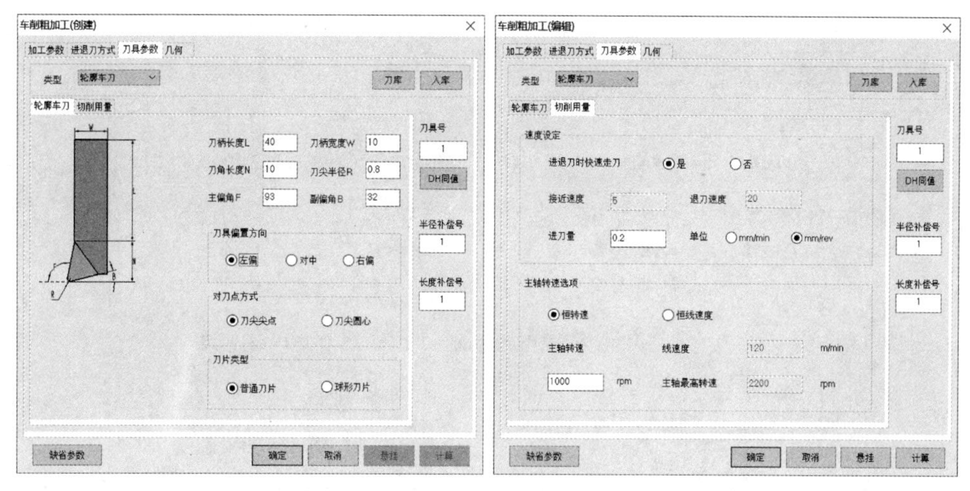

图 2-22　“刀具参数”选项卡

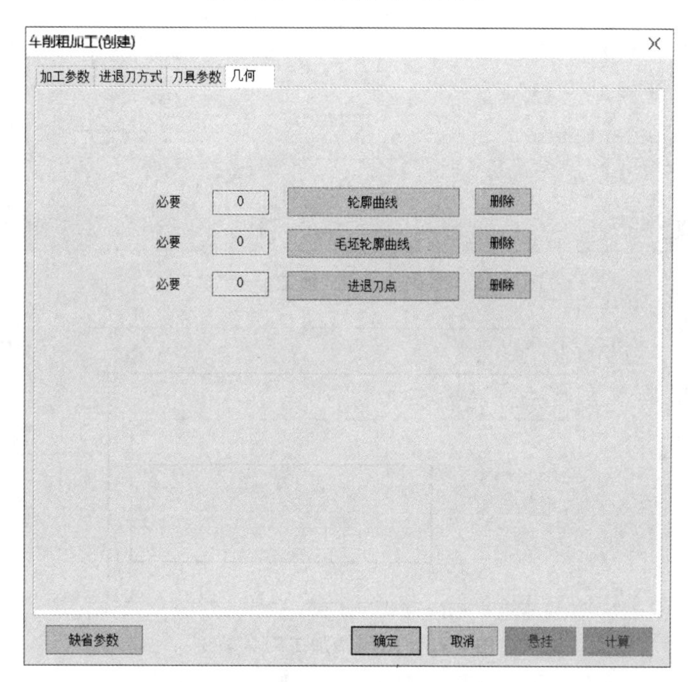

图 2-23　车削粗加工“几何”选项卡

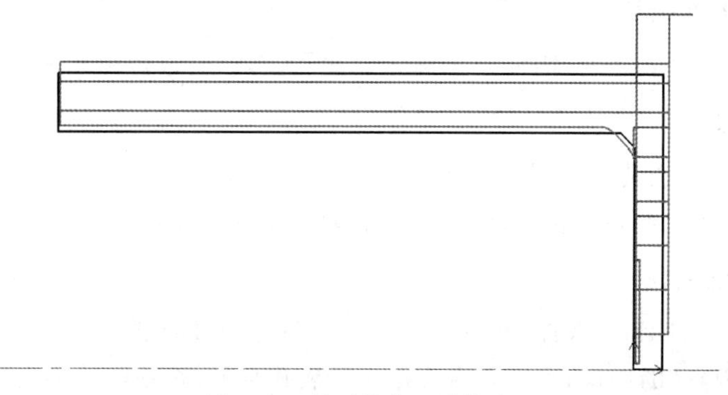

图 2-24　外形轮廓刀具轨迹

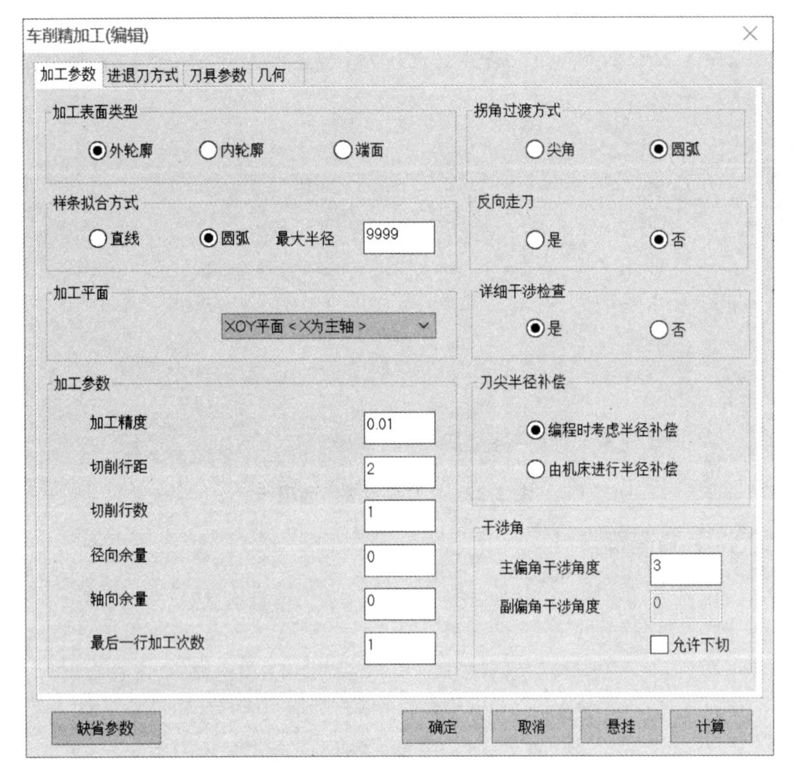

图 2-25 车削精加工“加工参数”选项卡

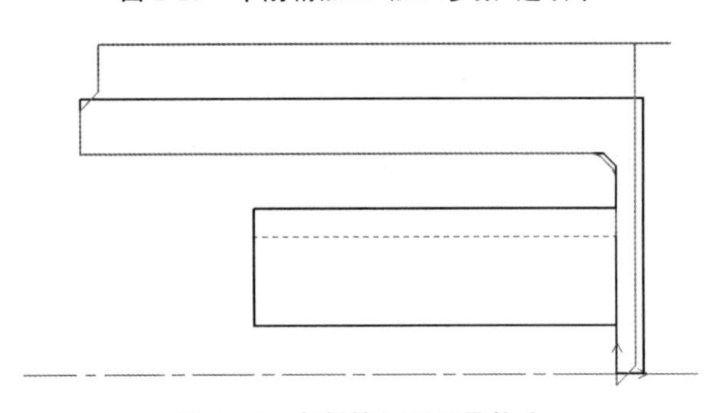

图 2-26 车削精加工刀具轨迹

2.11.2 右侧ϕ12 内孔编程

在加工右侧 ϕ12 孔之前，需要先预钻 ϕ10 底孔，然后再进行镗孔加工。

1. CAD 建模

如图 2-27 所示，首先绘制零件外形轮廓及毛坯轮廓，即定义加工区域。模型原点应与编程原点保持一致，CAD 软件当前版本不支持自定义加工原点。

2. CAM 自动编程

在设置内轮廓加工时，选择的轮廓曲线应在交点处打断，否则会导致选取失败。另外，需要注意“进退刀点”的选择，进退刀点应略小于底孔直径，否则可能会出现刀具与工件发生干涉的情况。

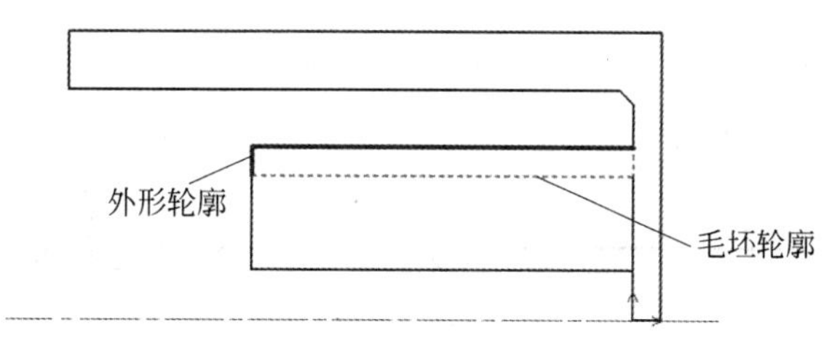

图 2-27　外形轮廓及毛坯轮廓

如图 2-28 所示，设置精加工参数，“几何”选项卡无须选择“毛坯轮廓曲线”，只需选择“轮廓曲线”即可，然后生成刀具轨迹，如图 2-29 所示。

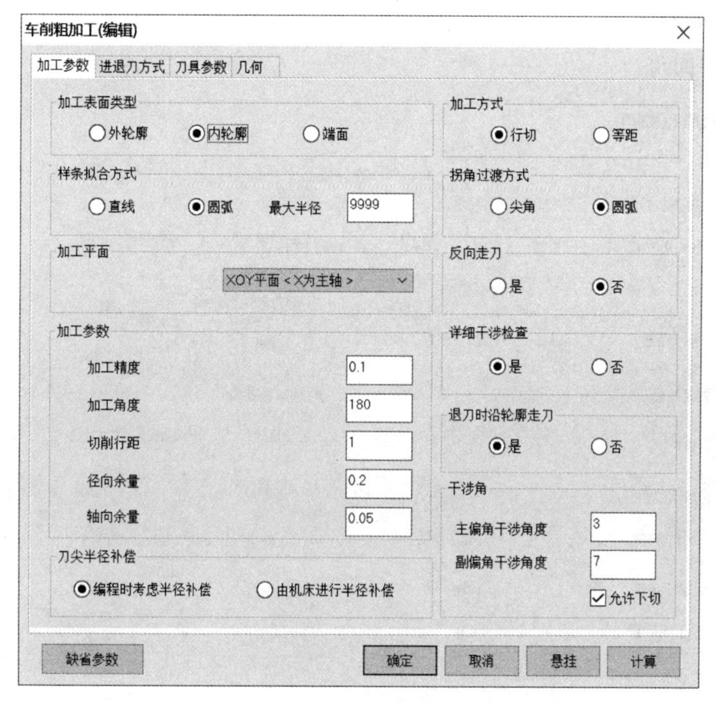

图 2-28　加工右侧 ϕ12 内孔参数配置

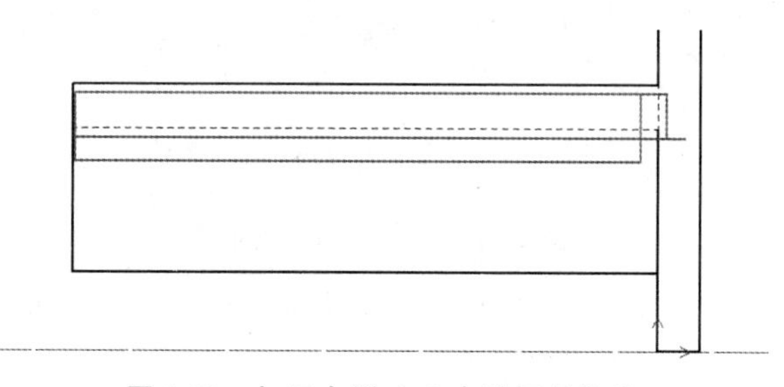

图 2-29　加工右侧 ϕ12 内孔刀具轨迹

精加工方法与粗加工方法基本一致，但不需要选择“毛坯轮廓曲线”。

2.11.3　圆弧槽编程

1. CAD 建模

如图 2-30 所示，首先绘制零件外形轮廓及毛坯轮廓，即定义加工区域。该圆弧槽在进行自动编程之前需要作辅助轮廓线，加工常见的矩形凹槽无须该步骤，直接进行选择即可。

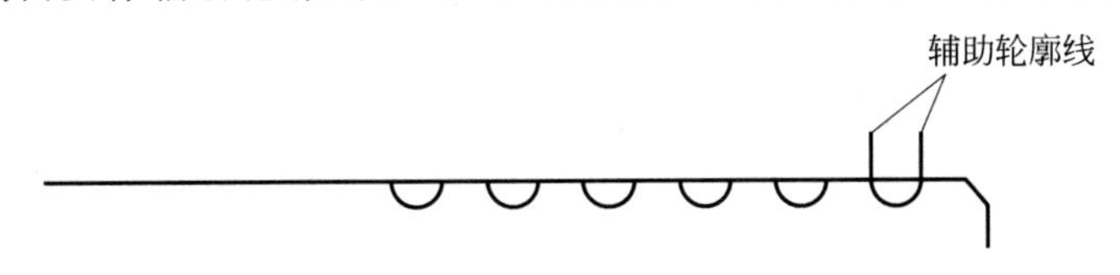

图 2-30　圆弧槽轮廓

2. CAM 自动编程

单击“车削槽加工”按钮，弹出“车削槽加工（编辑）”对话框，在其中的“加工参数”选项卡设置加工参数，如图 2-31 所示。该加工步骤选用的刀具为成型车刀，所以无须进行粗加工，直接进行精加工即可。

图 2-31　“加工参数”选项卡

如图 2-32 所示，设置刀具参数，CAM 软件默认刀具形状为普通切槽车刀，可以通过修改刀尖半径，将刀具设置为圆弧切槽车刀。

如图 2-33 所示，选择第一个凹槽和辅助线作为轮廓曲线。

生成的刀具轨迹如图 2-34 所示。

其余的凹槽加工可以通过复制的方式实现，如图 2-35 所示，选中左侧“管理树栏”中需要复制的内容，右击后再单击“带基点复制”命令。然后根据提示完成复制刀路的操作，复制结果如图 2-36 所示。

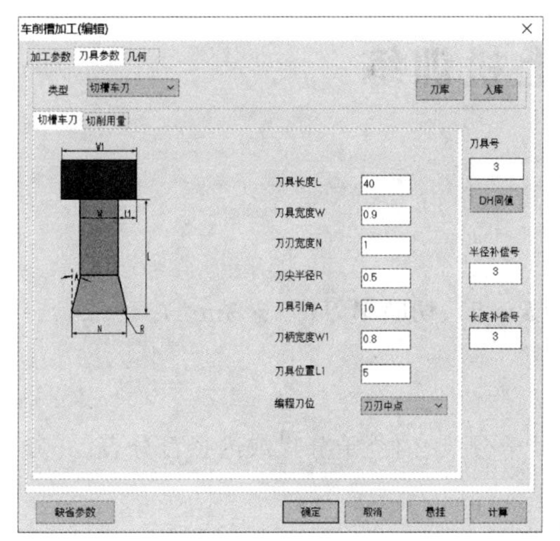

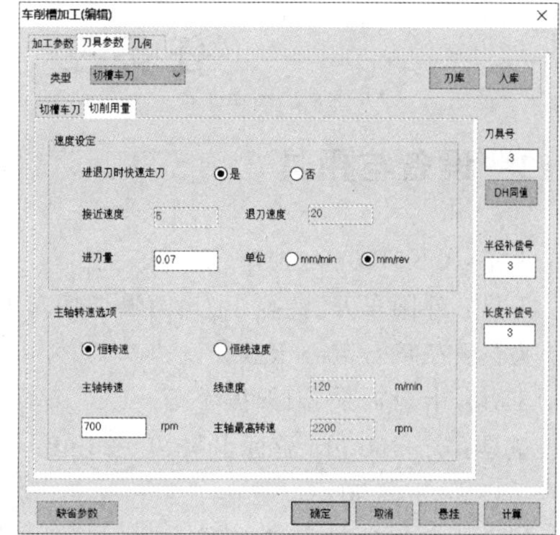

图 2-32　车削槽加工“刀具参数”选项卡

图 2-33　凹槽轮廓曲线

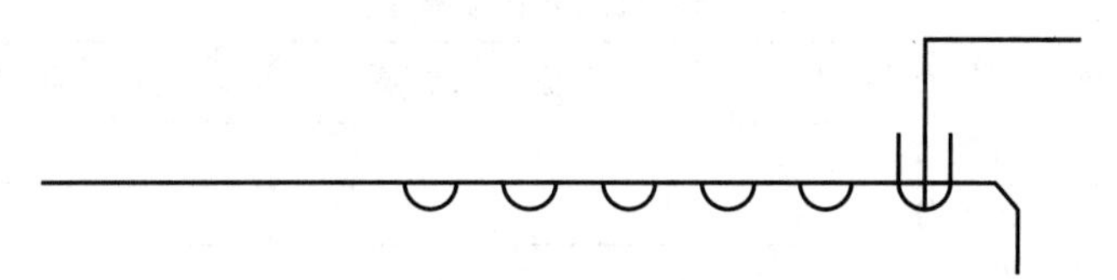

图 2-34　凹槽加工刀具轨迹

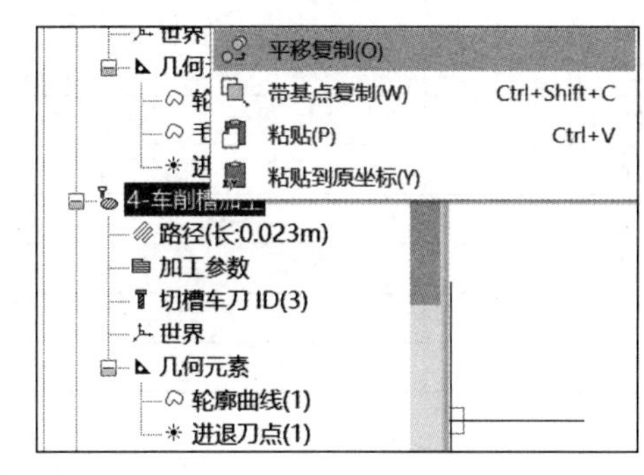

图 2-35　刀具轨迹平移复制

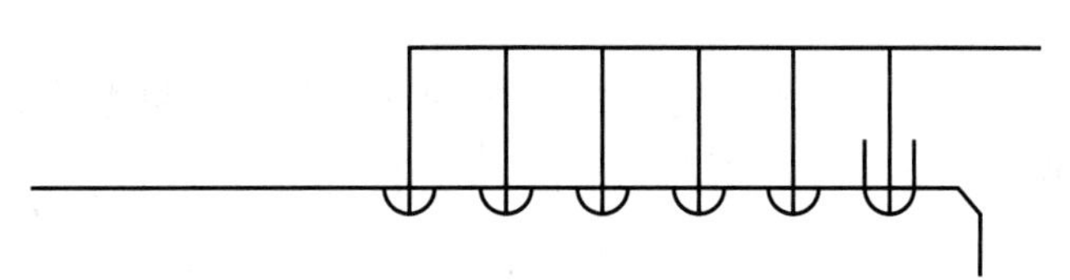

图 2-36　圆弧槽刀具轨迹

任务三 上机训练

2.12 设备与用具

设备：CK6150e 数控车床。

刀具：外圆车刀、镗孔车刀、圆弧切槽车刀($R0.5$)、切断车刀(刀宽 3mm)。

夹具：三爪自定心卡盘。

工具：卡盘扳手、刀架扳手等。

量具：0～150mm 游标卡尺、0～25mm 外径千分尺、内测千分尺或内径百分表。

毛坯：$\phi 20 \times 50$。

辅助用品：垫刀片、毛刷等。

2.13 开机前检查

可参考表 2-4 对机床状态进行点检。

表 2-4 机床开机准备卡片

检查项目		检查结果	异常描述
机械部分	主轴部分		
	进给部分		
	刀架		
	三爪自定心卡盘		
电器部分	主电源		
	冷却风扇		
数控系统	电气元件		
	控制部分		
	驱动部分		
辅助部分	冷却系统		
	压缩空气		
	润滑系统		

2.14 加工前准备

在加工前，应先将本任务所需刀具准备齐全，并安装正确。根据工艺要求设定工件原点，录入数控加工程序，并进行图形校验。

2.15 零件加工

在图形校验过程验证无问题后，即可进行零件加工。在零件加工前，应详细了解机床的安全操作要求，穿戴好劳动保护服装和用具。在进行零件加工时，应熟悉数控车床各操作按键的功能和位置，了解紧急状况的处置方法。在加工过程中，尤其是在即将切削之前，应对照显示屏“剩余移动量”栏显示的剩余移动量，观察刀具与工件之间的实际距离。若实际距离与剩余移动量相差过大，则应果断停机检查，以免发生撞机事故。若有异常，则应及时停止机床运动。

2.16　零件检测

在零件加工完成后，应当认真清理工件，并按照质量管理的相关要求，对加工完成的零件进行相关检验，保证生产质量。机械加工零件“三级”检验卡片如表 2-5 所示。

表 2-5　机械加工零件“三级”检验卡片

零部件图号		零部件名称		工　序　号	
材料		送检日期		工序名称	
检验项目	自检结果	互检结果	专业检验	备注	
检验结论	□合格　□不合格　□返修　□让步接收 检验签章： 年　月　日				
不符合项描述					

项 目 总 结

做功活塞作为数控车床的典型加工零件，在生产和生活中应用广泛。根据设备情况和精度的要求，其加工工艺也存在一些差别。编程人员及操作人员需要结合加工条件，合理制定加工工艺，以提高零件的加工精度和生产效率。

课 后 习 题

一、选择题

1. 常用 G 代码的功能是(　　)。

A. 刀具功能　　B. 准备功能
C. 坐标地址　　D. 进给速度

2. 圆弧编程中的 ***I***、***K*** 值是指（　　）的矢量值。
A. 起点到圆心　　B. 终点到圆心
C. 圆心到起点　　D. 圆心到终点

3. 在逆时针圆弧插补指令 G03 X_Y_R_中，R 后的数值表示圆弧的（　　）。
A. 半径　　B. 终点坐标值
C. 起点坐标值　　D. 圆心角

4. 以下叙述中错误的是（　　）。
A. 每一个存储在系统中的数控程序必须指定一个程序号
B. 程序段由 1 个或多个指令构成，表示数控机床的全部动作
C. 在大部分系统中，程序段号仅作为“跳转”或“程序检索”的目标位置指示
D. FANUC 系统的程序注释用“()”括起来

5. 下列哪项不属于数控机床的加工特点（　　）。
A. 加工精度高　　B. 适应性强
C. 生产效率高　　D. 适合大批量生产

6. 在测量工件凸肩厚度时，应选用（　　）。
A. 正弦规　　B. 外径千分尺
C. 三角板　　D. 块规

二、判断题

1. 在数控车床编程时，应采用直径尺寸编程。（　　）
2. 刀片圆弧半径一般适宜选取进给量的 2～3 倍。（　　）
3. 当圆弧插补用 R 代码编程，圆弧所对应的圆心角等于 180°时，*R* 取负值。（　　）
4. 在用倒圆角过渡指令编制圆弧程序时，不可以描述整圆。（　　）

三、填空题

1. 数控编程一般分为________和________。
2. 刀尖圆弧半径补偿的过程分为________、________和________。
3. 数控车床的刀具补偿有________和________两种。
4. 将输入计算机的零件设计和加工信息自动转换为数控装置能够读取和执行的指令（或信息）的过程就是________。

四、简答题

1. 简述圆弧车刀的作用。
2. 在加工圆弧槽零件时，入刀点该怎么选择？
3. CAXA 数控车二维加工主要包括哪些功能？
4. 镗孔加工有哪些注意事项？
5. 常用的数控加工工艺文件有哪些？

五、分析题

分析零件图（见图 2-37），确定工艺过程，填写工艺过程卡片，如表 2-6 所示。

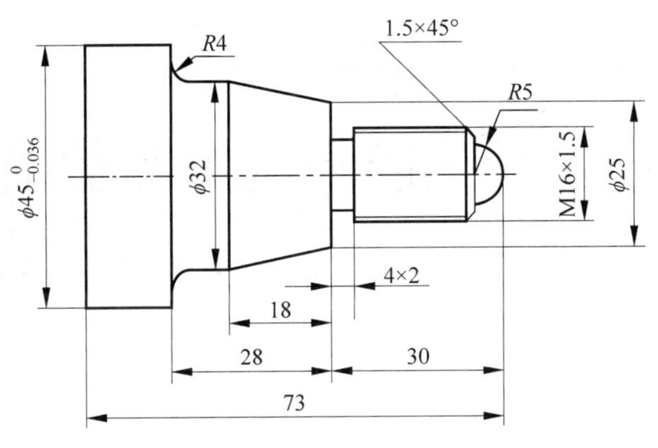

图 2-37　题五图

表 2-6　工艺过程卡片

序　　号	加工步骤及内容	刀　具　号	刀具类型	切削用量说明

自我学习检测评分表如表 2-7 所示。

表 2-7　自我学习检测评分表

项　　目	目标要求	分　　值	评分细则	得　　分	备　　注
学习关键知识点	(1) 了解 CAXA 数控车自动编程软件的基本组成 (2) 熟悉 CAXA 数控车自动编程软件的常用按键 (3) 掌握 CAXA 数控车自动编程软件的基本操作 (4) 掌握 CAXA 数控车加工的几个基本概念 (5) 掌握 CAXA 数控车刀库的设置	20	理解与掌握		
工艺准备	(1) 能够正确识读零件图 (2) 能够根据零件图分析、确定工艺过程 (3) 能够根据工序加工工艺、编写正确的加工程序	30	理解与掌握		
上机训练	(1) 会正确选择相应的设备与用具 (2) 能够正确操作数控车床，并根据加工情况调整加工参数	50	(1) 理解与掌握 (2) 操作流程		

思政小课堂

项目三　加热腔车削编程加工训练

➤思维导图

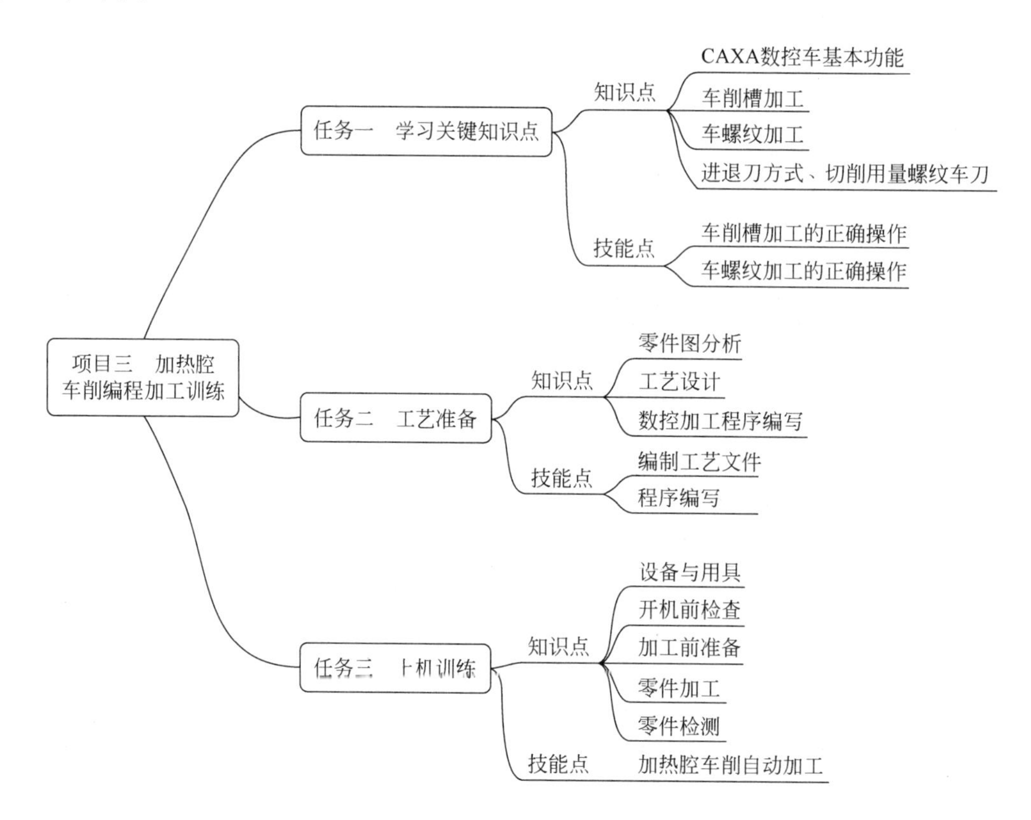

➤学习目标

知识目标

(1) 了解加热腔零件的加工特点。

(2) 理解孔加工循环指令各参数的含义。

能力目标

(1) 能够独立确定加工工艺路线,并正确填写工艺文件。

(2) 能够正确操作数控车床,并根据加工情况调整加工参数。

(3) 能够根据零件结构特点和精度合理选用量具,并正确、规范地测量相关尺寸。

素养目标

(1) 培养学生的科学探究精神和态度。

(2) 培养学生的工程意识。

(3) 培养学生的团队合作能力。

➢任务引入

根据零件图(见图 3-1)要求,制定加工工艺、编写数控加工程序,并完成加热腔零件的加工。该零件毛坯材料为 45 钢,调质处理,要求表面光整,无划伤。

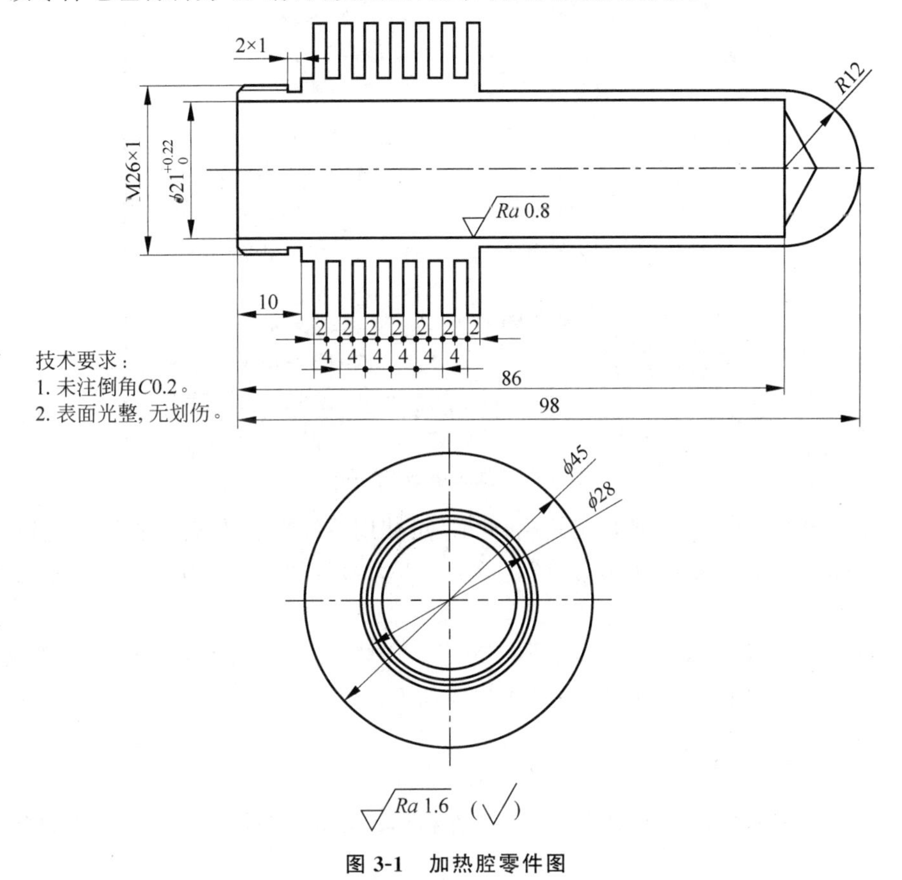

图 3-1　加热腔零件图

任务一　学习关键知识点

3.1　CAXA 数控车基本功能

3.1.1　“车削槽加工(创建)”对话框

车削槽加工用于在工件外轮廓表面、内轮廓表面和端面切槽。

切槽时要确定被加工轮廓,被加工轮廓就是加工结束后的工件表面轮廓,被加工轮廓不能闭合或自相交。车削槽加工的操作步骤如下。

（1）在菜单栏中单击“数控车”标签，再单击“车削槽加工”按钮，弹出“车削槽加工（创建）”对话框，其中“加工参数”选项卡如图 3-2 所示。在“加工参数”选项卡中，首先要确定被加工的是外轮廓表面，还是内轮廓表面或端面；接着按加工要求确定其他各加工参数。

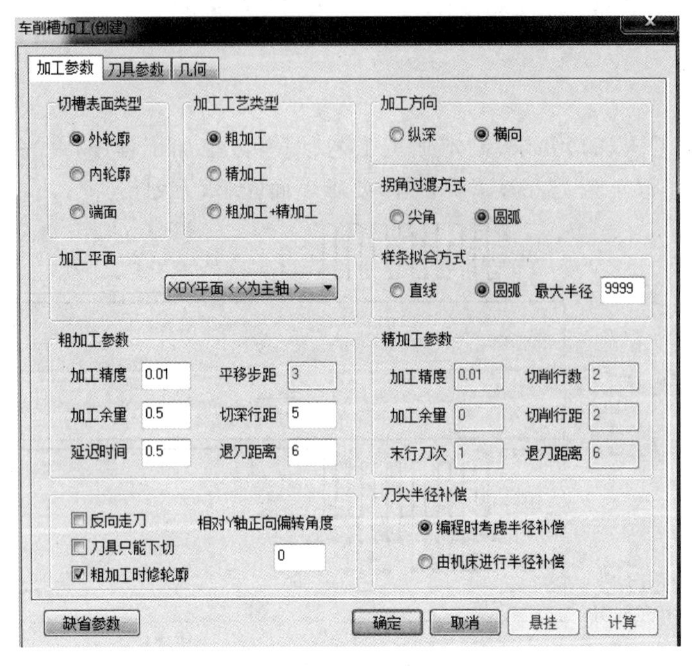

图 3-2 “加工参数”选项卡

（2）在确定参数后，拾取被加工轮廓，此时，可使用系统提供的“轮廓拾取”工具。

（3）选择完轮廓后确定“进退刀点”。指定一点为刀具加工前和加工后所在的位置。右击可忽略该点的输入。

在完成上述步骤后，即可生成切槽加工轨迹。单击“数控车”标签，再单击“后置处理”按钮，拾取刚生成的切槽加工刀具轨迹，即可生成加工指令。

3.1.2 “加工参数”选项卡

加工参数主要对切槽加工中各种工艺条件和加工方式进行限定。各加工参数含义说明如下。

1. “切槽表面类型”选项组

（1）“外轮廓”：外轮廓切槽，或者用切槽车刀加工外轮廓。

（2）“内轮廓”：内轮廓切槽，或者用切槽车刀加工内轮廓。

（3）“端面”：端面切槽，或者用切槽车刀加工端面。

2. “加工工艺类型”选项组

（1）“粗加工”：对槽只进行粗加工。

（2）“精加工”：对槽只进行精加工。

（3）“粗加工＋精加工”：对槽进行粗加工之后，接着进行精加工。

3. “拐角过渡方式”选项组

(1)“圆弧”：当切削过程遇到拐角时，在刀具从轮廓的一边到另一边的过程中，以圆弧的方式过渡。

(2)“尖角”：当切削过程遇到拐角时，在刀具从轮廓的一边到另一边的过程中，以尖角的方式过渡。

4. “粗加工参数”选项组

(1)“延迟时间”：在粗加工槽时，刀具在槽的底部停留的时间。

(2)“切深行距”：在粗加工槽时，刀具每一次纵向切槽的切入量(机床 X 向)。

(3)“平移步距”：在粗加工槽时，刀具在切到指定的切深平移量后，进行下一次切削前的水平平移量(机床 Z 向)。

(4)“退刀距离”：在粗加工槽中，进行下一行切削前，退刀到槽外的距离。

(5)“加工余量”：在粗加工槽时，被加工表面未加工部分的预留量。

5. “精加工参数”选项组

(1)“切削行距”：精加工槽行与行之间的距离。

(2)“切削行数”：精加工槽刀具轨迹的加工行数，不包括最后一行的重复次数。

(3)“退刀距离”：在精加工槽中切削完一行之后，进行下一行切削前，退刀的距离。

(4)“加工余量”：在精加工槽时，被加工表面未加工部分的预留量。

(5)“末行刀次”：在精加工槽时，为提高加工表面的质量，最后一行常常在相同进给量的情况下，进行多次车削，该处定义最后一行多次切削的次数。

3.1.3 “切槽车刀”选项卡

单击“刀具参数”标签可进入“切槽车刀”选项卡(见图 3-3、图 3-4)。该选项卡用于对加工中所用的切槽刀具参数进行设置。具体参数说明请参考 2.7.2 节中的说明。

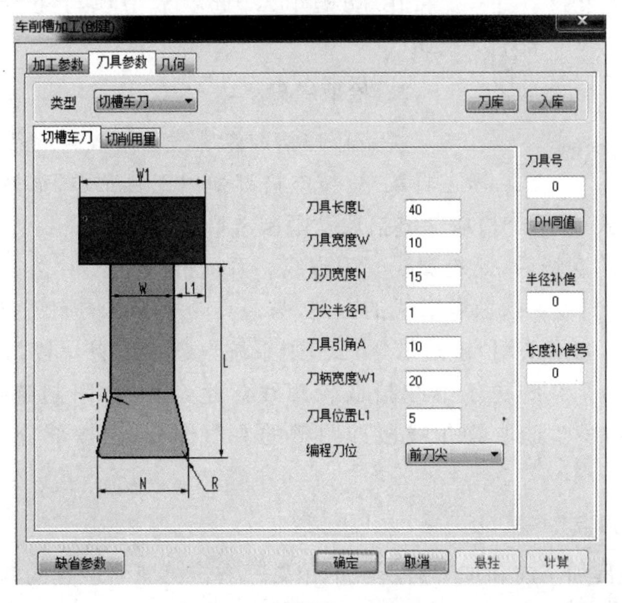

图 3-3　“切槽车刀”选项卡

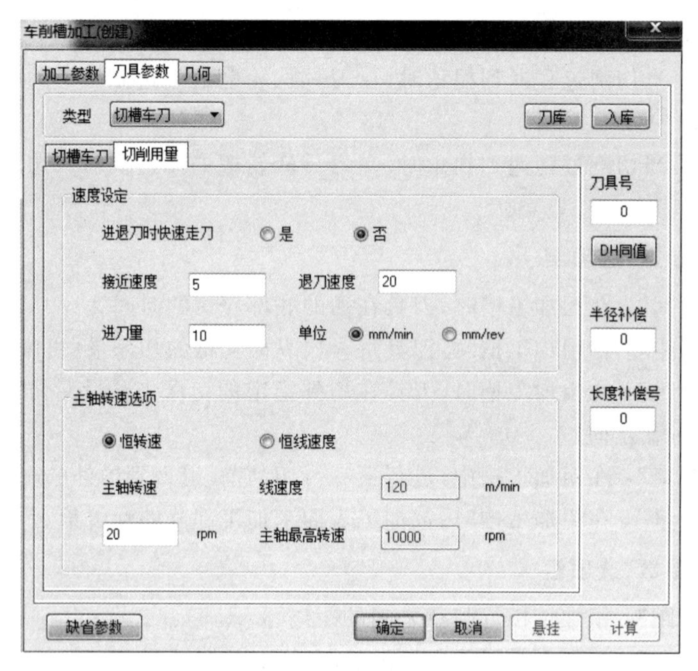

图 3-4 “切削用量”选项卡

3.1.4 车削槽加工实例

车削槽加工实例的步骤具体如下。

1. 确定加工轮廓

如图 3-5 所示,螺纹退刀槽凹槽部分为要加工出的轮廓。

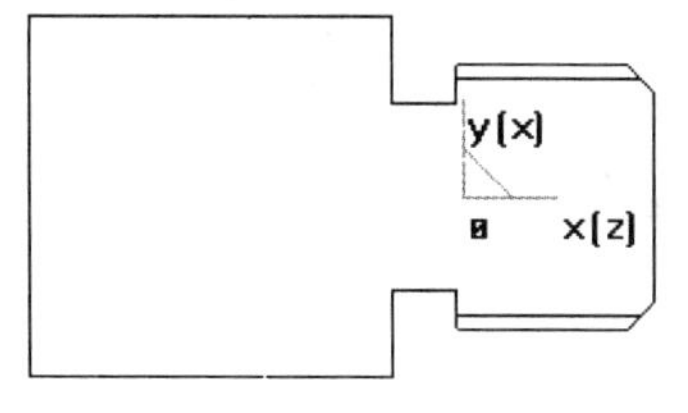

图 3-5 工件图纸

2. 填写参数表

在“切槽车刀”选项卡中填写完参数后,单击“确认”按钮。

3. 拾取轮廓

提示用户拾取轮廓线。拾取轮廓线可以利用曲线拾取工具菜单,按空格键弹出工具菜单,如图 3-6 所示。工具菜单提供 3 种拾取方式:“单个拾取”“链拾取”“限制链拾取”。

在拾取第一条轮廓线后,此轮廓线变为虚线。系统给出提示:“选择方向”,要求用户选择一个方向,此方向只表示拾取轮廓线的方向,与刀具的加工方向无关,如图 3-7 所示。

在选择方向后,如果采用“链拾取”方式,则系统自动拾取首尾连接的轮廓线;如果采用“单个拾取”方式,则系统提示继续拾取轮廓线。此处采用“限制链拾取”方式,系统继续提示:“选取限制线”,选取终止线段即凹槽的左边部分,凹槽部分变为虚线,如图 3-8 所示。

4. 确定“进退刀点”

指定一点为刀具加工前和加工后所在的位置。右击可忽略该点的输入。

图 3-6 拾取方式　　　　图 3-7 选择轮廓线方向

5. 生成刀具轨迹

在确定进退刀点之后,系统生成刀具轨迹,如图 3-9 所示。

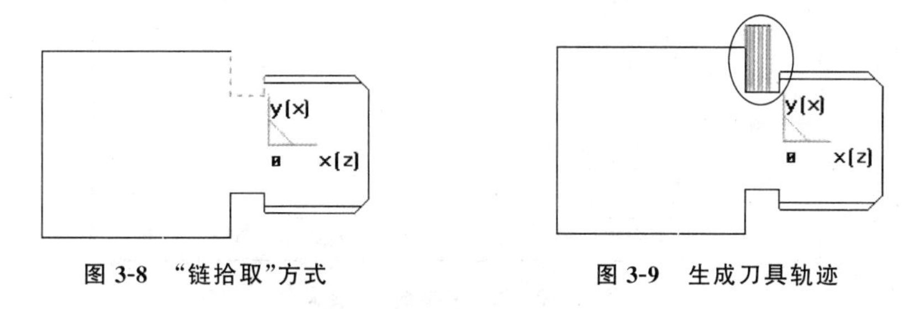

图 3-8 “链拾取”方式　　　　图 3-9 生成刀具轨迹

注意:被加工轮廓不能闭合或自相交。生成刀具轨迹与切槽车刀刀角半径、刀刃宽度等参数密切相关。可按实际需要只绘出退刀槽的上半部分。

3.1.5 “车螺纹加工(创建)”对话框

车螺纹加工为非固定循环方式加工螺纹,可对螺纹加工中的各种工艺条件、加工方式进行更为灵活的控制。车螺纹加工的操作步骤如下。

(1) 单击“数控车”标签,再单击“车螺纹加工”按钮,弹出“车螺纹加工(创建)”对话框,如图 3-10 所示。用户可在该对话框中确定各加工参数。

(2) 拾取螺纹“起点”“终点”“进退刀点”。

(3) 参数填写完毕,单击“确认”按钮,即生成螺纹车削刀具轨迹。

(4) 单击“数控车”标签,再单击“后置处理”按钮,拾取刚生成的刀具轨迹,即可生成螺纹加工指令。

3.1.6 “螺纹参数”选项卡

单击“车螺纹加工(创建)”对话框中的“螺纹参数”标签,进入“螺纹参数”选项卡,“螺纹参数”选项卡主要包括与螺纹性质相关的参数,如“螺纹类型”“螺纹节距”“螺纹头数”等。螺纹“起点”和“终点”坐标来自前一步的拾取结果,用户也可以进行修改。各螺纹参数含义说明如下。

(1) “起点”坐标:车螺纹的起始点坐标,单位为 mm。

(2) “终点”坐标:车螺纹的终止点坐标,单位为 mm。

(3) “进退刀点”坐标:车螺纹加工进刀与退刀点的坐标,单位为 mm。

(4) “螺纹牙高”:螺纹牙的高度。

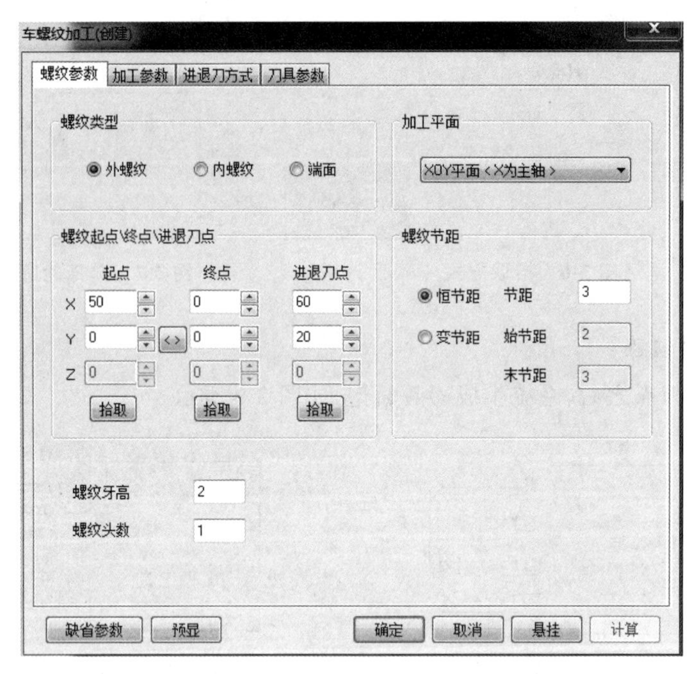

图 3-10 “螺纹参数”选项卡

(5)“螺纹头数”:螺纹起始点到终止点之间的牙数。

(6)“螺纹节距”:包括5种,“恒节距”是指两个相邻螺纹轮廓上对应点之间的距离为恒定值;“节距”是指恒定节距值;“变节距”是指两个相邻螺纹轮廓上对应点之间的距离为变化值;“始节距”是指起始端螺纹的节距;“末节距”是指终止端螺纹的节距。

3.1.7 “加工参数”选项卡

“加工参数”选项卡用于对螺纹加工中的工艺条件和加工方式进行设置,如图3-11所示。

各螺纹加工参数含义说明如下。

1. “加工工艺”选项组

(1)“粗加工”:直接采用粗加工方式加工螺纹。

(2)“粗加工+精加工”方式:根据指定的粗加工深度进行粗切后,再采用精加工方式(如采用更小的行距)切除剩余余量(精加工深度)。

2. “参数”选项组

(1)“末行走刀次数”:为提高加工表面的质量,最后一行常常在相同进给量的情况下,进行多次车削,该处定义最后一行多次切削的次数。

(2)“螺纹总深”:螺纹粗加工和精加工总的切深量。

(3)“粗加工深度”:螺纹粗加工的切深量。

(4)“精加工深度”:螺纹精加工的切深量。

3. “每行切削用量”下拉列表框

“每行切削用量”包括以下内容。

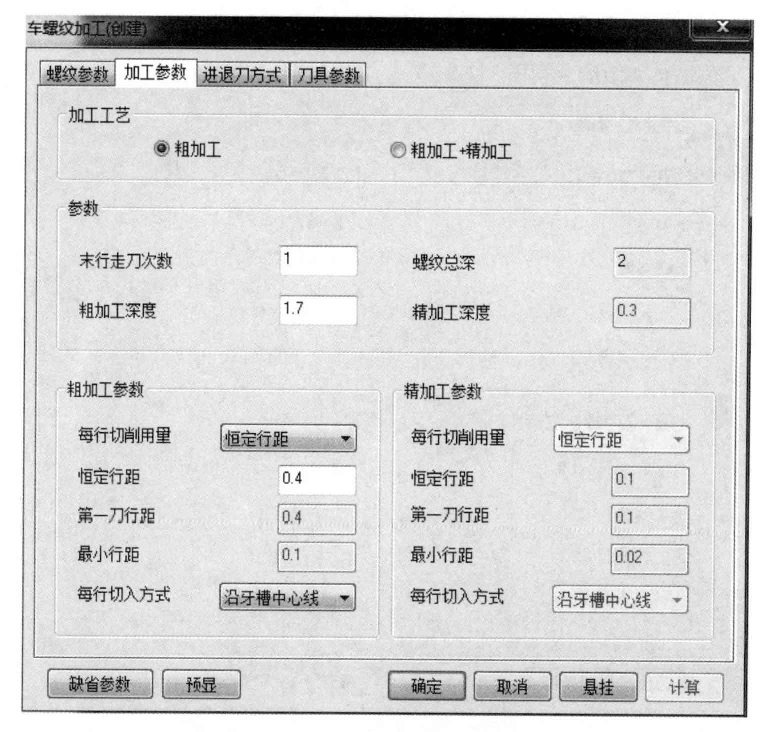

图 3-11　“加工参数”选项卡

(1)“恒定行距”：定义在沿恒定的行距进行加工时的行距。

(2)“恒定切削面积”：为保证每次切削的切削面积恒定，各次切削深度将逐步减小，直至等于最小行距。用户需指定第一刀行距及最小行距。吃刀深度规定为第 n 刀的吃刀深度为第一刀的吃刀深度的 $\sqrt{n}$ 倍。

(3)“变节距”：两个相邻螺纹轮廓上对应点之间的距离为变化值。

(4)“始节距”：起始端螺纹的节距。

(5)“末节距”：终止端螺纹的节距。

4. “每行切入方式”下拉列表框

“每行切入方式”：刀具在螺纹始端切入时的切入方式。刀具在螺纹末端的退出方式与切入方式相同。

(1)“沿牙槽中心线”：切入时沿牙槽中心线。

(2)“沿牙槽右侧”：切入时沿牙槽右侧。

(3)“左右交替”：切入时沿牙槽左右交替。

3.1.8　“进退刀方式”选项卡

单击“进退刀方式”标签，进入“进退刀方式”选项卡。如图 3-12 所示，该选项卡用于对加工中的进、退刀方式进行参数设定。

1. 进刀方式

(1)“垂直”：刀具直接进刀到每一切削行的起始点。

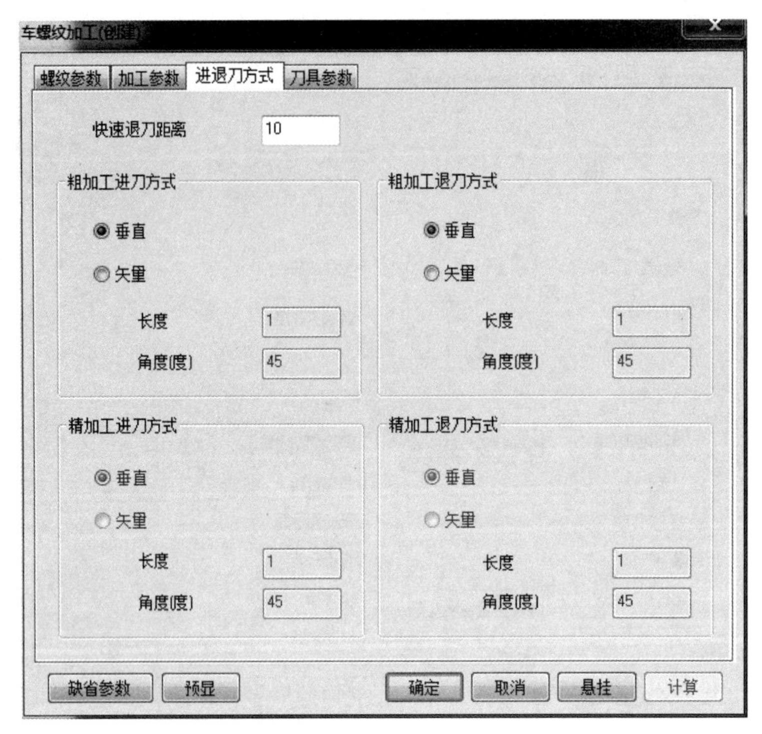

图 3-12 “进退刀方式”选项卡

(2) “矢量”：在每一切削行前，加入一段与系统 X 轴(机床 Z 轴)正方向成一定夹角的进刀段，刀具进刀到该进刀段的起点，再沿该进刀段进刀至切削行。

① “长度”：定义矢量(进刀段)的长度。

② “角度”：定义矢量(进刀段)与系统 X 轴正方向的夹角。

2. 退刀方式

(1) “垂直”：刀具直接退刀到每一切削行的起始点。

(2) “矢量”：在每一切削行后，加入一段与系统 X 轴(机床 Z 轴)正方向成一定夹角的退刀段，刀具先沿该退刀段退刀，再从该退刀段的末点开始垂直退刀。

① “长度”：定义矢量(退刀段)的长度。

② “角度”：定义矢量(退刀段)与系统 X 轴正方向的夹角。

3. 退刀距离

以给定的退刀速度回退的距离(相对值)，在此距离上，以机床允许的最大进给速度退刀。

3.1.9 “切削用量”选项卡

“切削用量”选项卡的说明与 2.11.1 节中介绍的“车削粗加工”过程类似，具体设置如图 3-13 所示。

3.1.10 “螺纹车刀”选项卡

单击“螺纹车刀”标签可进入“螺纹车刀”选项卡。如图 3-14 所示，该选项卡用于对加工中所用的螺纹车刀参数进行设置。具体参数说明请参考 2.7.3 节中的说明。

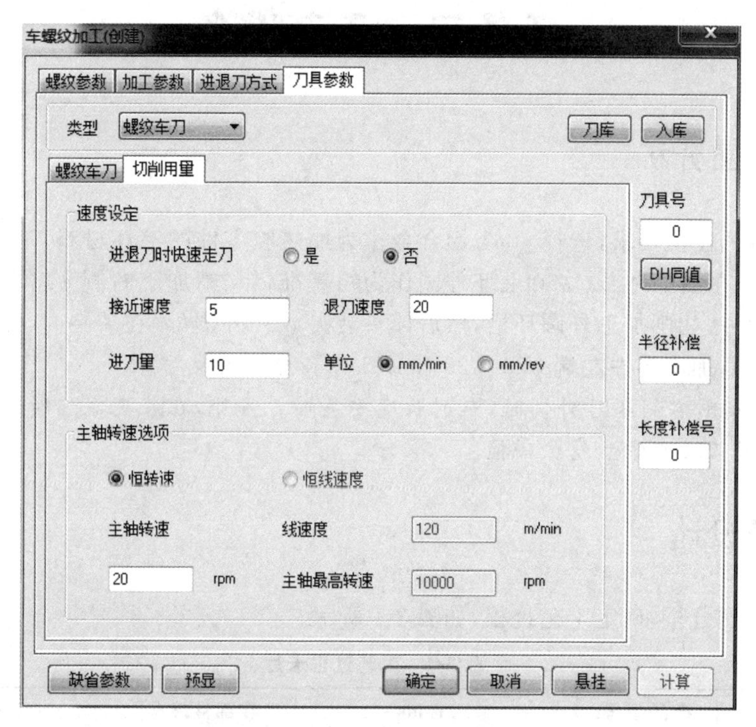

图 3-13　“刀具参数”选项卡

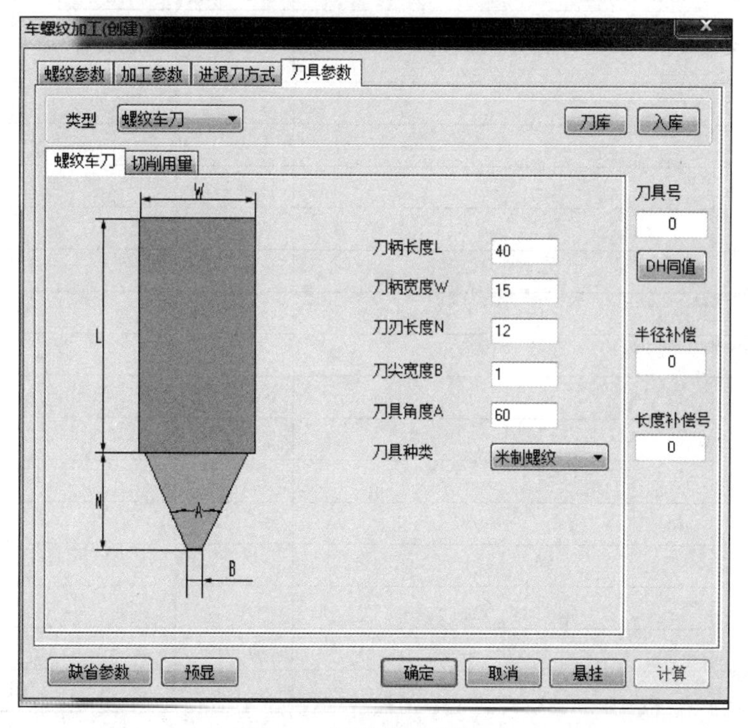

图 3-14　“螺纹车刀”选项卡

任务二　工艺准备

3.2　零件图分析

根据零件的使用要求，选择6061铝合金作为加热腔零件的毛坯材料，毛坯下料尺寸定为 $\phi 50 \times 102$。在加工时，以 $\phi 50$ 毛坯外圆作为粗基准，粗、精加工右侧 $\phi 24$ 外圆、$\phi 45$ 圆柱表面至要求尺寸，切削加工外圆环槽，然后掉头装夹 $\phi 24$ 外圆处（在装夹时，注意做好保护，以防表面夹伤），加工零件左端 $\phi 26$、$\phi 28$ 和 $\phi 21$ 内孔至要求尺寸。

注意：在车削右侧 $\phi 45$ 外圆时，车削长度要足够。另外，在装夹毛坯时，应注意棒料伸出的长度，以免刀具与卡盘发生碰撞。

3.3　工艺设计

根据零件图分析，确定工艺过程，如表3-1所示。

表3-1　工艺过程卡片

<table>
<tr><td colspan="2" rowspan="2">机械加工
工艺过程卡片</td><td>产品型号</td><td>STL-00</td><td colspan="2">零部件序号</td><td colspan="4">STL-03</td><td>第1页</td></tr>
<tr><td>产品名称</td><td>斯特林发动机模型</td><td colspan="2">零部件名称</td><td colspan="4">加热腔</td><td>共1页</td></tr>
<tr><td>材料牌号</td><td>6061</td><td>毛坯规格</td><td>φ50×102</td><td>毛坯质量</td><td colspan="2">kg</td><td colspan="2">数量</td><td colspan="2">1</td></tr>
<tr><td rowspan="2">工序号</td><td rowspan="2">工序名</td><td colspan="2" rowspan="2">工序内容</td><td rowspan="2">工段</td><td colspan="3" rowspan="2">工艺装备</td><td colspan="3">工时/min</td></tr>
<tr><td colspan="2">准结</td><td>单件</td></tr>
<tr><td></td><td></td><td colspan="2"></td><td></td><td colspan="3"></td><td colspan="2"></td><td></td></tr>
<tr><td></td><td></td><td colspan="2"></td><td></td><td colspan="3"></td><td colspan="2"></td><td></td></tr>
<tr><td></td><td></td><td colspan="2"></td><td></td><td colspan="3"></td><td colspan="2"></td><td></td></tr>
<tr><td></td><td></td><td colspan="2"></td><td></td><td colspan="3"></td><td colspan="2"></td><td></td></tr>
<tr><td></td><td></td><td colspan="2"></td><td></td><td colspan="3"></td><td colspan="2"></td><td></td></tr>
<tr><td></td><td></td><td colspan="2"></td><td></td><td colspan="3"></td><td colspan="2"></td><td></td></tr>
<tr><td></td><td></td><td colspan="2"></td><td></td><td colspan="3"></td><td colspan="2"></td><td></td></tr>
<tr><td></td><td></td><td colspan="2"></td><td></td><td colspan="3"></td><td colspan="2"></td><td></td></tr>
<tr><td></td><td></td><td colspan="2"></td><td></td><td colspan="3"></td><td colspan="2"></td><td></td></tr>
<tr><td></td><td></td><td colspan="2"></td><td></td><td colspan="3"></td><td colspan="2"></td><td></td></tr>
<tr><td></td><td></td><td colspan="2"></td><td></td><td colspan="3"></td><td colspan="2"></td><td></td></tr>
<tr><td></td><td></td><td colspan="2"></td><td></td><td colspan="3"></td><td colspan="2"></td><td></td></tr>
<tr><td></td><td></td><td colspan="2"></td><td></td><td colspan="3"></td><td colspan="2"></td><td></td></tr>
<tr><td></td><td></td><td colspan="2"></td><td></td><td colspan="3"></td><td colspan="2"></td><td></td></tr>
<tr><td></td><td></td><td colspan="2"></td><td></td><td colspan="3"></td><td colspan="2"></td><td></td></tr>
</table>

本训练任务针对加热腔零件进行工序设计，制订工序卡片，如表3-2所示。

表 3-2 车削加工工序卡片

机械加工工序卡片	产品型号	STL-00	零部件序号	STL-03	第 1 页
	产品名称	斯特林发动机模型	零部件名称	加热腔	共 1 页

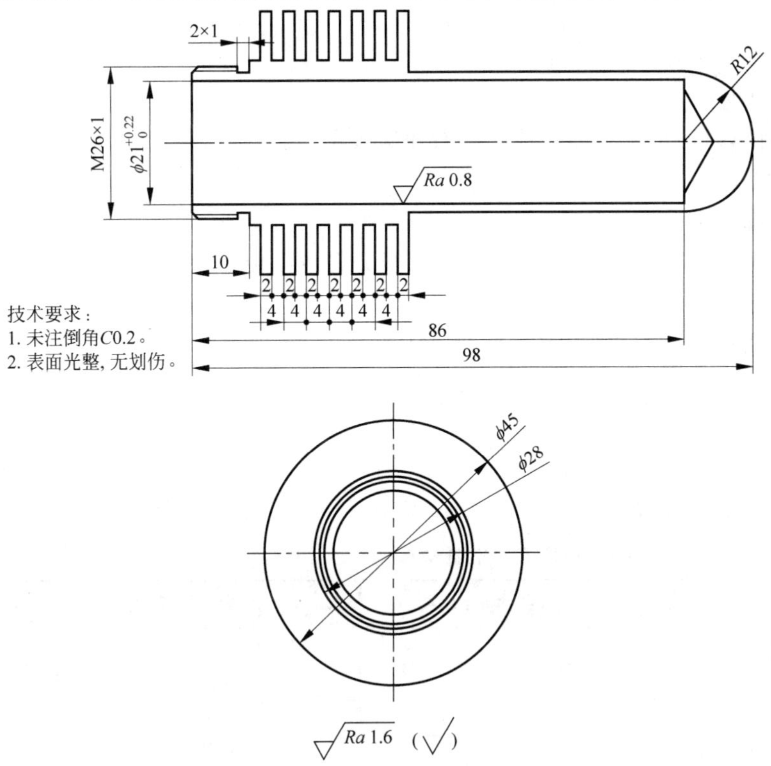

2×1
R12
M26×1
$\phi21^{+0.22}_{0}$
Ra 0.8
10
2
4
86
98
技术要求：
1. 未注倒角C0.2。
2. 表面光整，无划伤。
φ45
φ28
Ra 1.6 (√)

工序号	15
工序名	车加工
材料	6061
设备	数控车床
设备型号	CK6150e
夹具	三爪自定心卡盘
量具	游标卡尺
	千分尺
	内径百分表
准结工时	
单件工时	

工步	工步内容	刀具	S/(r/min)	F/(mm/r)	a_p/mm	工步工时/min	
						机动	辅助

3.4 数控加工程序编写

根据工序加工工艺，利用 CAXA 数控车分别对右端和左端创建轮廓模型，生成刀具轨迹，以及数控加工程序。

任务三　上机训练

3.5　设备与用具

设备：CK6150e 数控车床。
刀具：外圆车刀、切断车刀（刀宽 2mm）、镗孔车刀和螺纹车刀。
夹具：三爪自定心卡盘。
工具：卡盘扳手、刀架扳手等。
量具：0～150mm 游标卡尺、0～25mm 千分尺、内测千分尺或内径百分表。
毛坯：$\phi 50 \times 102$。
辅助用品：垫刀片、毛刷等。

3.6　开机前检查

可参考表 3-3 对机床状态进行点检。

表 3-3　机床开机准备卡片

检查项目		检查结果	异常描述
机械部分	主轴部分		
	进给部分		
	刀架		
	三爪自定心卡盘		
电器部分	主电源		
	冷却风扇		
数控系统	电气元件		
	控制部分		
	驱动部分		
辅助部分	冷却系统		
	压缩空气		
	润滑系统		

3.7　加工前准备

在加工前，应先将本任务所需刀具准备齐全，并安装正确。根据工艺要求设定工件原点，录入数控加工程序，并进行图形校验。

3.8　零件加工

在图形校验过程验证无问题后，即可进行零件加工。在零件加工前，应详细了解机床的安全操作要求，穿戴好劳动保护服装和用具。在进行零件加工时，应熟悉数控车床各操作按键的功能和位置，了解紧急状况的处置方法。在加工过程中，尤其是在即将切削之前，应对照显示屏“剩余移动量”栏显示的剩余移动量，观察刀具与工件之间的实际距离。若实际距离与剩余移动量相差过大，则应果断停机检查，以免发生撞机事故。若有异常，则应及时停止机床运动。

3.9　零件检测

在零件加工完成后，应当认真清理工件，并按照质量管理的相关要求，对加工完成的零件进行相关检验，保证生产质量。机械加工零件“三级”检验卡片如表 3-4 所示。

表 3-4　机械加工零件“三级”检验卡片

零部件图号		零部件名称		工　序　号	
材料		送检日期		工序名称	
检验项目	自检结果	互检结果	专业检验	备注	
检验结论	□合格　□不合格　□返修　□让步接收 检验签章： 年　月　日				
不符合项描述					

项目总结

加热腔作为数控车床的典型加工零件，在生产和生活中应用广泛。根据设备情况和精度的要求，其加工工艺也存在一些差别。编程人员及操作人员需要结合加工条件，合理制定加工工艺，以提高零件的加工精度和生产效率。

课后习题

一、选择题

1. 固定循环指令：(　　)。

A. 只需 1 个指令，便可完成某项加工　　B. 只能循环 1 次

C. 不能用其他指令代替　　D. 只能循环 2 次

2. 在进行孔类零件加工时，钻孔→平底钻扩孔→倒角→精镗孔的方法适用于(　　)。

A. 阶梯孔　　B. 小孔径的盲孔

C. 大孔径的盲孔　　　　D. 较大孔径的平底孔

3. 在数控系统中，(　　)指令在加工过程中是非模态的。

A. G01　　B. G04　　C. G17　　D. G81

4. 当数控机床主轴以 800r/min 转速顺时针转时，其指令应是(　　)。

A. S800 M03；　　B. S800 M04；　　C. S800 M05；　　D. S800 M06；

5. 数控机床空运行主要用于检查(　　)。

A. 程序编制的正确性　　　　B. 刀具轨迹的正确性

C. 机床运行的稳定性　　　　D. 加工精度的正确性

6. 在测量孔内径时，应选用(　　)。

A. 正弦规　　B. 内测千分尺　　C. 三角板　　D. 块规

二、判断题

1. 在 FANUC 系统中，00 组的 G 代码都是非模态指令。　(　　)

2. G04 指令为模态代码。　(　　)

3. 非模态代码，只在该代码的程序段中才有效。　(　　)

4. FANUC 系统数控车床的 G73 指令中不能含有宏程序加工指令。　(　　)

5. 钻工件内孔表面的 IT 值为 5.9。　(　　)

三、填空题

1. FANUC 系统数控车床内、外圆切削单一固定循环用________指令来指定，而端面切削循环则采用________指令来指定。

2. FANUC 系统数控车床中的径向切槽固定循环用________指令来实现，而端面切槽固定循环用________指令来实现。

3. 孔加工循环指令为________，一旦某个孔加工循环指令有效，在接着所有的位置均采用该孔加工循环指令进行孔加工，直到用________指令取消孔加工循环为止。

4. 如果槽的宽度小于深度，则使用________；如果宽度大于深度，则使用________；在加工细长工件时，可以使用________。

四、简答题

1. 简述凹槽加工的方法和注意事项。

2. 完成加热腔的手工编程。

3. 什么是模态代码和非模态代码？请分别举例说明。

自我学习检测评分表如表 3-5 所示。

表 3-5　自我学习检测评分表

项　目	目标要求	分　值	评分细则	得　分	备　注
学习关键知识点	(1) 掌握车削槽加工的操作步骤及加工参数的设定 (2) 掌握车螺纹加工的操作步骤及加工参数的设定 (3) 掌握进、退刀方式，切削用量，螺纹车刀的设置方法	20	理解与掌握		

续表

项　　目	目 标 要 求	分　　值	评 分 细 则	得　　分	备　　注
工艺准备	(1) 能够正确识读零件图 (2) 能够根据零件图分析、确定工艺过程 (3) 能够根据工序加工工艺，编写正确的加工程序	30	理解与掌握		
上机训练	(1) 会正确选择相应的设备与用具 (2) 能够正确操作数控车床，并根据加工情况调整加工参数	50	(1) 理解与掌握 (2) 操作流程		

思政小课堂

项目四　飞轮铣削编程加工训练

➤思维导图

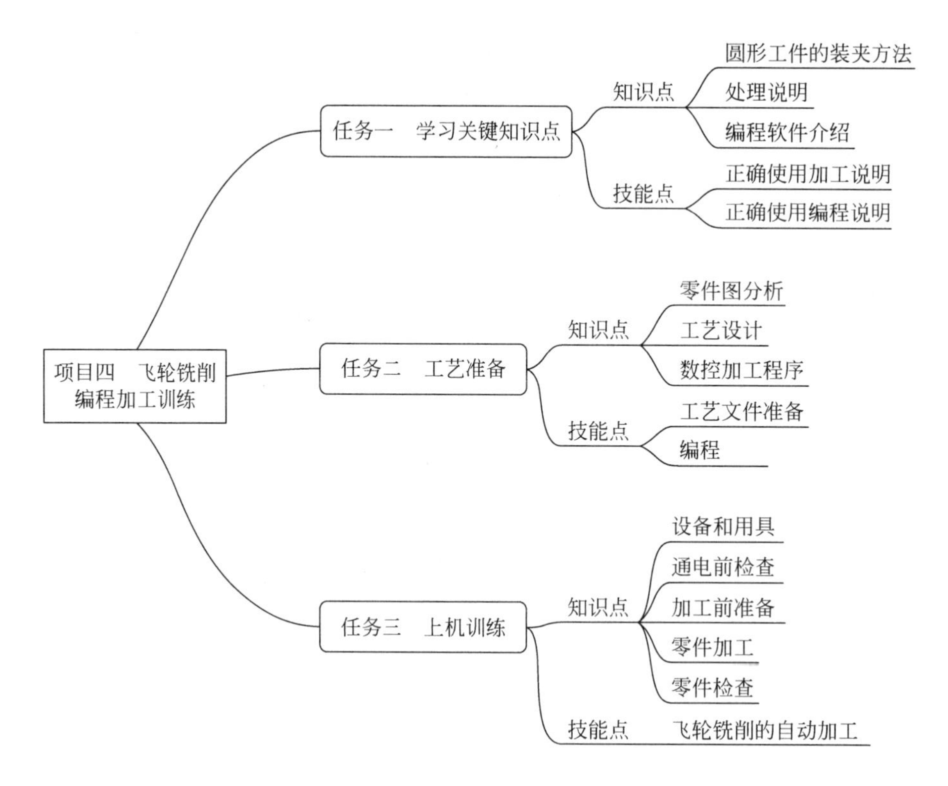

➤学习目标

知识目标

(1) 具备盘类零件的识图能力。

(2) 了解飞轮的用途和特点。

(3) 了解 CAXA 软件的绘图与编程方法。

能力目标

(1) 掌握圆形工件的铣削加工方法。

(2) 能够独立确定加工工艺路线,并正确填写工艺文件。

(3) 能够正确操作数控加工中心,并根据加工情况调整加工参数。

(4) 能够根据零件结构特点和精度合理选用量具,并正确、规范地测量相关尺寸。

素养目标

(1) 培养学生的科学探究精神和态度。

(2) 培养学生的工程意识。

(3) 培养学生的团队合作能力。

➢任务引入

飞轮是斯特林发动机的重要组成零件。斯特林发动机通过气缸内工作介质(氢气或氦气)经过冷却、压缩、吸热、膨胀为一个周期的循环来输出动力,只有膨胀过程对外输出正功。因此,为保证机器工作周期的正常循环,需要飞轮提供足够的转动惯性。同时,飞轮的存在也使斯特林发动机的转动更加平稳。

根据其工作特点,在设计过程中,应尽量使大部分的质量远离工件的回转中心位置,以保证足够的惯性,所以零件辐板比较薄、刚性比较低,为加工带来一定的难度;在加工过程中,应尽量保证装夹和切削加工内外同心,以保证斯特林发动机在运转过程中保持平稳,并减少能量损失和噪声。

根据零件图(如图 4-1)要求,制定加工工艺、编写数控加工程序,并完成飞轮零件的加工。该零件毛坯材料为 C45 钢。

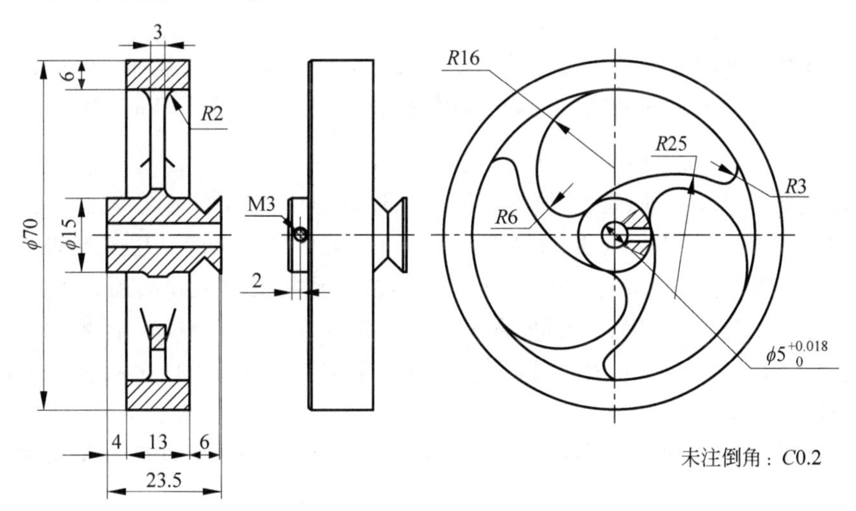

图 4-1　飞轮零件图

任务一　学习关键知识点

4.1　圆形工件的装夹方法

本任务介绍在铣削机床上加工圆形工件的方法。通常,圆形工件装夹可通过三爪自定心卡盘装夹或通过 V 形块装夹两种方式。

1. 三爪自定心卡盘

三爪自定心卡盘由卡盘体、活动卡爪和卡爪驱动机构组成。三爪自定心卡盘上 3 个卡爪导向部分的下面，有螺纹与碟形伞齿轮背面的平面螺纹相啮合，当用扳手通过四方孔转动小伞齿轮时，碟形齿轮转动，背面的平面螺纹同时带动3个卡爪向中心靠近或退出，用以夹紧不同直径的工件。将3个卡爪换成3个反爪，可用来装夹直径较大的工件。三爪自定心卡盘的自行对中精确度为0.05～0.15mm。用三爪自定心卡盘加工工件的精度受到卡盘制造精度和使用后磨损情况的影响。为保证定位精度，通常使用软三爪铣削出与所装夹零件安装面相同的形状，如图4-2所示，这样一方面大幅提高了定准精度，另一方面减少了装夹对工件的损伤。

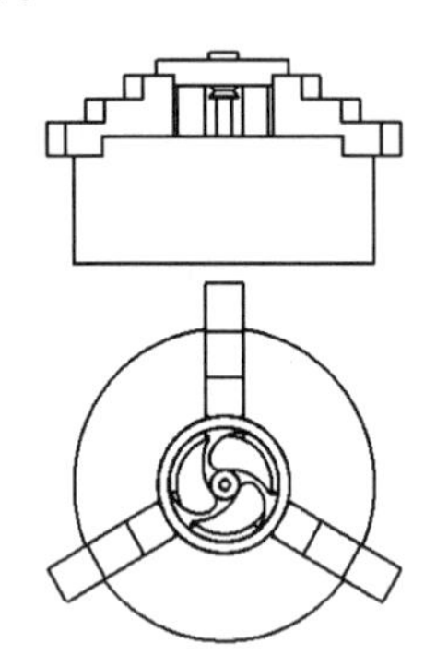

图4-2　三爪自定心卡盘装夹示意图

飞轮零件在装夹时要综合考虑零件的刚性，因加工后辐板比较薄，而且有镂空，所以要尽量将轮辐置于3个卡爪的夹紧位置，并合理控制夹紧力，以减少工件的形变。

2. V形块

V形块主要用来装夹轴、套筒、圆盘等圆形工件，以便圆形工件定位圆心。一般V形块都是一副两块，且两块的平面与V形槽都是在同一次安装中磨出的。精密V形块的尺寸相互表面间的平行度、垂直度误差均在0.01mm之内，V形槽的中心线必须在V形架的对称平面内并与底面平行，同心度、平行度的误差也均在0.01mm之内，V形槽半角误差在±(0.5°～1°)。精密V形块也可做画线，带有夹持弓架的V形块，可以把圆柱形工件牢固地夹持在V形块上，翻转到各个位置画线。

在铣削加工中，V形块配合精密平口钳使用，有的平口钳在出厂时自带V形块，如图4-3所示。

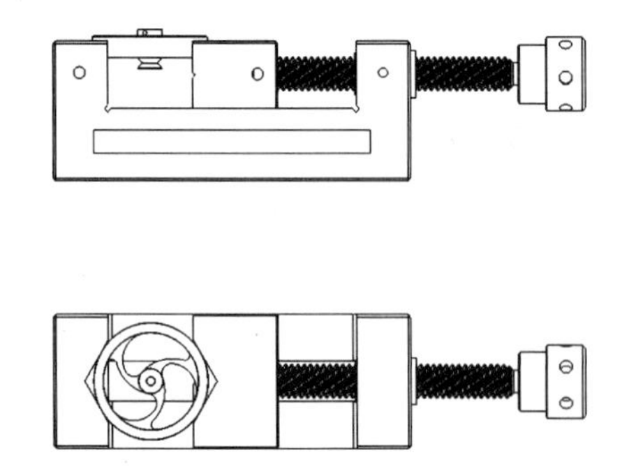

图4-3　平口钳自带V形块装夹示意图

通常，平口钳包括一个定钳口和一个动钳口，所以无自定心功能，真正起作用的V形块为固定钳口一侧的V形块，因此，在安装定位中，工件外圆的公差对定心误差影响较大，图4-4所示为在V形块定位时，定位误差计算的示意图。在加工中需要考虑由于飞轮外圆的加工误差δ_d所造成的定位中心偏移

$$\overline{O_1O_2}=\frac{d}{2\sin\frac{\alpha}{2}}-\frac{d-\delta_d}{2\sin\frac{\alpha}{2}}=\frac{\delta_d}{2\sin\frac{\alpha}{2}} \tag{4-1}$$

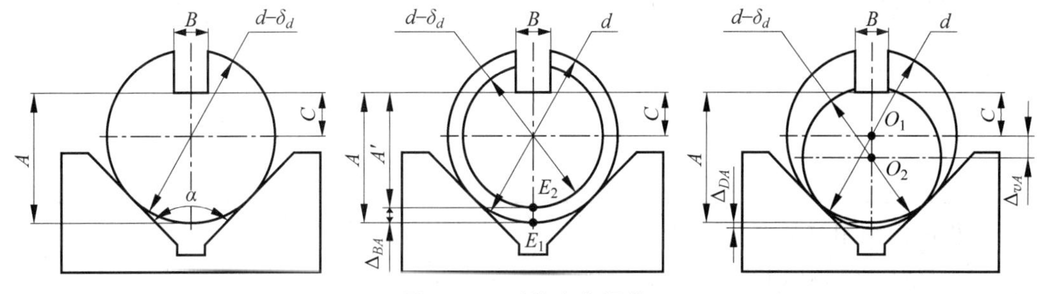

图4-4　V形块定位误差

当V形块装夹圆形工件时，受力部分为“线”接触；当三爪自定心卡盘使用软爪装夹时，受力部分为“面”接触，所以使用三爪自定心卡盘配合软爪装夹更好一些。在条件允许的情况下，推荐在实训过程中使用三爪自定心卡盘进行工件的装夹。

4.2　加工指令

1. 子程序指令(M98、M99)

在一个加工程序中，如果其中有些加工内容完全相同或相似，如图4-5所示，为了简化程序，可以把这些重复的程序段单独列出，并按一定的格式编写成子程序。主程序在执行过程中如果需要某子程序，则通过调用指令来调用该子程序。在子程序执行完毕后，又返回到主程序，主程序继续执行后面的程序段。

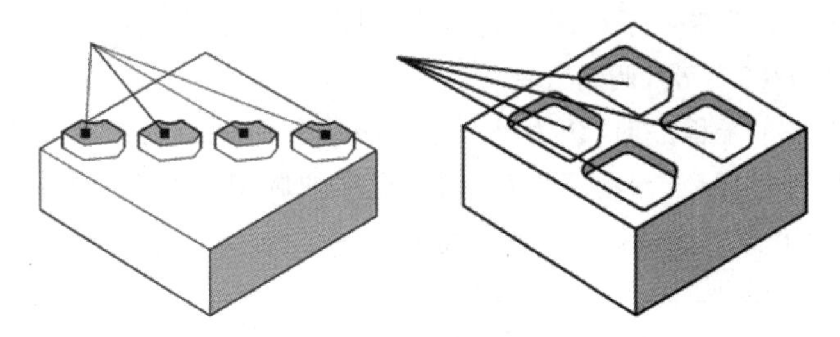

图4-5　子程序的应用

子程序的主要功能如下。

(1) 零件上若干处具有相同的轮廓形状，在这种情况下，只要编写一个加工该轮廓形状的子程序，然后通过用主程序多次调用该子程序的方法，即可完成对工件的加工。

(2) 在加工中，反复出现具有相同轨迹的走刀路线，如果这种相同轨迹的走刀路线出现在某个加工区域，或者在这个区域的各个层面上，则采用子程序编写加工程序比较方便。在程序中，常用增量值确定切入深度。

(3) 在加工较复杂的零件时，往往包含许多独立的工序，有时工序之间需要适当的调整。为了优化加工程序，把每一个独立的工序编成一个子程序，这样就形成了模块式的程序结构，便于对加工顺序进行调整，而在主程序中只有换刀和调用子程序等指令。

FANUC 0*i* MF 系统子程序调用可以嵌套 15 级，其程序构成如表 4-1 所示。

表 4-1　子程序的构成

子　程　序	解 释 说 明
O◇◇◇◇；	子程序名称，命名与主程序要求相似
…	
…	子程序内容
…	
M99；	子程序返回指令，M99 指令可不必构成一个独立的程序段

子程序的调用有如下几种方法。

(1) 调用 4 位数以下程序号的子程序的指令格式为：

　　M98　P◆◆◆◆◇◇◇◇；

或　M98　P◇◇◇◇　L◆◆◆◆◆◆◆◆◆；

其中，◆◆◆◆为调用次数；◇◇◇◇为子程序号。

(2) 调用 5 位数以上程序号的子程序的指令格式为：

　　M98　P◇◇◇◇◇◇◇◇◇　L◆◆◆◆◆◆◆◆◆；

其中，◆◆◆◆◆◆◆◆◆为调用次数；◇◇◇◇◇◇◇◇◇为子程序号。

(3) 以程序名来调用子程序的指令格式为：

　　M98　〈○○○○〉　L◆◆◆◆◆◆◆◆◆；

其中，◆◆◆◆◆◆◆◆◆为调用次数；○○○○为子程序名。

在常规编程中，可能应用的子程序调用基本都是调用 4 位数以下程序号的子程序，所以本课程只需掌握第一种调用方法，其他两种了解即可。

在第一种调用方法中，调用次数为 0001～9999，当调用次数为 1 次时，"1"可以省略，而且次数前边的 0 也可以省略；子程序号应为 4 位数字，子程序号不足 4 位的前边用 0 补齐。例如，调用子程序"O365"5 次，调用方法为：M98 P50365。

当主程序调用一个子程序时，认为是一个 1 级子程序调用，FANUC 0*i* MF 系统最多可以进行 10 级调用。如果配合宏程序指令，则子程序的调用与宏程序的调用一起可以嵌套多达 15 级。图 4-6 所示为子程序调用分级示意图。

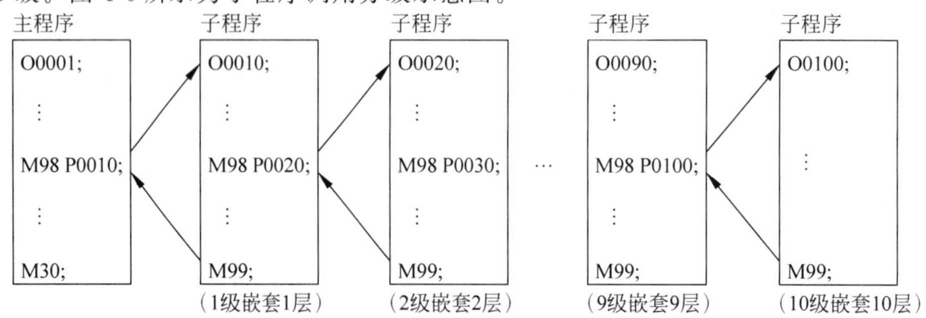

图 4-6　子程序调用分级

2. 坐标系旋转指令(G68、G69)

坐标系旋转指令(G68、G69)用于使编程形状旋转。通过使用这一功能,当安装的工件处在相对于机床旋转的位置上时,即可通过坐标系旋转指令来进行补偿,如图 4-7 所示。此外,当存在一个形状旋转的图形时,通过编写一个形状子程序,在使其旋转后调用该子程序,就可以缩短编程所需的时间和程序长度。

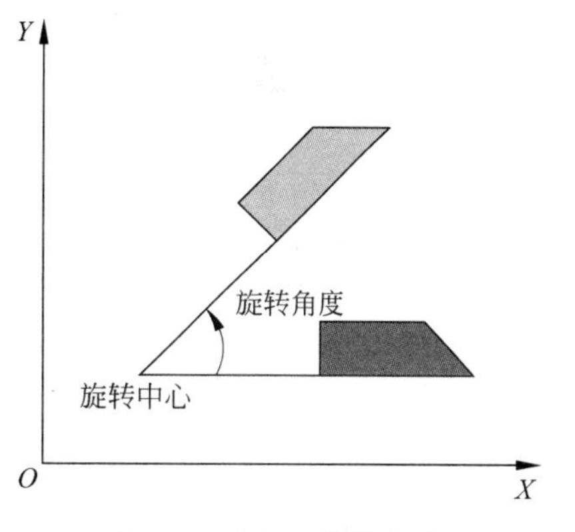

图 4-7　坐标系旋转指令

在 FANUC 0*i* MF 系统中,坐标旋转编程的格式为:

子　程　序	解释说明
G68 X_ Y_ R_;	在 *XOY* 平面内,形状绕旋转中心(x,y)旋转角度 *R*
…	
G69;	取消坐标旋转

在有些版本的系统中,G68 指令可以配合相对值(增量值)编程功能进行相对旋转某一指令角度。角度的单位为(°)。

可通过修改系统参数 RIN(No. 5400#0)来设定坐标旋转是绝对值还是增量值。

3. 坐标变换与子程序配合编程示例

本节以图 4-8 所示零件为例,练习使用子程序指令、坐标系旋转指令编写数控加工程序的基本方法。

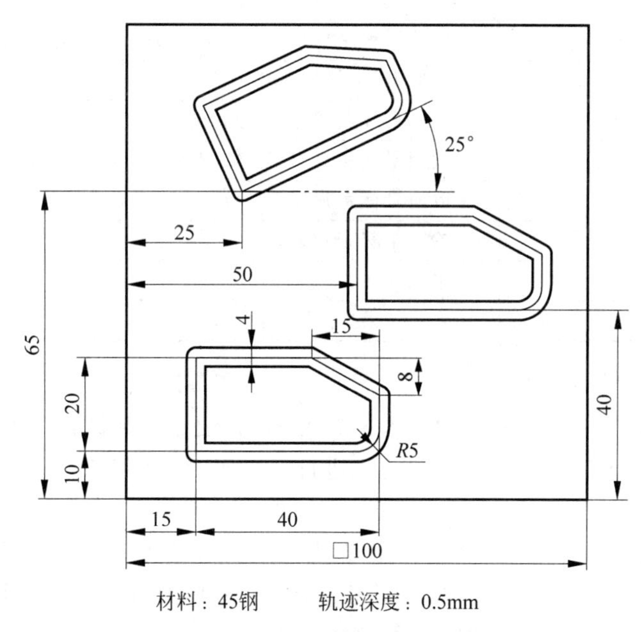

图 4-8　子程序指令、坐标系旋转指令编程练习

首先按照图 4-8 所示中左下角图形,使用相对坐标指令编写子程序,如表 4-2 所示。

表 4-2　子程序

子　程　序	程序注释
O1234；	命名子程序名为"1234"
G1 G90 Z-0.5 F50；	以 F50 速度下刀到 Z-0.5 位置
G1 G91 X40 R5 F500；	沿 X 正方向切削 40mm，并在终点处倒圆角 R5
Y12；	沿 Y 正方向切削 12mm
X-15 Y8；	同时向 X 负方向切削 15mm，Y 正方向切削 8mm
X-25；	沿 X 负方向切削 25mm
Y-20；	沿 Y 负方向切削 20mm，回到起点
G1 G90 Z10；	加工结束，抬刀至 Z10 位置
M99；	子程序结束，返回调用位置

在加工零件时，只需在坐标(15,10)、(50,40)位置直接调用一次子程序即可。在(25,65)位置先将坐标系旋转 25°，再调用一次子程序即可完成形状加工。编写主程序如表 4-3 所示。

表 4-3　主程序

子　程　序	程序注释
O0001；	命名主程序名为"0001"，"0"可以省略
T1 M6；	调用 T1 刀具：ϕ4 键槽立铣刀
S3000 M3；	设置主轴转速为 3000r/min，并使主轴正转
G0 G90 G54 X15 Y10；	在 G54 坐标系下快速定位到绝对坐标(15,10)位置
G43 Z10 H1；	调用 H1 刀具长度补偿，并快速定位到安全高度 Z10 位置
M98 P0011234；	调用子程序"1234"1 次，P 参数可简写为"P1234"
G0 X50 Y40；	快速定位到(50,40)处
M98 P0011234；	调用子程序"1234"1 次，P 参数可简写为"P1234"
G0 X25 Y65；	快速定位到(25,65)处
G68 X25 Y65 R25；	坐标系绕点(25,65)旋转 25°
M98 P0011234；	调用子程序"1234"1 次，P 参数可简写为"P1234"
G69；	取消坐标系旋转
G0 Z100；	抬刀至 Z100 位置
M5；	主轴停转
M30；	程序结束，并返回程序头

4. 铰孔加工指令(G84)

如图 4-9 所示，铰孔加工基本动作为先沿 X 轴、Y 轴定位到铰孔位置，刀具快速进给到点 R 平面；然后从点 R 平面到点 Z 设定点进行铰孔加工；在到达 Z 轴指定位置后，刀具以进给速度 F 回切到点 R 平面。

如要铰多个孔，则在以 G98 指令指定时，加工完一个孔后刀具抬高到初始平面再向下一个孔移动；在以 G99 指令指定时，加工完一个孔后刀具抬高到点 R 平面再向下一个孔移动。所以，在编程时要注意工件加工表面上方有无干涉。

G85 X_Y_Z_R_F_K_;
X_Y_ ：孔位置数据
Z_ ：从点R到孔底的距离
R_ ：从基准平面到点R的距离
F_ ：切削进给速度
K_ ：重复次数（仅限需要重复时）

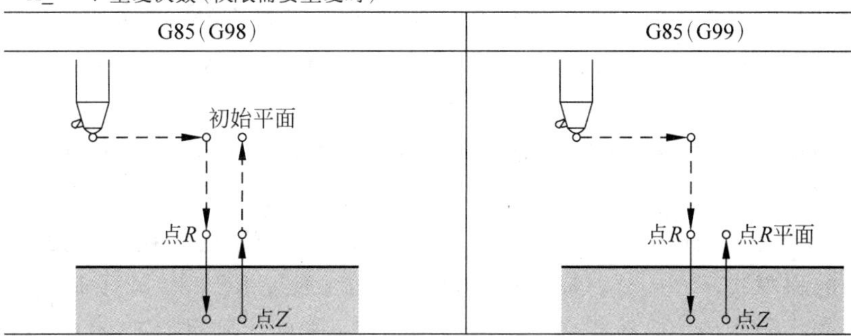

图 4-9　铰孔加工指令

4.3　编程软件简介

CAM 制造工程师软件是北京数码大方科技股份有限公司（以下简称数码大方）（CAXA）开发的用于数控铣削加工编程的软件，数码大方（CAXA）是中国自主的工业软件和工业互联网公司。该公司是“智能化协同制造技术及应用”国家工程实验室承建单位，也是高新技术企业、中关村高新技术企业、中关村示范区信用五星级企业、北京市高新技术成果转化示范企业、北京市专利示范单位、工信部首批智能制造系统供应商。

数码大方始终坚持技术创新，自主研发数字化设计（Computer Aided Design，CAD）、产品全生命周期管理（product lifecycle management，PLM）、数字化制造（manufacturing execution system，MES）软件，是中国早期从事此领域的软件公司。其研发团队拥有多年专业经验积累，具备国际先进技术水平，在北京市、南京市和美国亚特兰大设有 3 个研发中心，目前已拥有 330 余项商标、专利、专利申请及著作权。该公司是国家智能制造标准化总体组成员单位、大数据标准工作组全权成员单位、中国信息技术标准化技术委员会委员、北京标准化协会单位会员，牵头或参与国家智能制造标准、工业云、工业大数据、增材制造等标准体系的建设。

数码大方主要为装备、汽车、电子电器、航空航天、教育等行业提供工业软件、智能制造解决方案、工业云平台等产品和服务。数码大方产品线完整，包括数字化设计、产品全生命周期管理、数字化制造环节，贯通企业设计制造核心环节数字化。

在造型方向，CAM 制造工程师软件秉承实体和曲面混合造型的方法，以及可视化设计理念。实体造型主要有拉伸、旋转、导动、放样、倒角、圆角、打孔、筋板、拔模、分模等特征造型方式，可以将二维的草图轮廓快速生成三维实体模型；还提供多种构建基准平面的功能，用户可以根据已知条件构建各种基准平面。曲面造型提供多种非均匀有理 B 样条（non-uniform rational B-splines，NURBS）曲面造型手段：可通过扫描、放样、旋转、导动、等距、边界和网格等多种形式生成复杂曲面；并提供曲面线裁剪和面裁剪、曲面延伸、按照平均切矢或选定曲面切矢的曲面缝合功能、多张曲面之间的拼接功能；另外，提供强大的曲面过渡功能，可以实现两面、三面、系列面等曲面过渡方式，还可以实现等半径或变半径过渡。系统支持实体与复杂曲面混合的造型方法，应用于复杂零件设计或模具设计。提供曲面裁剪实体功能、曲面加厚成实体、闭合曲面填充生成实体功能。另外，系统还允许将实体的表面生成曲面供用户直接引用。曲面和实体造型方法的完美结合，是CAM制造工程师软件在CAD软件上的一个突出特点。每一个操作步骤，软件的提示区都有操作提示功能，不管是初学者还是具有丰富CAD软件经验的工程师，都可以根据软件的提示迅速掌握诀窍，设计出自己想要的零件模型。

在数控编程方面，数控铣加工编程模块具有方便的代码编辑功能，简单易学，非常适合手工编程使用。同时，支持自动导入代码和手工编写代码，其中包括宏程序代码的轨迹仿真，能够有效验证代码的正确性。支持多种系统代码的相互后置转换，实现加工程序在不同数控系统上的程序共享。此外，还具有通信传输功能，通过RS 232口可实现数控系统与编程软件之间的代码互传。系统提供7种粗加工方式：平面区域粗加工(2D)、区域粗加工、等高粗加工、扫描线、摆线、插铣、导动线(2.5轴)；提供14种精加工方式：平面轮廓、轮廓导动、曲面轮廓、曲面区域、曲面参数线、轮廓线、投影线、等高线、导动、扫描线、限制线、浅平面、三维偏置、深腔侧壁；提供3种补加工方式：等高线补加工、笔式清根、区域补加工；提供2种槽加工方式：曲线式铣槽、扫描式铣槽。在多轴加工方面，不仅可以进行四轴曲线、四轴平切面加工，而且支持五轴等参数线、五轴侧铣、五轴曲线、五轴曲面区域、五轴G01钻孔、五轴定向、转四轴轨迹等加工对叶轮、叶片类零件。除以上这些加工方法以外，系统还提供专用的叶轮粗加工及叶轮精加工功能，可以实现对叶轮和叶片的整体加工。

CAM制造工程师软件采用Windows操作系统，基于微机平台，采用原创Windows菜单和交互，多语言用户界面，可以轻松流畅地学习和操作。

在数据接口方面，CAM制造工程师软件提供了丰富的数据接口，包括直接读取市场上流行的三维CAD软件，如CATIA、Pro/ENGINEER的数据接口；基于曲面的DXF和IGES标准图形接口；基于实体的STEP标准数据接口；Parasolid几何核心的x-T、x-B格式文件；ACIS几何核心的SAT格式文件；面向快速成型设备的STL，以及面向Internet和虚拟现实的VRML等接口。这些接口保证了与世界流行的CAD软件进行双向数据交换，使企业可以跨平台和跨地域地与合作伙伴实现虚拟产品的开发和生产。

4.3.1 基本界面介绍

CAXA 3D实体设计的设计环境是完成各种设计任务的窗口，提供了各种工具及条件。如图4-10所示为CAXA CAM制造工程师软件实体设计环境。

CAXA 3D实体设计环境最上方为快速启动栏、软件名称和当前文件名称。其下方是按照功能划分的菜单栏。中间是设计工作显示区域。工作显示区域上方为多文档标签页，左边显示“设计树”“属性”等，右边是可以自动隐藏的设计元素库。最下方是状态栏，这里主要有操作提示、“视图尺寸”、单位、视向设置、设计模式选择、配置设置等内容。

4.3.2 CAD造型

通过CAXA 3D对飞轮零件进行3D建模，建模过程坐标系与加工尽量相同，以便在加工时直接选用 Z 轴方向为主轴方向。图4-11所示为进行飞轮零件3D建模的环境。

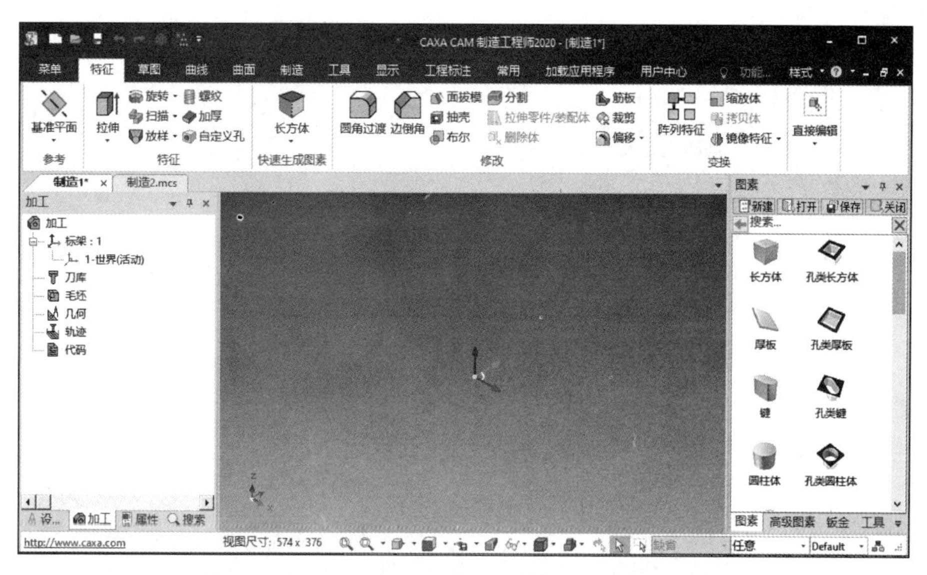

图 4-10　CAXA CAM 制造工程师软件实体设计环境

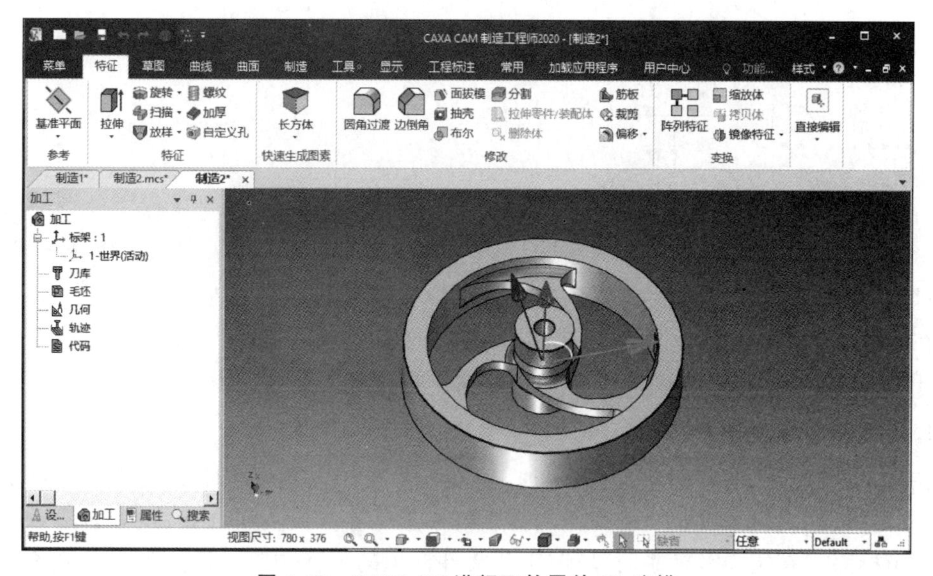

图 4-11　CAXA 3D 进行飞轮零件 3D 建模

本章主要讲述数控编程技术，详细的 3D 建模技术可参考相关资料，此处不再赘述。

4.3.3　CAM 编程

CAM 编程的基本流程如图 4-12 所示。

在进行编程前，应先根据不同加工工艺过程设置合适的加工坐标系，推荐在进行零件建模时，将设计坐标系与加工工艺的加工坐标系重合，以保证加工精度。

在进行刀具定义时，应严格按照工艺要求进行，尽量选择通用刀具，当在加工过程中需要使用多种规格的刀具时，刀具尺寸规格应尽量少，以减少换刀时间。

当零件形状不复杂时，可以不设置加工几何体，直接在模型上选择加工元素。

在生成刀具加工轨迹后，应利用刀具加工轨迹的仿真功能对切削过程进行模拟，观察是

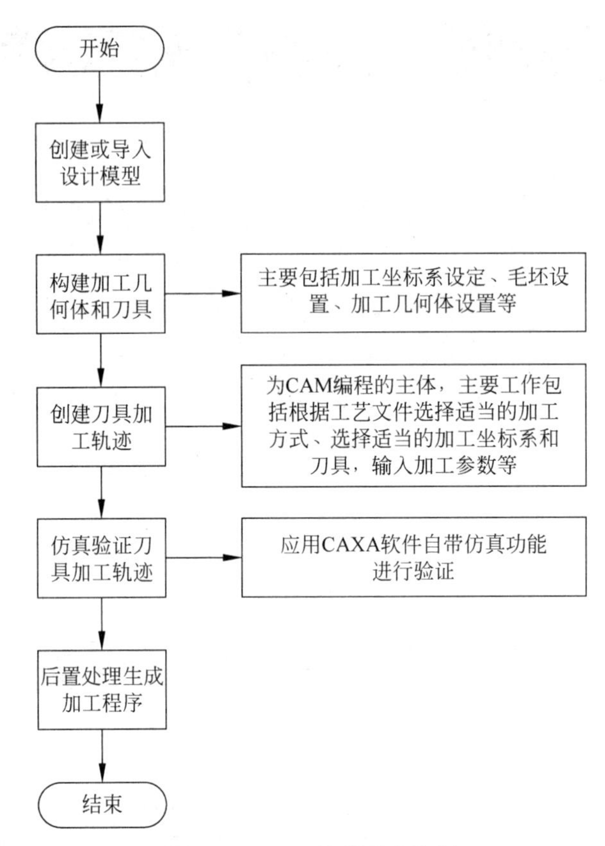

图 4-12　CAM 编程基本流程

否有过切、快速定位过程是否有碰撞、余量过大位置是否需要安排清角等。

在模拟完成后，可后置处理生成机床识别的 G 代码，通过存储介质或分布式数控(distributed numberical control，DNC)将 G 代码程序输入数控机床中。

1. 建立加工坐标系

在窗口左侧导航器界面单击 加工 加工图标，进入“加工”下拉列表，右击坐标系 坐标架：1 图标，单击“创建坐标系”命令，在弹出的对话框中进行参数设置和选择，建立如图 4-13 所示的加工坐标系。在定义坐标系时，应与工艺文件标识的坐标方向一致。在编程时，使用哪个加工坐标系就将哪个坐标系激活。

2. 建立刀具

右击“加工”下拉列表上的 刀库 图标，单击“创建刀具”命令，弹出“创建刀具”对话框，如图 4-14 所示。

刀具的选择和创建要与工艺文件一致，还要注意选择通用刀具。

3. 创建毛坯

为仿真需要，应为加工过程创建毛坯。在“加工”下拉列表中单击 毛坯：1 图标，单击“创建毛坯”命令。如零件比较简单、加工部位较少，则可选择自动生成的毛坯；如零件复杂、潜在的碰撞危险较多，则可运用模型导入法建立毛坯。

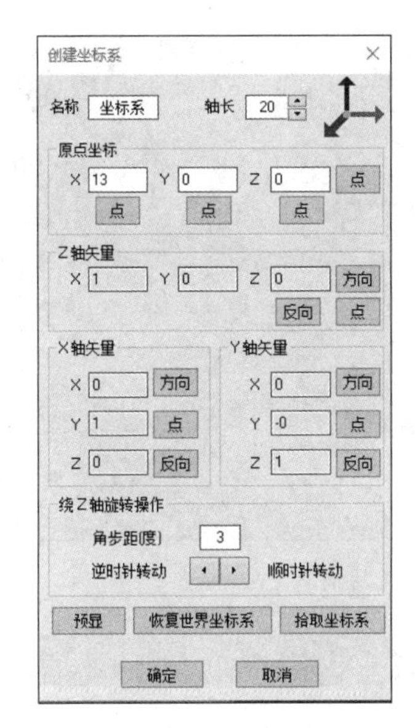

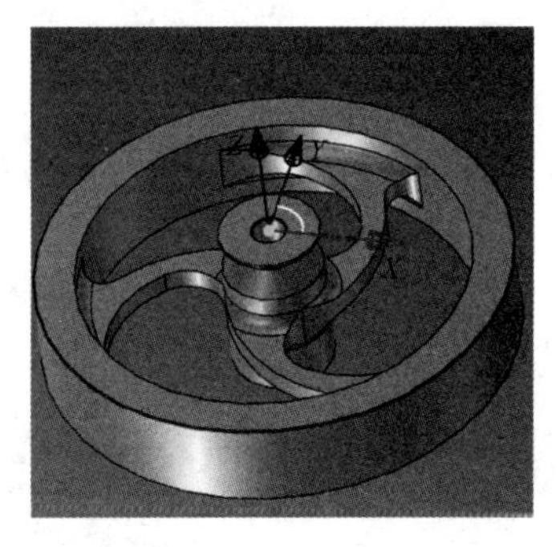

图 4-13　建立加工坐标系

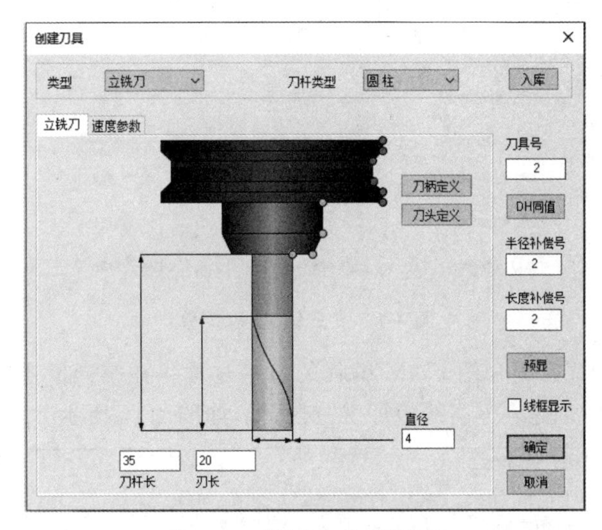

图 4-14　"创建刀具"对话框

4. 建立几何

建立几何的目的是精确指定加工位置。形状简单的零件可直接选择零件的边界线；形状较复杂的零件可通过草图绘制的方法，精确绘制加工边界。图 4-15 所示为其中一个轮辐镂空区域加工时所需要的几何边界。

5. 轨迹生成

通过单击"制造"标签，进入"制造"选项卡，如图 4-16 所示。

本节使用 CAXA CAM 制造工程师软件的三轴加工功能，包含的加工方式如图 4-17 所示。

三轴加工功能包含 2 个粗加工方式和 13 个精加工方式。其中，"等高线粗加工"为常用的粗加工方式，"自适应粗加工"常用于高速铣削。

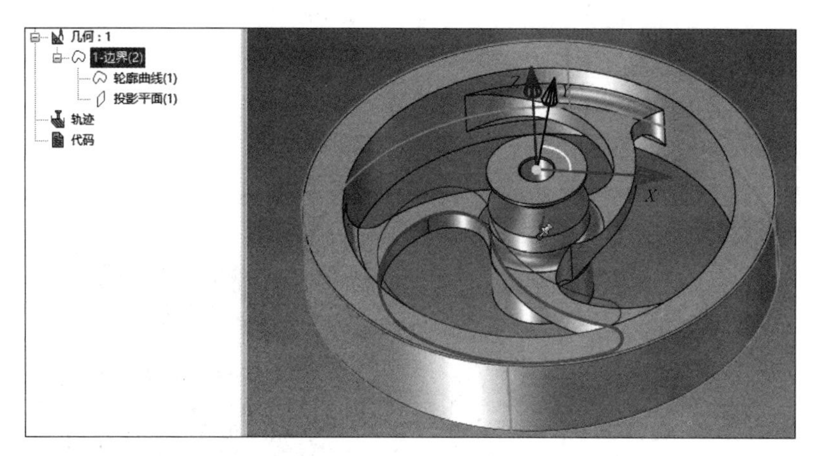

图 4-15　建立几何

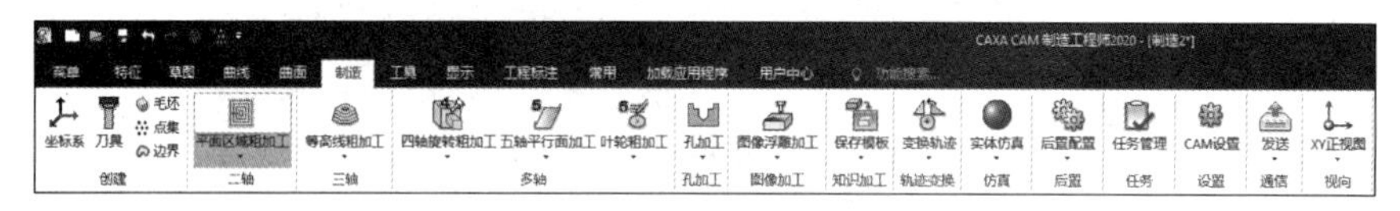

图 4-16　“制造”选项卡

图 4-17　“三轴”下拉菜单

以“等高线粗加工”为例介绍 CAM 编程的基本过程。在“三轴”下拉菜单中单击“等高线粗加工”命令，弹出“创建：等高线粗加工”对话框，如图 4-18 所示。

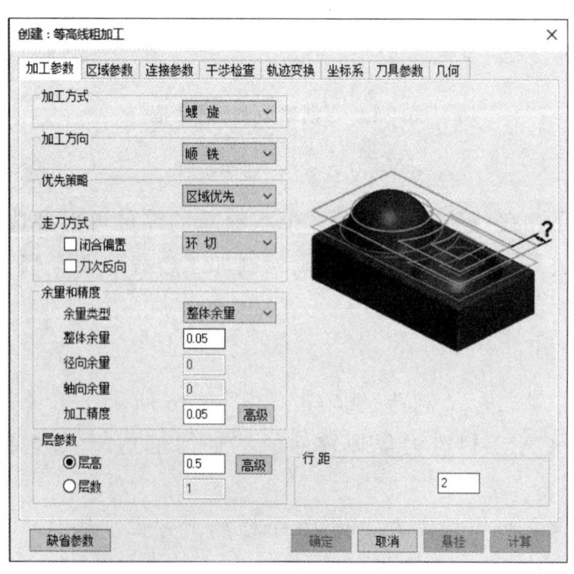

图 4-18　“创建：等高线粗加工”对话框

“加工参数”选项卡用于定义“加工方式”“加工方向”“优先策略”“走刀方式”、加工“余量和精度”、切削“层参数”、切削“行距”等。

“区域参数”选项卡中用于定义加工“高度范围”“起始点”“加工边界”“工件边界”“补加工”等，图 4-19 所示为通过拾取方式定义加工起始高度和终止高度。

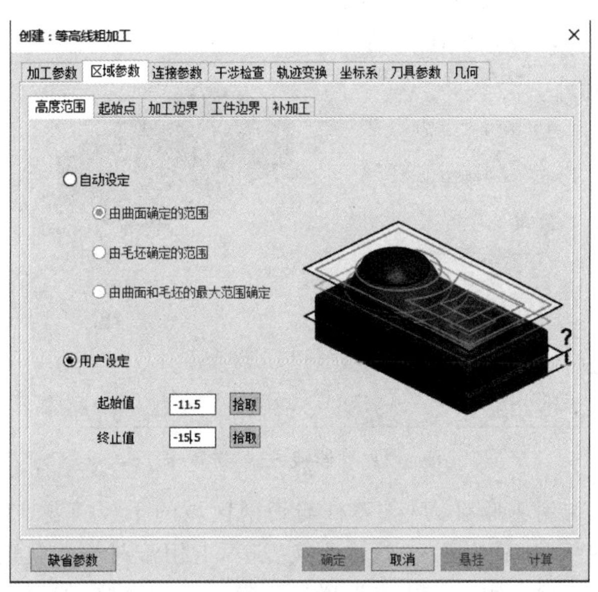

图 4-19　“区域参数”选项卡

利用“加工边界”选项卡确定加工几何边界，如图 4-20 所示。在生成刀具轨迹时，只加工所选择加工边界的内部。

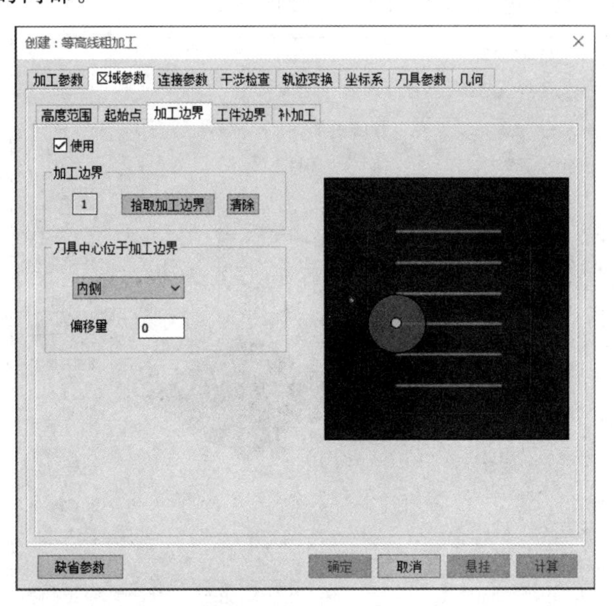

图 4-20　“加工边界”选项卡

“连接参数”选项卡用于设置加工过程各区域的“连接方式”“下刀方式”“空切区域”“空切距离”“光滑”等参数。图 4-21 所示为区域内及区域外“连接方式”选项卡的参数设置。

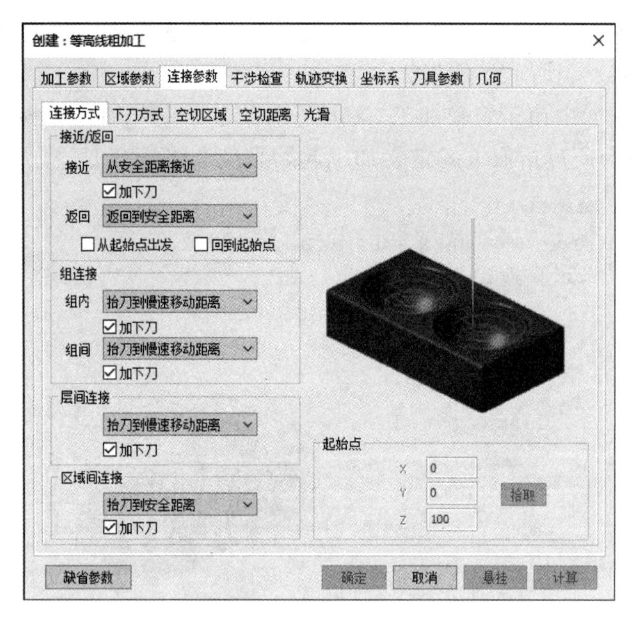

图 4-21 “连接方式”选项卡

“干涉检查”选项卡用于验证刀柄、刀杆与切削区域的干涉问题，在多轴加工中应用较多，三轴及以下加工中可以不设置。“轨迹变换”选项卡用于对刀具轨迹进行平移、旋转、镜像等操作。“坐标系”选项卡用于前边未设置加工坐标系的情况，如第一步已设置并激活会直接带入进来。“几何”选项卡用于选择加工部件和毛坯，可直接选择零件模型和毛坯。

“刀具参数”选项卡用于选择刀具和定义切削要素，一般在进行刀具定义时已经完成，如有需要，则可在速度参数中进行修改，如图 4-22 所示。

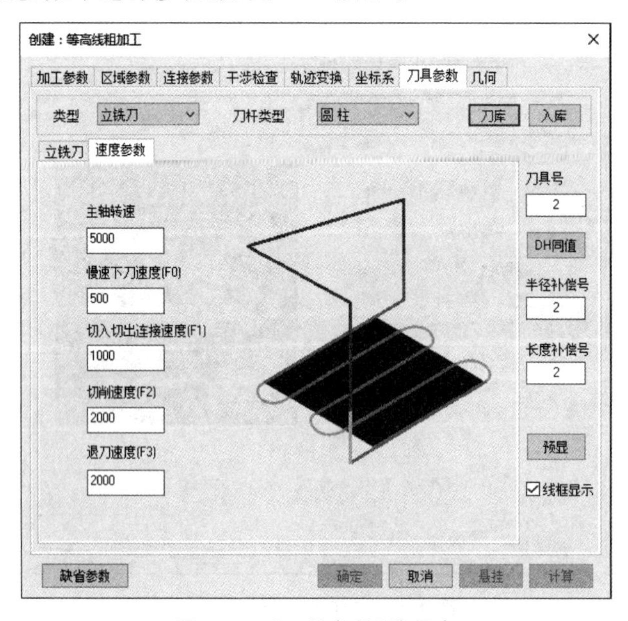

图 4-22 “刀具参数”选项卡

在设置完成后，单击“确定”按钮，生成的刀具轨迹如图 4-23 所示，如需修改“加工参数”，则要单击“计算”按钮，进行刀具轨迹重新生成。

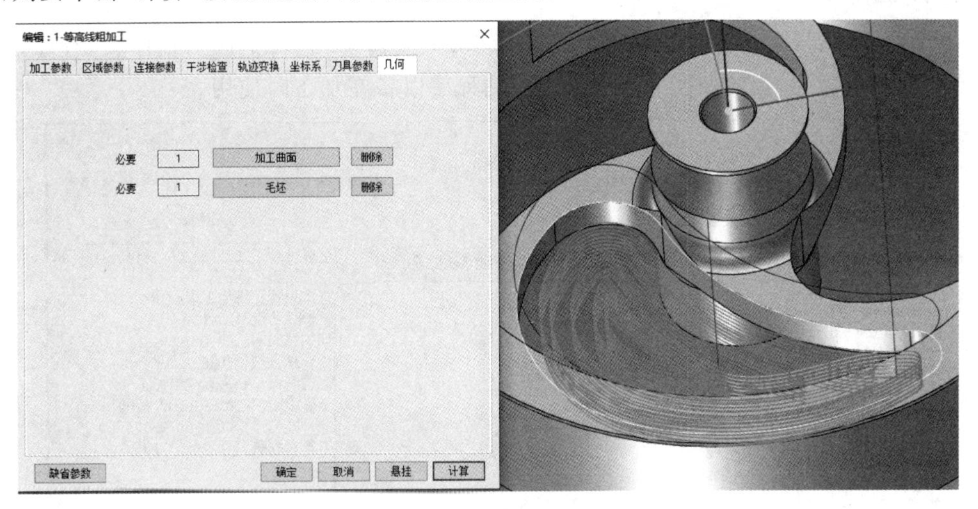

图 4-23　刀具轨迹

6. 仿真加工

右击刀具轨迹，单击“实体仿真”命令，在弹出的对话框中选择用于仿真的刀具轨迹名称，并拾取毛坯和零件用于验证材料清除情况，观察有无过切。

7. 后置处理

右击刀具轨迹，单击“后置处理”命令，在“控制系统”列表框中选择 Fanuc，在“机床配置”中选择“铣加工中心_3X”，拾取要进行后置处理的刀具轨迹，定义加工时的起止高度，单击“后置”按钮，如图 4-24 所示。

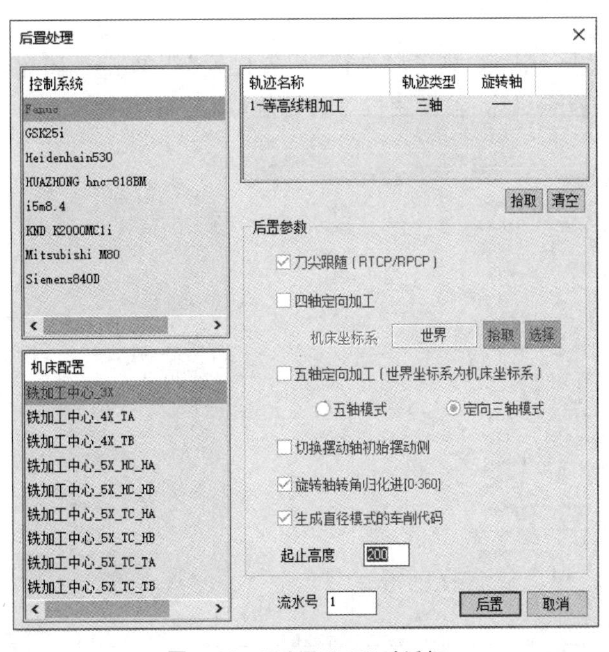

图 4-24　“后置处理”对话框

8. 编辑代码

在后置处理后，自动弹出“编辑代码”对话框，可按需要，编辑生成的程序头、程序尾、起始位置、切削参数等内容，如图 4-25 所示，存储程序以备在加工时使用。

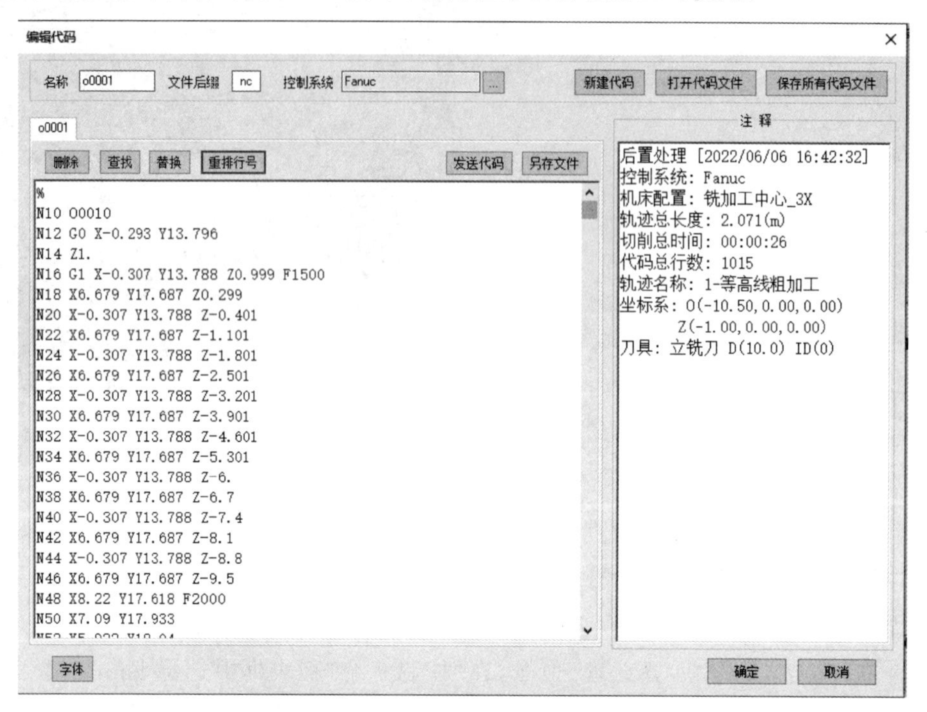

图 4-25 “编辑代码”对话框

9. 后置设置

若对生成的 G 代码不满意，则可以进行“后置设置”。单击 后置配置 图标，通过修改后置处理文件参数，进行程序输出、机床、输出变量等的控制，如图 4-26 所示。

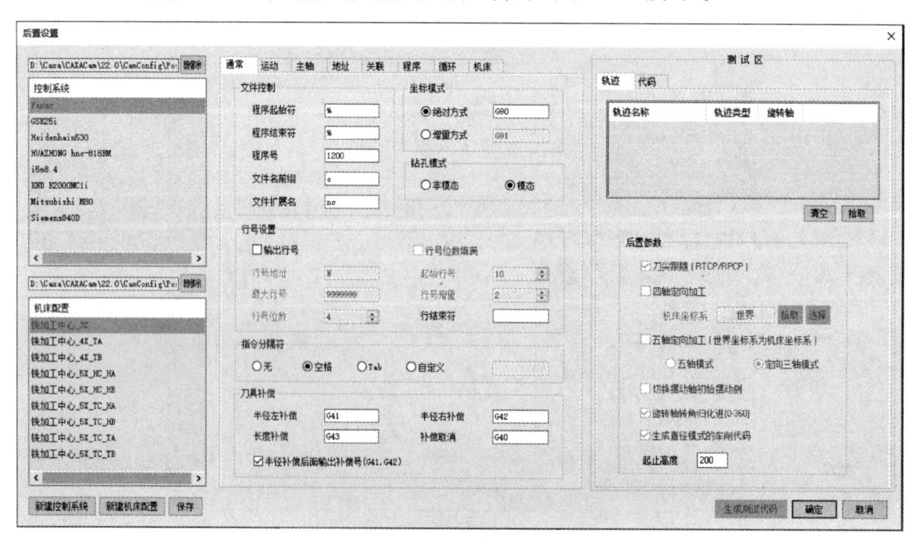

图 4-26 “后置设置”对话框

任务二　工艺准备

4.4　零件图分析

根据零件的使用要求，可以选择45钢作为飞轮零件的毛坯材料，毛坯下料尺寸定为$\phi72\times25$。在实际生产中，可结合自身情况，灵活掌握下料尺寸，便于进行二次加工。

如图4-1所示，采用三爪自定心卡盘装夹，工件高于三爪自定心卡盘顶面的有效高度不应小于17mm，使用$\phi12$立铣刀粗、精加工顶面、外圆；使用$\phi12R2$圆角立铣刀加工辐板左侧。在翻转后，使用三爪自定心卡盘采用软爪装夹，加工零件右侧部分，并钻、铰中心孔。最后钳工加工M3顶丝孔，M3孔钳工加工本书不再介绍。

4.5　工艺设计

根据零件图分析，确定工艺过程，如表4-4所示。

表4-4　工艺过程卡片

<table>
<tr><td colspan="2" rowspan="2">机械加工
工艺过程卡片</td><td>产品型号</td><td>STL-01</td><td colspan="2">零部件序号</td><td colspan="4">FL-01</td><td>第1页</td></tr>
<tr><td>产品名称</td><td>斯特林发动机</td><td colspan="2">零部件名称</td><td colspan="4">飞轮</td><td>共1页</td></tr>
<tr><td>材料牌号</td><td>C45</td><td>毛坯规格</td><td>$\phi72\times25$</td><td>毛坯质量</td><td colspan="2">0.8kg</td><td colspan="2">数量</td><td colspan="2">1</td></tr>
<tr><td rowspan="2">工序号</td><td rowspan="2">工序名</td><td colspan="2" rowspan="2">工序内容</td><td rowspan="2">工段</td><td colspan="3" rowspan="2">工艺装备</td><td colspan="3">工时/min</td></tr>
<tr><td colspan="2">准结</td><td>单件</td></tr>
<tr><td>5</td><td>备料</td><td colspan="2">按$\phi72\times25$尺寸备料</td><td>外购</td><td colspan="3">锯床</td><td colspan="2"></td><td></td></tr>
<tr><td>10</td><td>铣加工</td><td colspan="2">使用三爪自定心卡盘装夹$\phi72$圆柱，伸出有效长度大于17mm；铣加工$\phi15$顶面、$\phi70$外圆及侧面、辐板侧面，使用成型铣刀，保证$R2$尺寸同时加工完成</td><td>铣</td><td colspan="3">加工中心
游标卡尺</td><td colspan="2">60</td><td>45</td></tr>
<tr><td>15</td><td>铣加工</td><td colspan="2">使用三爪自定心卡盘采用软三爪装夹$\phi70$外圆，铣削零件顶面，保证尺寸为23.5mm，铣削轮辐保证尺寸为3mm、7mm；使用$\phi4$立铣刀，加工辐板镂空；采用成形铣刀加工90°槽口；最后钻铰$\phi5$孔</td><td>铣</td><td colspan="3">加工中心
游标卡尺</td><td colspan="2">90</td><td>60</td></tr>
<tr><td>20</td><td>钳加工</td><td colspan="2">钻床加工M3底孔到$\phi2.5$；攻丝完成M3紧固螺纹孔加工</td><td>钳</td><td colspan="3">加工中心
游标卡尺</td><td colspan="2">15</td><td>10</td></tr>
<tr><td>25</td><td>清理</td><td colspan="2">清理工件，锐角倒钝</td><td>钳</td><td colspan="3"></td><td colspan="2">15</td><td>5</td></tr>
<tr><td>30</td><td>检验</td><td colspan="2">检验工件尺寸</td><td>检</td><td colspan="3"></td><td colspan="2">15</td><td>5</td></tr>
<tr><td></td><td></td><td colspan="2"></td><td></td><td colspan="3"></td><td colspan="2"></td><td></td></tr>
<tr><td></td><td></td><td colspan="2"></td><td></td><td colspan="3"></td><td colspan="2"></td><td></td></tr>
<tr><td></td><td></td><td colspan="2"></td><td></td><td colspan="3"></td><td colspan="2"></td><td></td></tr>
</table>

本训练任务针对第10工序铣加工，进行工序设计，制订工序卡片，如表4-5所示。

表 4-5 第 10 工序铣加工工序卡片

机械加工工序卡片	产品型号	STL-01	零部件序号	FL-01	第 1 页
	产品名称	斯特林发动机	零部件名称	飞轮	共 1 页

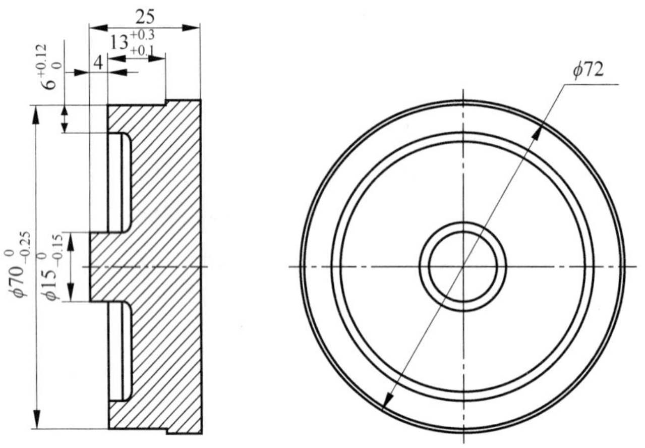

项目	内容
工序号	10
工序名	铣加工
材料	C45
设备	加工中心
设备型号	VAL650e
夹具	三爪自定心卡盘
量具	游标卡尺
准结工时	60min
单件工时	45min

工步	工步内容	刀具	S/(r/min)	F/(mm/r)	a_p/mm	a_e/mm	工步工时/min 机动	辅助
1	工件安装，保证有效伸出≥17.3mm							5
2	铣削 ϕ72 上表面，保证 ϕ15 圆柱端面平整	ϕ12 立铣刀	1500	500	0.5	6	5	
3	粗铣 ϕ70 端面，深度为 4mm，底面留 0.1mm，侧面留 0.1mm	ϕ12 立铣刀	1500	500	2	6	10	
4	精铣 ϕ70 端面，深度为 4mm，底面留 0mm，侧面留 0.1mm	ϕ12 立铣刀	2500	500	0.1	6	5	
5	粗铣轮辐左侧，底面留 0.1mm，侧面留 0.2mm	ϕ10R2 立铣刀	1500	500	0.5	5	10	
6	精铣轮辐左侧，保证尺寸为 6mm、ϕ15	ϕ10R2 立铣刀	2500	500	0.1	5	5	
7	锐角倒钝							2
8	拆卸、清理工件							3
9								
10								

根据第 10 工序，本训练任务针对第 15 工序铣加工，进行工序设计，制订工序卡片，如表 4-6 所示（表头部分已完成）。

提示：辐板镂空（3 处）采用 CAXA CAM 制造工程师软件进行编程。

表 4-6　第 15 工序铣加工工序卡片

机械加工工序卡片	产品型号	STL-01	零部件序号	FL-01	第 1 页
	产品名称	斯特林发动机	零部件名称	飞轮	共 1 页

3　6　R2　$\phi70$　$\phi15$　13　6　23.5　R16　R25　R3　R6　$\phi5^{+0.018}_{0}$	工序号	15
未注倒角：C0.2	工序名	铣加工
	材料	C45
	设备	加工中心
	设备型号	VAL650e
	夹具	三爪自定心卡盘
	量具	游标卡尺
	准结工时	90min
	单件工时	60min

工步	工步内容	刀具	S/(r/min)	F/(mm/r)	a_p/mm	a_e/mm	工步工时/min	
							机动	辅助
1								
2								
3								
4								
5								
6								
7								
8								
9								
10								

4.6　数控加工程序编写

根据第 10 工序铣加工工序卡片，编写第 10 工序铣加工数控程序，如表 4-7 所示。

表 4-7　第 10 工序铣加工数控程序

段　号	程序语句	注　释
	O0001;	
	T1 M6;	
	S1500 M3;	
	G0 G90 G54 X45 Y0;	
	G43 Z10 H1;	
N1	G1 Z0.1 F500;	粗加工下刀

续表

段　号	程序语句	注　释
	G1 X36；	第 1 刀
	G2 I-36 J0；	第 1 刀铣平面，ϕ72 圆
	G1 X30；	第 2 刀
	G2 I-30 J0；	第 2 刀铣平面，ϕ60 圆
	G1 X24；	第 3 刀
	G2 I-24 J0；	第 4 刀铣平面，ϕ48 圆
	G1 X18；	第 5 刀
	G2 I-18 J0；	第 5 刀铣平面，ϕ36 圆
	G1 X12；	第 6 刀
	G2 I-12 J0；	第 6 刀铣平面，ϕ24 圆
	G1 X6；	第 7 刀
	G2 I-6 J0；	第 7 刀铣平面，ϕ12 圆
	G1 X0；	加工中心孔
	G1 X45；	返回铣平面起点
	G1 Z0 F500；	精加工下刀
	G1 X36；	第 1 刀
	G2 I-36 J0；	第 1 刀铣平面，ϕ72 圆
	G1 X30；	第 2 刀
	G2 I-30 J0；	第 2 刀铣平面，ϕ60 圆
	G1 X24；	第 3 刀
	G2 I-24 J0；	第 4 刀铣平面，ϕ48 圆
	G1 X18；	第 5 刀
	G2 I-18 J0；	第 5 刀铣平面，ϕ36 圆
	G1 X12；	第 6 刀
	G2 I-12 J0；	第 6 刀铣平面，ϕ24 圆
	G1 X6；	第 7 刀
	G2 I-6 J0；	第 7 刀铣平面，ϕ12 圆
	G1 X0；	加工中心
	G1 Z10；	精加工完成，抬刀至 Z10 位置
N2	G0 X45 Y0；	加工 13 侧面
	G1 Z0.1；	
	M98 P0080010；	调用子程序“0010”4 次，粗加工 13 侧面
	G1 Z-3.5；	
	M98 P0010010；	调用子程序“0010”4 次，精加工 13 侧面
N3	G1 Z-17；	加工 ϕ70 外圆
	G1 G41 X45.1 Y10 D1；	粗加工阶段，调用刀具半径补偿指令
	G3 X35.1 Y0 R10；	圆弧进刀
	G2 I-35.1 J0；	粗加工切削外圆
	G3 X45.1 Y-10 R10；	圆弧退刀
	G1 G40 X45 Y0；	取消刀具半径补偿指令
	G1 G41 X44.95 Y10 D1；	精加工阶段，调用刀具半径补偿指令
	G3 X34.95 Y0 R10；	圆弧进刀

续表

段　号	程序语句	注　释
	G2 I-34.95 J0；	精加工切削外圆
	G3 X44.95 Y-10 R10；	圆弧退刀
	G1 G40 X45 Y0；	取消刀具半径补偿指令
	Z10；	抬刀
	G0 Z100；	
	M5；	
	T2 M6；	调用 $\phi 10R2$ 圆弧立铣刀
	S1500 M3；	
	G0 G90 G54 X18 Y0；	定准到加工起点
	G43 Z10 H2；	
N4		粗精加工辐板左侧
	G1Z-4 F500；	运动到切削起点
	M98 P0100011；	调用切削子程序“0011”10 次，进行粗、精加工
	G1 Z10；	
	G0 Z100；	
	M5；	
	M30；	
	%	
	%	
	O0010；	加工 13 左侧面子程序
	G1 G91 Z-0.5 F500；	
	G1 X36；	第 1 次进刀
	G2 I-36 J0；	
	G1 X30；	第 2 次进刀
	G2 I-30 J0；	
	G1 X24；	第 3 次进刀
	G2 I-24 J0；	
	G1 X18；	第 4 次进刀
	G2 I-18 J0；	
	G1 X13.6；	加工最后一圈，单面留 0.1mm 加工余量
	G2 I-13.6 J0；	
	G1 X45 Y0；	返回起点
	M99；	子程序返回
	%	
	O0011；	加工辐板左侧面
	G3 G91 Z-0.5 I-18 J0；	螺旋下刀，每层为 0.5mm
	G3 G90 I-18 J0；	铣平螺旋下刀层
	G1 G41 X18.94 Y-10 D2；	调用刀具半径补偿指令
	G3 X28.94 Y0 R10；	圆弧进刀
	I-28.94 J0；	加工外沿
	X18.94 Y10 R10；	圆弧退刀
	G1 G40 X18 Y0；	取消刀具半径补偿指令

续表

段　　号	程 序 语 句	注　　释
	G1 G41 X17.5375 Y10；	调用刀具半径补偿指令
	G3 X7.5375 Y0 R10；	圆弧进刀
	G2 I-7.5375 J0；	切削加工 ϕ15
	G3 X7.5375 Y-10 R10；	圆弧退刀
	G1 G40 X18 Y0；	取消刀具半径补偿指令
	M99；	子程序返回
	%	

根据第 15 工序铣加工工序卡片，编写第 15 工序铣加工数控程序，如表 4-8 所示。本程序为手工编程与自动编程结合生成，在实际生产中，可以根据需要进行改动。

表 4-8　第 15 工序铣加工数控程序

段　　号	程 序 语 句	注　　释
	O0002；	
	T1 M6；	建议对刀时自 ϕ70 侧面（定位面）上返 19mm 设置 Z0
	S1500 M3；	
	G0 G90 G54 X45 Y0；	
	G43 Z10 H1；	
N1	G1 Z0.1 F500；	粗加工下刀
	G1 X36；	第 1 刀
	G2 I-36 J0；	第 1 刀铣平面，ϕ72 圆
	G1 X30；	第 2 刀
	G2 I-30 J0；	第 2 刀铣平面，ϕ60 圆
	G1 X24；	第 3 刀
	G2 I-24 J0；	第 4 刀铣平面，ϕ48 圆
	G1 X18；	第 5 刀
	G2 I-18 J0；	第 5 刀铣平面，ϕ36 圆
	G1 X12；	第 6 刀
	G2 I-12 J0；	第 6 刀铣平面，ϕ24 圆
	G1 X6；	第 7 刀
	G2 I-6 J0；	第 7 刀铣平面，ϕ12 圆
	G1 X0；	加工中心孔
	G1 X45；	返回铣平面起点
	G1 Z0 F500；	精加工下刀
	G1 X36；	第 1 刀
	G2 I-36 J0；	第 1 刀铣平面，ϕ72 圆
	G1 X30；	第 2 刀
	G2 I-30 J0；	第 2 刀铣平面，ϕ60 圆
	G1 X24；	第 3 刀
	G2 I-24 J0；	第 4 刀铣平面，ϕ48 圆
	G1 X18；	第 5 刀
	G2 I-18 J0；	第 5 刀铣平面，ϕ36 圆

续表

段　号	程序语句	注　释
	G1 X12;	第 6 刀
	G2 I-12 J0;	第 6 刀铣平面，ϕ24 圆
	G1 X6;	第 7 刀
	G2 I-6 J0;	第 7 刀铣平面，ϕ12 圆
	G1 X0;	加工中心孔
	G1 Z10;	精加工完成，抬刀至 Z10 位置
N2	G0 X45 Y0;	加工 13 侧面
	G1 Z0.1;	
	M98 P0130010;	调用子程序“0010”4 次，粗加工 13 右侧面
	G1 Z-6;	
	M98 P0010010;	调用子程序“0010”4 次，精加工 13 右侧面
	T2 M6;	调用 $\phi 10R2$ 圆弧立铣刀
	S1500 M3;	
	G0 G90 G54 X18 Y0;	定准到加工起点
	G43 Z10 H2;	
N3		粗精加工辐板左侧
	G1 Z-6 F500;	运动到切削起点
	M98 P0100011;	调用切削子程序“0011”10 次，进行粗、精加工
	G1 Z10;	
	G0 Z100;	
	M5;	
	T3 M6;	调用 ϕ20 正 90°片铣刀加工尖槽
	S800 M3;	
	G0 G90 G54 X20 Y0;	
	G43 Z10 H3;	
N4	G1 Z-3.5 F200;	下刀，90°刀尖对齐尖槽
	G1 G41 X17.5 Y0 D3;	横向进刀
	G2 I-17.5 J0;	加工外层圆周
	G1 X17;	第 2 次进刀
	G2 I-17 J0;	
	G1 X16.5;	第 3 次进刀
	G2 I-16.5 J0;	
	G1 X16;	第 4 次进刀
	G2 I-16 J0;	
	G1 X15.5;	第 5 次进刀
	G2 I-15.5 J0;	
	G1 X15;	第 6 次进刀
	G2 I-15 J0;	
	G1 X14.5;	第 7 次进刀
	G2 I-14.5 J0;	
	G2 I-14.5 J0;	精修圆周
	G3 X24.5 Y-10 R10;	圆弧切出

续表

段　号	程序语句	注　释
	G1 G40 Y0；	取消刀具半径补偿指令
	Z10；	
	G0 Z100；	
	M5；	
	T4 M6；	调用 ϕ4 立铣刀加工辐板镂空
	S5000 M3；	
	G0 G90 G54 X0 Y0；	
	G43 Z10 H4；	
N5	P0010012；	调用由 CAXA 生成的加工第 1 处镂空子程序
	G68 X0 Y0 R120；	坐标系旋转 120°
	P0010012；	调用由 CAXA 生成的加工第 2 处镂空子程序
	G69；	取消坐标系旋转
	G68 X0 Y0 R240；	坐标系旋转 120°
	P0010012；	调用由 CAXA 生成的加工第 3 处镂空子程序
	G69；	取消坐标系旋转指令
	G0 Z100；	
	M5；	
	T5 M6；	调用 90°中心钻
	S2000 M3；	
	G0 G90 G54 X0 Y0；	
	G43 Z10 H5；	
N6	G83 Z-1 R0.5 Q0.5 F50；	加工 ϕ5 中心孔
	G0 G80 Z100；	
	M5；	
	T6 M6；	调用 ϕ4.8 钻头
	S2000 M3；	
	G0 G90 G54 X0 Y0；	
	G43 Z10 H6；	
N7	G83 Z-27 R0.5 Q1.5 F100；	加工 ϕ5 底孔至 ϕ4.8
	G0 G80 Z100；	
	M5；	
	T7 M6；	调用 ϕ5 铰刀
	S800 M3；	
	G0 G90 G54 X0 Y0；	
	G43 Z10 H7；	
N8	G85 Z-27 R0.5 F50；	加工 ϕ5 孔
	G0 G80 Z100；	
	M5；	
	%	
	O0010；	加工 13 左侧面子程序
	G1 G91 Z-0.5 F500；	
	G1 X36；	第 1 次进刀

续表

段　　号	程序语句	注　　释
	G2 I-36 J0；	
	G1 X30；	第 2 次进刀
	G2 I-30 J0；	
	G1 X24；	第 3 次进刀
	G2 I-24 J0；	
	G1 X18；	第 4 次进刀
	G2 I-18 J0；	
	G1 X13.6；	加工最后一圈，单面留 0.1mm 加工余量
	G2 I-13.6 J0；	
	G1 X45 Y0；	返回起点
	M99；	子程序返回
	%	
	O0011；	加工辐板左侧面
	G3 G91 Z-0.5 I-18 J0；	螺旋下刀，每层为 0.5mm
	G3 G90 I-18 J0；	铣平螺旋下刀层
	G1 G41 X18.94 Y-10 D2；	调用刀具半径补偿指令
	G3 X28.94 Y0 R10；	圆弧进刀
	I-28.94 J0；	加工外沿
	X18.94 Y10 R10；	圆弧退刀
	G1 G40 X18 Y0；	取消刀具半径补偿指令
	G1 G41 X17.5375 Y10；	调用刀具半径补偿指令
	G3 X7.5375 Y0 R10；	圆弧进刀
	G2 I-7.5375 J0；	切削加工 $\phi 15$
	G3 X7.5375 Y-10 R10；	圆弧退刀
	G1 G40 X18 Y0；	取消刀具半径补偿指令
	M99；	子程序返回
	%	
	O0012；	加工辐板镂空子程序，由 CAXA CAM 制造工程师软件自动编程生成，编辑程序头为定位到加工起点，在程序尾加工完抬刀后，加上“M99”指令

任务三　上机训练

4.7　设备与用具

设备：AVL650e 立式加工中心。

刀具：$\phi 12$ 立铣刀、$\phi 10R2$ 圆角铣刀、$\phi 20$ 正 90°片铣刀、90°中心钻、$\phi 4.8$ 麻花钻、$\phi 5$H8 铰刀。

夹具：K11-320 三爪自定心卡盘(配软爪)。

工具：什锦锉刀。

量具：0～150mm 游标卡尺、0.02mm 杠杆百分表(配安装杆)。

毛坯：$\phi72\times25$。

辅助用品：卡盘扳手、橡胶锤、毛刷等。

4.8 开机检查

检查机床外观各部位(如防护罩、脚踏板等部位)是否存在异常；检查机床润滑油、冷却液是否充足；检查刀架、夹具、导轨护板上是否有异物；检查机床面板各旋钮状态是否正常；在开机后，检查机床是否存在报警等。可参考表 4-9 对机床状态进行点检。

表 4-9 机床开机准备卡片

检查项目		检查结果	异常描述
机械部分	主轴部分		
	进给部分		
	换刀机构		
	夹具系统		
电器部分	主电源		
	冷却风扇		
数控系统	电气元件		
	控制部分		
	驱动部分		
辅助部分	冷却系统		
	压缩空气		
	润滑系统		

4.9 加工前准备

(1) 在机床启动后，各坐标轴按操作说明书返回机床原点。

(2) 安装工件，工件伸出三爪自定心卡盘的有效长度须大于 17mm。

(3) 用分中法找正工件中心，并在 G54 坐标系中设定为 $X0$、$Y0$。

(4) 安装刀具。

(5) 对刀。建议以最高的一处卡爪顶部为基准对刀，以确保在加工时刀具与卡爪不会发生干涉。

(6) 程序的编辑。因程序比较长，建议先在微机上进行程序的编辑，在确认检查无误后，再使用存储卡将程序导入数控机床中。

4.10 零件加工

在程序自动运行前，应当在数控机床上进行一次试运行，以确保程序的正确性，防止加工事故的发生。

在图形校验过程验证无问题后，即可进行零件加工。在零件加工前，应详细了解机床的安全操作要求，穿戴好劳动保护服装和用具。在进行零件加工时，应熟悉加工中心各操作按键的功能和位置，了解紧急状况的处置方法。

注意：如已进行图形校验操作，则在操作完成后，必须执行机床回归机械零点操作，然后再进行其他的相关操作。如果未回零点而进行了相关操作，则会造成坐标偏移，甚至出现撞机、程序乱跑等异常现象，从而导致危险。

在操作面板上选择“记忆”模式(MEM)，即程序的自动执行模式，调入需要进行加工的数控加工程序，按循环启动功能按键进行自动加工。

数控加工程序的首次自动运行应在调试状态下进行。机床进给倍率旋钮先调低至0%，按下操作面板上功能按键，通过单击显示屏底部“程序检查”按钮切换到运行状态检查窗口，如图 4-27 所示。

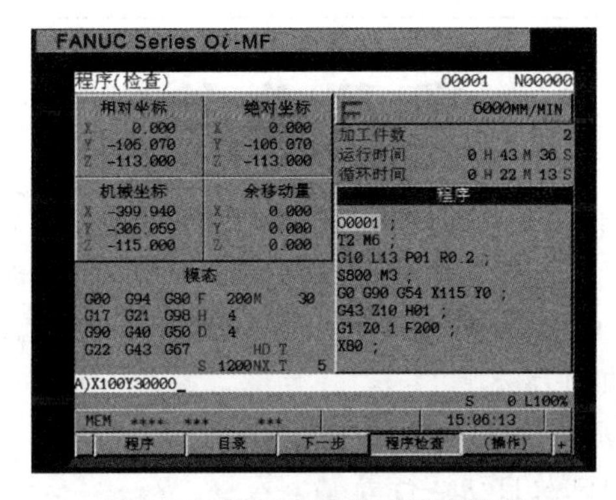

图 4-27　程序自动运行状态检查

在此状态下，每按一次循环启动功能按键程序都只自动执行光标所在的一行。在按循环启动功能按键前，应观察刀具与工件之间的距离是否安全；在按下循环启动功能按键后，通过进给倍率旋钮控制机床的运动速度，同时对照显示屏“剩余移动量”栏显示的剩余移动量，观察刀具与工件之间的实际距离。若实际距离与剩余移动量相差过大，则应果断停机检查，以免发生撞机事故。在程序调试过程中，还应密切注意显示屏的“模态”状态显示，确保主轴转速、进给速度、工件坐标系号、补偿状态及补偿号等无异常发生。

4.11　零件检测

在零件加工完成后，应当认真清理工件，并按照质量管理的相关要求，对加工完成的零件进行相关检验，保证生产质量。机械加工零件“三级”检验卡片如表 4-10 所示。

表 4-10　机械加工零件“三级”检验卡片

零部件图号		零部件名称		工　序　号	
材料		送检日期		工序名称	
检验项目	自检结果	互检结果	专业检验	备注	

续表

零部件图号		零部件名称		工　序　号	
检验结论	□合格　□不合格　□返修　□让步接收 检验签章： 年　月　日				
不符合项描述					

项目总结

通过飞轮零件数控铣削加工，需要掌握数控铣削程序的基本格式和基本切削加工指令的使用方法，能够应用基本切削加工指令、刀具半径补偿指令、子程序，以及坐标变换指令等进行零件切削加工。掌握 G8 * 系列指令在较高精度孔加工中的应用方法。掌握 CAXA CAM 制造工程师软件的使用方法。

掌握立式加工中心的基本操作方法，包括开关机、刀具安装、工件找正、对刀、程序编辑、图形校验、数控加工程序调试及自动运行等。

通过任务训练，养成良好的职业素养，培养正确的加工中心安全操作规范，养成基本的机械加工质量意识。

课后习题

、选择题

1. M98 指令表示(　　)。

A. 返回子程序　B. 调用子程序　C. 冷却液打开　D. 冷却液关闭

2. 利用一般计算工具，运用各种数学方法，人工进行刀具轨迹的运算并进行指令编程称为(　　)。

A. 机械编程　B. 手工编程　C. CAD 编程　D. CAM 编程

3. 子程序返回主程序的指令为(　　)指令。

A. P98　B. M99　C. M08　D. M09

4. V 形架用于工件外圆定位，其中，短 V 形架限制(　　)个自由度。

A. 10　B. 12　C. 2　D. 9

5. 加工中心按照功能特征分类，可分为复合加工中心、(　　)加工中心和钻削加工中心。

A. 刀库＋主轴换刀　B. 卧式　C. 镗铣　D. 三轴

6. 车床上的三爪自定心卡盘和铣床上的平口钳属于(　　)。

A. 通用夹具　　B. 专用夹具　　C. 组合夹具　　D. 随行夹具

7. V形块的主要作用是轴类零件(　　)。

A. 夹紧　　B. 测量　　C. 定位　　D. 导向

二、判断题

1. FANUC系统指令“M98 P0050012”和“M98 P50012”调用子程序的次数是相同的。(　　)

2. 一般规定加工中心的宏编程采用A类宏指令,数控铣床宏编程采用B类宏指令。(　　)

3. 在固定循环功能中的K指令指重复加工次数,一般在增量方式下使用。(　　)

4. 采用通用的三爪自定心卡盘对小型的盘类零件进行工件的装夹,四爪卡盘或花盘适用于中、大型盘类零件的装夹。(　　)

三、填空题

1. 若所有坐标点的坐标值均从某一固定的坐标原点计量,则称为________坐标编程;若运动轨迹的终点坐标相对于线段的起点来计量,则称为________坐标编程。

2. G代码按照在程序段中的续效性可分为________代码和________代码两种。

3. 盘类零件主要是由端面、外圆、内孔、台阶面、槽四轴排列孔等组成,属于同轴会转体,其主要特征是________。

4. 采用CAXA数控车进行零件加工造型的方法可以分为________、________、________三类。

5. 常见的铣削加工形式有________、________、________、________、________、________、________、________、________。

四、简答题

1. 简述数控刀具半径补偿指令的使用方法。

2. 数控加工中心上圆形工件的装夹方法有哪些?各有什么特点?

3. 简述铰孔加工指令G85与钻孔指令G83的区别。

4. 零件加工工序的安排原则有哪些?

自我学习检测评分表如表4-11所示。

表4-11　自我学习检测评分表

项　目	目标要求	分　值	评分细则	得　分	备　注
学习关键知识点	(1) 掌握圆形工件的装夹方法 (2) 掌握子程序指令的使用 (3) 掌握坐标系旋转指令的使用 (4) 能够将坐标变换与子程序配合进行编程 (5) 掌握铰孔加工指令的使用 (6) 了解CAM编程软件,并熟悉CAM编程的基本流程	20	理解与掌握		

续表

项　目	目标要求	分　值	评分细则	得　分	备　注
工艺准备	(1) 能够正确识读零件图 (2) 能够根据零件图分析、确定工艺过程 (3) 能够根据工序加工工艺，编写正确的加工程序	30	理解与掌握		
上机训练	(1) 会正确选择相应的设备与用具 (2) 能够正确操作数控车床，并根据加工情况调整加工参数	50	(1) 理解与掌握 (2) 操作流程		

思政小课堂

项目五　缸体安装座铣削编程加工训练

➢思维导图

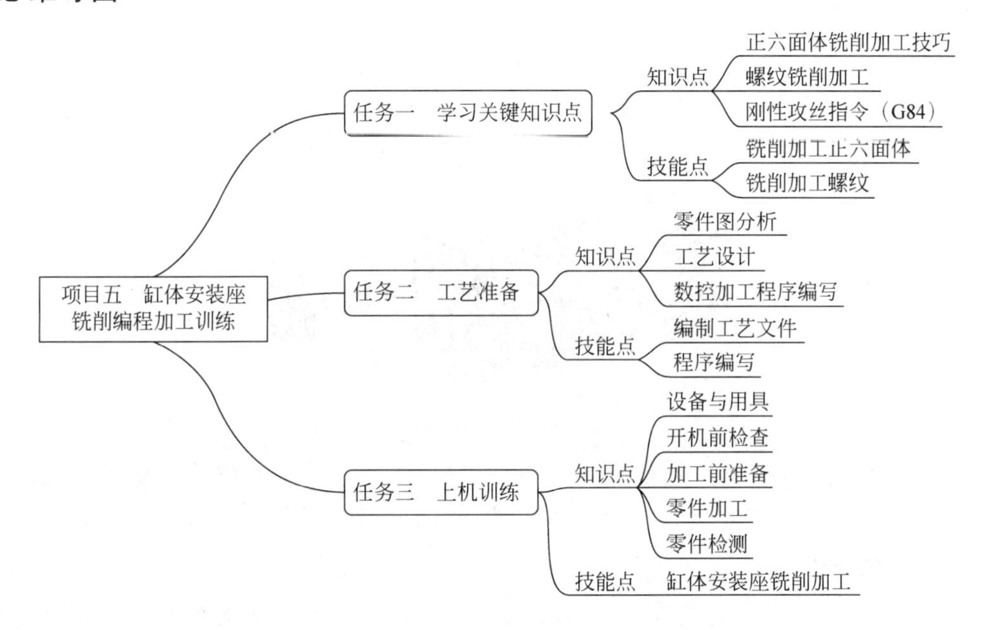

➢学习目标

知识目标

（1）了解缸体安装座的工作原理。

（2）了解缸体安装座在本机构中的用途和关键要求。

能力目标

（1）掌握平板类零件的加工技巧。

（2）掌握螺纹铣削刀具的选择与使用方法。

（3）掌握螺纹铣削加工的特点和编程方法。

（4）能够独立确定加工工艺路线，并正确填写工艺文件。

（5）能够正确操作数控加工中心，并根据加工情况调整加工参数。

（6）能够根据零件结构特点和精度合理选用量具，并正确、规范地测量相关尺寸。

素养目标

（1）培养学生的科学探究精神和态度。

（2）培养学生的工程意识。

（3）培养学生的团队合作能力。

任务一 学习关键知识点

5.1 正六面体铣削加工技巧

正六面体形状看似简单,但往往都有较高的尺寸精度要求和相对位置精度要求。在加工过程中,若加工方法不当,则会造成在产品装配后形状扭曲或无法装配,所以掌握其加工技巧相当重要。

5.1.1 夹具选择

铣削加工正六面体使用的夹具通常为平口钳,如图 5-1 所示,一部精密平口钳包括钳体、固定钳口、活动钳口、夹紧螺杆等主要部分。平口钳组成简练、结构紧凑。用扳手转动螺杆,通过丝杠螺母带动活动钳身移动,达到对工件夹紧与松开的目的。

图 5-1 精密平口钳

在加工时,尤其是在加工互相关联的表面时(如本项目缸体安装座),应事先仔细校正平口钳在工作台的纵向、横向及水平位置,方可进行加工。完整的校正有如下 3 步。

1. 纵向位置的校正

在平口钳内夹一块平行垫铁,在刀架上装一个百分表,使百分表的触头与平行垫铁侧面接触,百分表的压缩量控制在 0.2mm 左右。然后移动滑枕,观察百分表指针是否摆动。如果表针不动,则说明平口钳的纵向位置正确;如果表针摆动,则松开平口钳底座螺母,然后转动平口钳再进行调整,直至表针不动。

2. 横向位置的校正

将平口钳的角度旋转 90°,使百分表仍然与平行垫铁的侧面接触,然后移动工作台,根据表针摆动情况进行调整。

3. 水平位置的校正

平口钳的水平位置也要在横向与纵向两个方向调整。在校正纵向水平时,在平口钳内夹一个角尺,使百分表触头与角尺上棱面接触,然后移动滑枕进行调整。在校正横向水平时,在平口钳钳身滑动面上放一个平行垫铁,然后将百分表触头与平行垫铁上平面接触,移动工作台,根据表针摆动情况进行调整。

在实际生产中,一般先进行水平校正,再选择 X 向、Y 向中的一个方向进行纵向或横向校正,就可以进行加工了。在使用平口钳时还应注意以下几个问题。

(1) 工件的被加工面必须高出钳口,否则就要用平行垫铁垫高工件。

(2) 为了能够装夹牢固,防止在加工时工件松动,必须把比较平整的平面贴紧在垫铁和钳口上。要使工件贴紧在垫铁上,应该一面夹紧、一面用手锤轻击工件的平面。光洁的平面要用铜棒敲击以防敲伤光洁表面。

(3) 在使用垫铁夹紧工件时,要用木锤或铜手锤轻击工件的上表面,使工件贴紧垫铁。在夹紧后,要用手抽动垫铁,如有松动,则说明工件与垫铁贴合不好,在切削时工件可能会移动。这时应松开平口钳重新夹紧。

(4) 在装夹刚性较差的工件时,为了防止工件形变,应先将工件的薄弱部分支撑起来或垫实。

(5) 如果工件按画线加工,则可用画线盘或内卡钳来校正工件。

本项目平口钳的安装如图 5-2 所示。平口钳的底面与机床 XOY 平面平行,固定钳口侧面与 X 轴方向平行。

图 5-2　平口钳的安装

5.1.2　加工顺序

在进行六面体加工时,通常选择面积较大、较平的面先加工,如图 5-3 所示中的①顶面。

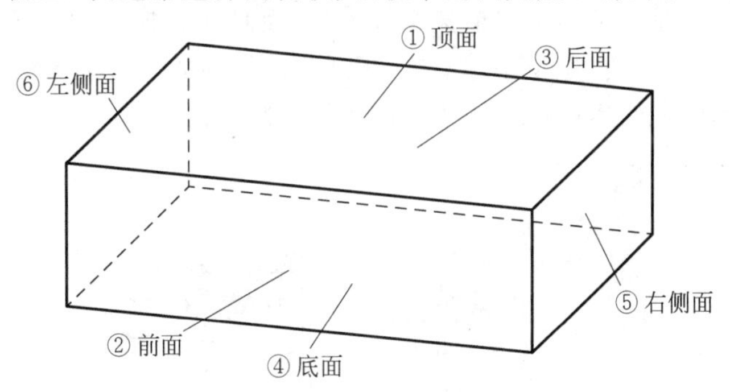

图 5-3　六面体铣削顺序

将铣完的①顶面转到平口钳的固定钳口侧,并与其贴紧,然后加工②前面。

在②前面加工完成后,仍以①顶面接触平口钳固定钳口侧,同时,使刚加工完成的②前面接触平口钳底面,加工③后面,保证零件宽度尺寸。

再以①顶面、②前面、③后面分别接触平口钳底面、固定钳口侧、活动钳口侧,加工零件④底面,保证零件厚度尺寸。

在加工⑤右侧面时有两种方法。当工件较厚时，选择以①顶面、④底面分别接触平口钳固定钳口侧和活动钳口侧，使用角尺或百分表使①顶面或③后面与平口钳底面垂直，铣削⑤右侧面；当工件不太厚时，可选择加工④底面在装夹时，零件侧面伸出平口钳一小段距离，用铣刀侧刃加工⑤右侧面。

最后选择以①顶面、④底面分别接触平口钳固定钳口侧和活动钳口侧，⑤右侧面接触平口钳底面，切削加工⑥左侧面，保证长度尺寸。

注意：当以橡胶锤、木锤或铜棒敲打工件时（工件底侧为铣过的平面），若工件底部有垫铁，则需敲打直到虎钳底部的垫铁不能抽动；若以百分表量测垂直度，则需以百分表的移动量来决定敲打力量，以最快的速度达到垂直度。

5.2 螺纹铣削加工

螺纹铣削加工的研究工作主要集中在理论和工艺两方面。螺纹铣削一般是多次走刀，分粗加工和精加工，但对于难加工的材料，在数控铣削时切削力较大，对加工精度和刀具寿命的影响较大，因此，一般在实际工程中采用的加工策略对于难加工材料的螺纹铣削加工缺乏指导意义。

5.2.1 螺纹铣削刀具

传统的螺纹加工方法主要为采用螺纹车刀车削螺纹或采用丝锥、板牙手工攻丝及套扣。随着数控加工技术的发展，尤其是三轴联动数控加工系统的出现，使更先进的螺纹加工方式——螺纹数控铣削加工得以实现。螺纹铣削加工与传统螺纹加工方式相比，在加工精度、加工效率方面具有极大的优势，且在加工时，不受螺纹结构和螺纹旋向的限制，如图 5-4 所示的螺纹铣刀可加工多种不同旋向的内、外螺纹。对于不允许有过渡扣或退刀槽结构的螺纹，采用传统的车削方法或丝锥、板牙很难加工，但采用数控铣削加工却十分容易实现。此外，螺纹铣刀的耐用度是丝锥的十多倍甚至数十倍，而且在数控铣削加工螺纹过程中，对螺纹直径尺寸的调整极为方便，这是采用丝锥、板牙难以做到的。由于螺纹铣削加工的诸多优势，发达国家的大批量螺纹生产已较广泛地采用了铣削工艺。

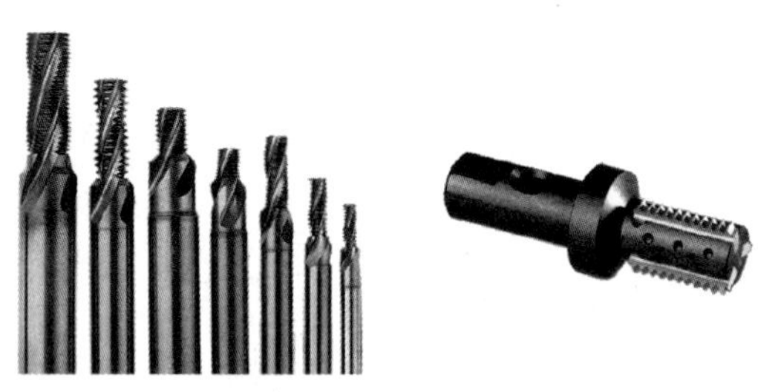

图 5-4 螺纹铣削刀具

5.2.2 螺纹铣削工艺

螺纹铣削加工工艺应注意以下几点。

（1）必须使用螺旋插补铣生成螺距。

（2）根据螺纹类型与加工方法，可使用顺时针或逆时针进给方向（右侧或左侧），生成外螺纹或内螺纹。

（3）推荐使用顺铣。

（4）推荐使用冷却液。在对淬硬材料进行螺纹加工时除外。

（5）同一把刀具可用于加工内、外螺纹与左、右螺纹。

铣削加工内、外螺纹如图 5-5 所示。

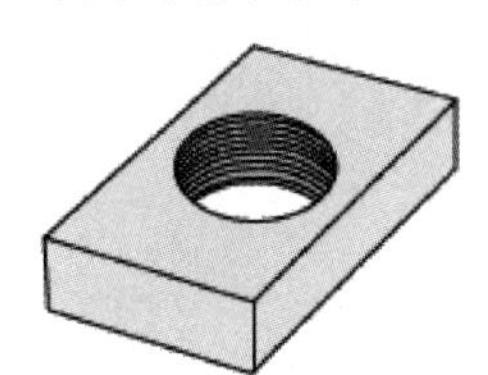

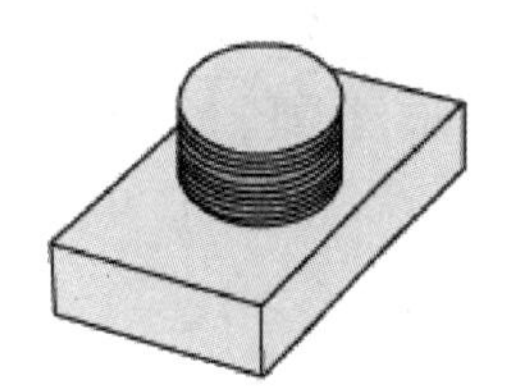

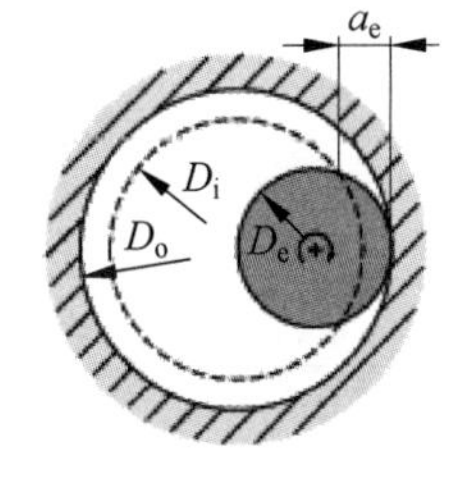

$D_i=D_o-2h$

螺纹	h
ISO	$0.60\times p$
UN	$0.60\times p$
W	$0.69\times p$
BSPT	$0.69\times p$
NPT	$0.78\times p$

径向进给值a_p：

$$a_e=\frac{D_o^2-D_i^2}{4(D_o-D_c)}$$

p=螺距（mm）
h=螺纹深度（mm）
D_c=刀具直径（mm）
D_o=大径（mm）
D_i=小径（mm）

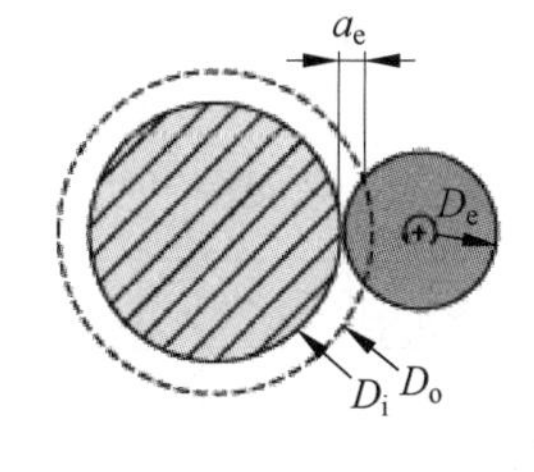

$D_i=D_o-2h$

螺纹	h
ISO	$0.65\times p$
UN	$0.65\times p$
W	$0.69\times p$
BSPT	$0.69\times p$
NPT	$0.78\times p$

径向进给值a_p：

$$a_e=\frac{D_o^2-D_i^2}{4(D_i+D_c)}$$

p=螺距（mm）
h=螺纹深度（mm）
D_c=刀具直径（mm）
D_o=大径（mm）
D_i=小径（mm）

图 5-5　螺纹铣削加工工艺示意图

5.2.3　螺纹铣削编程方法

螺纹铣削加工需要用到螺旋插补指令。螺旋线的形成是在刀具作圆弧插补运动的同时，与其同步地作轴向运动，其指令格式为：

G2/G3　X_Y_Z_I_J_F_；

走刀路线如图 5-6 所示。

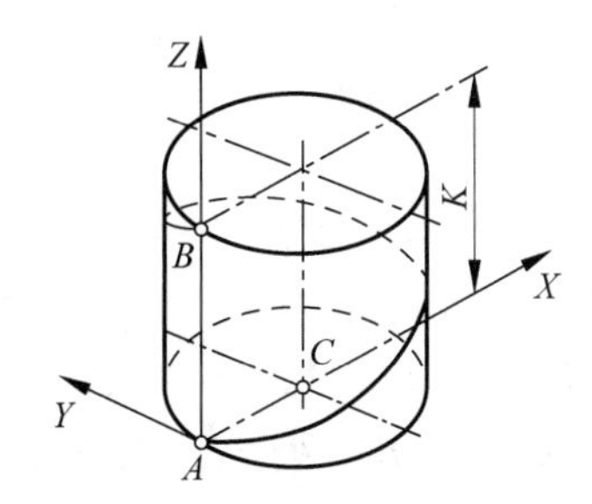

图 5-6　螺纹铣削加工时的走刀路线

5.3 刚性攻丝指令(G84)

刚性攻丝是指攻丝的刀柄是刚性的，且没有自动调整间隙；而柔性攻丝是指攻丝的刀柄是柔性的，且有调整间隙。指定刚性攻丝的指令有如下3种。

(1) 在与攻丝的指令相同的程序段中指令“M29 S_”的方法，其指令格式为：

G84 X_ Y_ Z_ R_ F_ M29 S_；

(2) 在攻丝循环之前指令“M29 S_；”的方法，其指令格式为：

M29 S_；

G84 X_ Y_ Z_ R_ F_；

(3) 不指令“M29 S_”的方法，而将参数 G84(No. 5200#0)设定为“1”，即将刚性攻丝设定主机床默认，其指令格式为：

G84 X_ Y_ Z_ R_ F_；

刚性攻丝指令的运动过程如图5-7所示。

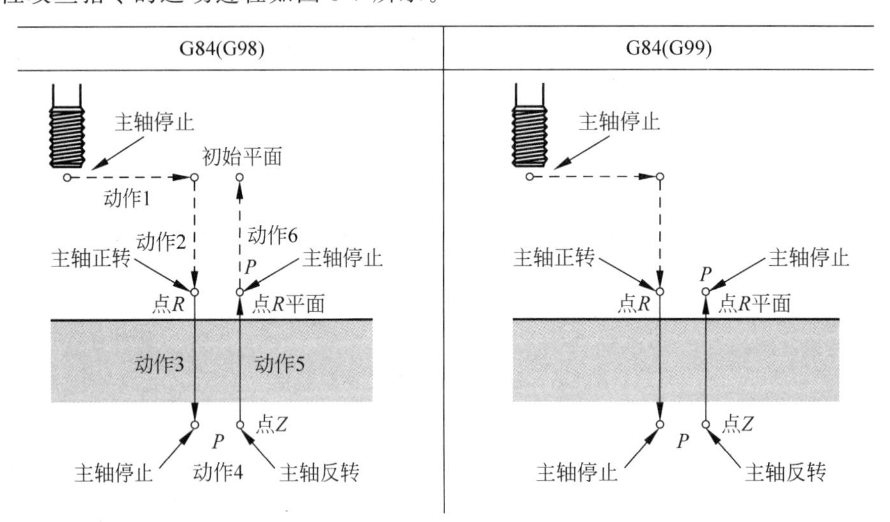

图5-7 刚性攻丝指令的运动过程

任务二 工艺准备

5.4 零件图分析

图5-8所示为缸体安装座零件图，加工过程中选择45钢作为缸体安装座零件的毛坯材料，毛坯下料尺寸为75mm×45mm×15mm。在加工过程中，先按六面体加工工艺要求进行外形加工，保证各面的平行度与垂直度，在加工70mm尺寸一端时同时完成$R4$圆角的加工；正面找正左下角，加工$\phi4.5$螺纹过孔2处、M8螺纹孔、铣削M26×1螺纹孔；反面加工M20×1螺纹孔；最后加工2处M4螺钉连接孔，底孔加工尺寸为$\phi3.3$。

注意：加工$R4$圆角时，注意零件伸出平口钳的距离，以防刀具与钳口发生干涉；如编程加工倒角，则要注意倒角铣刀刀尖位置，以免底部与钳口干涉。

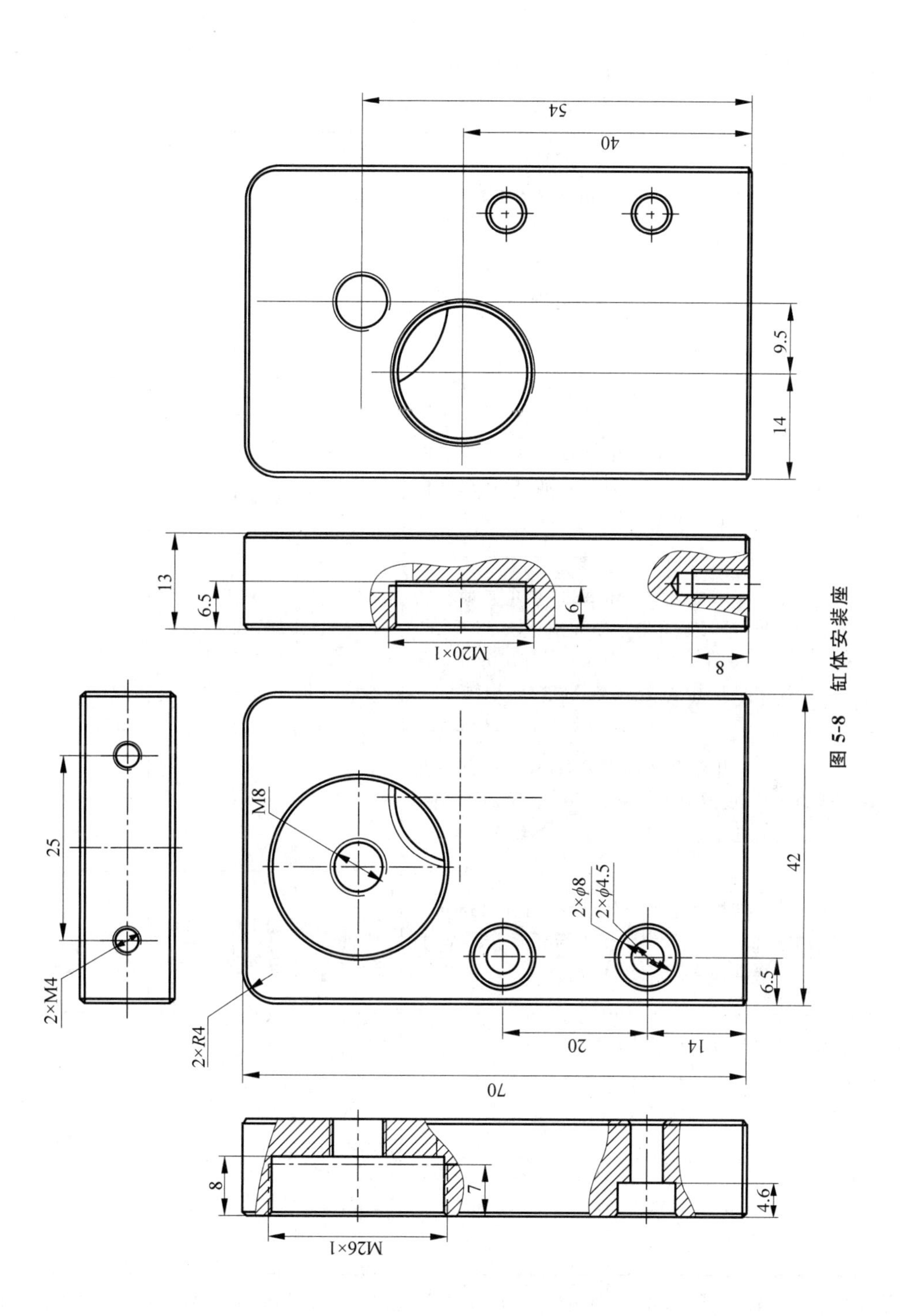
54
40
9.5
14
13
6.5
6
M20×1
8
25
2×M4
M8
2×φ8
2×φ4.5
42
6.5
20
14
2×R4
70
8
7
4.6
M26×1

图 5-8　缸体安装座

5.5 工艺设计

根据零件图分析，确定工艺过程，如表 5-1 所示。

表 5-1 工艺过程卡片

<table>
<tr><td colspan="2" rowspan="2">机械加工
工艺过程卡片</td><td>产品型号</td><td>STL-01</td><td colspan="2">零部件序号</td><td colspan="4">STL-02</td><td>第 1 页</td></tr>
<tr><td>产品名称</td><td>斯特林发动机</td><td colspan="2">零部件名称</td><td colspan="4">缸体安装座</td><td>共 1 页</td></tr>
<tr><td>材料牌号</td><td>C45</td><td>毛坯规格</td><td>75mm×45mm×15mm</td><td>毛坯质量</td><td colspan="2">0.4kg</td><td colspan="2">数量</td><td colspan="2">1</td></tr>
<tr><td rowspan="2">工序号</td><td rowspan="2">工序名</td><td colspan="2" rowspan="2">工序内容</td><td rowspan="2">工段</td><td colspan="3" rowspan="2">工艺装备</td><td colspan="3">工时/min</td></tr>
<tr><td colspan="2">准结</td><td>单件</td></tr>
<tr><td>5</td><td>备料</td><td colspan="2">按 75mm×45mm×15mm 尺寸备料</td><td>外购</td><td colspan="3">锯床</td><td colspan="2"></td><td></td></tr>
<tr><td>10</td><td>铣加工</td><td colspan="2">使用精密平口钳装夹，使用 ϕ63 平面铣刀、ϕ10 立铣刀加工六面，保证尺寸精度和形位公差</td><td>铣</td><td colspan="3">加工中心
游标卡尺</td><td colspan="2">60</td><td>45</td></tr>
<tr><td>15</td><td>铣加工</td><td colspan="2">使用精密平口钳装夹，找正底部左侧角，设定加工坐标系原点；钻 ϕ4.5 孔、ϕ6.8 孔，攻 M8 螺纹；铣 ϕ8 沉孔；铣 M26 底孔至 ϕ24；铣 M26×1 螺纹</td><td>铣</td><td colspan="3">加工中心
游标卡尺</td><td colspan="2">90</td><td>60</td></tr>
<tr><td>20</td><td>铣加工</td><td colspan="2">使用精密平口钳装夹，找正底部左侧角，设定加工坐标系原点；铣加工 M20 底孔至 ϕ18；铣 M20×1 螺纹</td><td>铣</td><td colspan="3">加工中心
游标卡尺</td><td colspan="2">30</td><td>10</td></tr>
<tr><td>25</td><td>铣加工</td><td colspan="2">使用精密平口钳装夹，分中 45mm×15mm 面，设定加工坐标系原点；钻 M4 底孔至 ϕ3.3，攻 M4 螺纹</td><td>铣</td><td colspan="3">加工中心
游标卡尺</td><td colspan="2">15</td><td>10</td></tr>
<tr><td>30</td><td>清理</td><td colspan="2">清理工件，锐角倒钝</td><td>钳</td><td colspan="3"></td><td colspan="2">15</td><td>5</td></tr>
<tr><td>35</td><td>检验</td><td colspan="2">检验工件尺寸</td><td>检</td><td colspan="3"></td><td colspan="2">15</td><td>5</td></tr>
<tr><td></td><td></td><td colspan="2"></td><td></td><td colspan="3"></td><td colspan="2"></td><td></td></tr>
<tr><td></td><td></td><td colspan="2"></td><td></td><td colspan="3"></td><td colspan="2"></td><td></td></tr>
</table>

第 10 工序为加工正六面体，在正六面体加工工艺介绍中已进行了说明，在此不做详细介绍。本训练任务针对第 15 工序铣加工，进行工序设计，制订工序卡片，如表 5-2 所示。

表 5-2 第 15 工序铣加工工序卡片

机械加工工序卡片	产品型号	STL-01	零部件序号	STL-02	第 2 页
	产品名称	斯特林发动机	零部件名称	缸体安装座	共 4 页

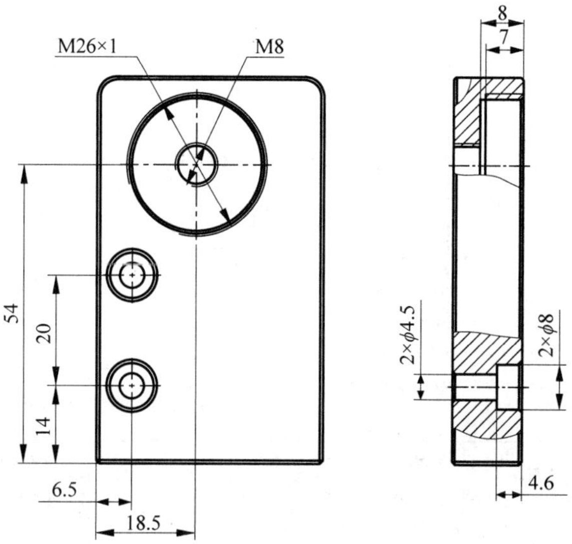

项目	内容
工序号	15
工序名	铣加工
材料	C45
设备	加工中心
设备型号	VAL650e
夹具	精密平口钳
量具	游标卡尺
准结工时	60min
单件工时	90min

工步	工步内容	刀具	S/(r/min)	F/(mm/r)	a_p/mm	a_e/mm	工步工时/min	
							机动	辅助
1	工件安装，顶面伸出钳口≥3mm，找正左上角							5
2	钻中心孔，2×ϕ4.5、M8 位置共 3 处	ϕ2 中心钻	1500	100	0.5	1	5	
3	钻 2×ϕ4.5 通孔	ϕ4.5 麻花钻	1500	200	0.1	2.25	5	
4	钻 M8 螺纹底孔至 ϕ6.8	ϕ6.8 麻花钻	1500	200	0.1	3.4	10	
5	攻 M8 螺纹	M8 机用丝锥	300	100	0.5	1	5	
6	粗、精铣 ϕ8×4.6 沉孔	ϕ4 立铣刀	2500	500	0.5	2	20	
7	铣 M26 螺纹底孔至 ϕ24	ϕ6 立铣刀	2500	500	0.5	3	20	
8	铣 M26×1 螺纹	M10×1 螺纹铣刀	2500	500	0.1	0.2	15	
9	拆卸、清理工件							5
10								

根据第 15 工序，本训练任务针对第 20 工序铣加工，进行工序设计，制订工序卡片，如表 5-3 所示(表头部分已完成)。

表 5-3　第 20 工序铣加工工序卡片

机械加工工序卡片	产品型号	STL-01	零部件序号	STL-02	第 3 页
	产品名称	斯特林发动机	零部件名称	缸体安装座	共 4 页

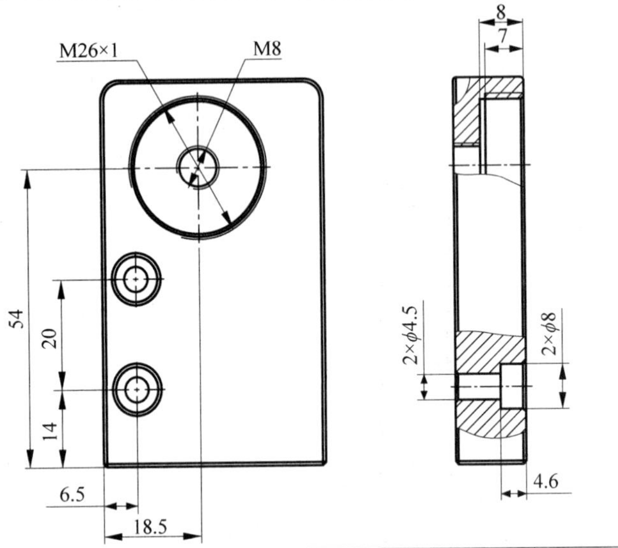

项目	内容
工序号	20
工序名	铣加工
材料	C45
设备	加工中心
设备型号	VAL650e
夹具	精密平口钳
量具	游标卡尺
准结工时	30min
单件工时	10min

工步	工步内容	刀具	S/(r/min)	F/(mm/r)	a_p/mm	a_e/mm	工步工时/min	
							机动	辅助
1								
2								
3								
4								
5								
6								
7								
8								
9								
10								

第 25 工序加工 2×M4 螺纹孔工艺相对比较简单，学生可以在课下进行练习。

5.6　数控加工程序编写

根据第 15 工序铣加工工序卡片，编写第 15 工序铣加工数控程序，如表 5-4 所示。

表 5-4　第 15 工序铣加工数控程序

段　号	程序语句	注　释
	O0001;	
N1	T1 M6;	调用 $\phi 2$ 中心钻
	S1500 M3;	
	G0 G90 G54 X6.5 Y14;	
	G43 Z10 H1;	

续表

段　号	程序语句	注　释
	G81 Z-1.5 R1 F100;	钻中心孔
	Y34;	
	X18.5 Y54;	
	G0 G80 Z100;	
	M5;	
N2	T2 M6;	调用 ϕ4.5 麻花钻
	S1500 M3;	
	G0 G90 G54 X6.5 Y14;	
	G43 Z10 H2;	
	G83 Z-15 R1 Q1 F200;	钻 ϕ4.5 孔
	Y34;	
	G0 G80 Z100;	
	M5;	
N3	T3 M6;	调用 ϕ6.8 麻花钻
	S1500 M3;	
	G0 G90 G54 X18.5 Y54;	
	G43 Z10 H3;	
	G83 Z-16 R1 Q1 F200;	钻 M8 螺纹底孔
	G0 G80 Z100;	
	M5;	
N4	T4 M6;	调用 M8 丝锥
	S300 M3;	
	G0 G90 G54 X18.5 Y54;	
	G43 Z10 H4;	
	M29 S300;	调用刚性攻丝指令
	G84 Z-15 R1 F100;	刚性攻丝 M8
	G0 G80 Z100;	
	M5;	
N5	T5 M6;	调用 ϕ4 立铣刀
	S2500 M3;	
	G0 G90 G54 X6.5 Y14;	
	G43 Z10 H5;	
	G1 Z0 F500;	
	M98 P0090010;	调用 ϕ8 沉孔粗加工子程序,加工第 1 个 ϕ8 沉孔
	G1 G90 Z-4.6;	精加工底面
	G1 G91 X1.95;	
	G3 I-1.95 J0;	
	G1 X-1.95;	精加工侧面
	G1 G41 X1 Y-1 D5;	
	G3 X3 Y3 R3;	
	I-4 J0;	
	X-3 Y3 R3;	

续表

段　号	程序语句	注　释
	G1 G40 X-1 Y-1;	
	G1 G90 Z10;	
	G0 X6.5 Y34;	
	G1 Z0 F500;	
	M98 P0090010;	调用 ϕ8 沉孔粗加工子程序,加工第 2 个 ϕ8 沉孔
	G1 G90 Z-4.6;	精加工底面
	G1 G91 X1.95;	
	G3 I-1.95 J0;	
	G1 X-1.95;	精加工侧面
	G1 G41 X1 Y-1 D5;	
	G3 X3 Y3 R3;	
	I-4 J0;	
	X-3 Y3 R3;	
	G1 G40 X-1 Y-1;	
	G1 G90 Z10;	
	G0 Z100;	
	M5;	
N6	T6 M6;	调用 ϕ6 立铣刀
	S2500 M3;	
	G0 G90 G54 X18.5 Y54;	
	G43 Z10 H6;	
	G1 Z0 F500;	
	M98 P0160011;	调用 M26 底孔加工程序,加工至 ϕ24
	G1 G90 Z10;	
	G0 Z100;	
	M5;	
N7	T7 M6;	调用螺纹铣刀 M10×1
	S2500 M3;	
	G52 X18.5 Y54;	
	G0 G90 G54 X0 Y0;	
	G43 Z10 H7;	
	G1 Z-7 F500;	铣螺纹起点
	G1 G41 X2.5 Y-10 D7;	第 1 刀进刀为 0.5mm
	G3 X12.5 Y0 R10;	
	Z-6 I-12.5 J0;	顺铣,逆时针从下向上,加工右旋螺纹
	X2.5 Y10 R10;	
	G1 G40 X0 Y0;	
	G1 Z-7 F500;	铣螺纹起点
	G1 G41 X2.7 Y-10 D7;	第 2 刀进刀为 0.2mm
	G3 X12.7 Y0 R10;	
	Z-6 I-12.7 J0;	顺铣,逆时针从下向上,加工右旋螺纹
	X2.7 Y10 R10;	

续表

段 号	程序语句	注 释
	G1 G40 X0 Y0;	
	G1 Z-7 F500;	铣螺纹起点
	G1 G41 X2.9 Y-10 D7;	第 3 刀进刀为 0.2mm
	G3 X12.9 Y0 R10;	
	Z-6 I-12.9 J0;	顺铣,逆时针从下向上,加工右旋螺纹
	X2.9 Y10 R10;	
	G1 G40 X0 Y0;	
	G1 Z-7 F500;	铣螺纹起点
	G1 G41 X3 Y-10 D7;	第 4 刀进刀为 0.1mm
	G3 X13 Y0 R10;	
	Z-6 I-13 J0;	顺铣,逆时针从下向上,加工右旋螺纹
	X3 Y10 R10;	
	G1 G40 X0 Y0;	
	Z10;	
	G52 X0 Y0;	
	G0 Z100;	
	M5;	
	M30;	
	%	
	O0010;	ϕ8×4.6 沉孔粗加工子程序
	G1 G91 Z-0.5 F500;	
	X1.95;	
	G3 I-1.95 J0;	
	G1 X-1.95;	
	M99;	
	%	
	O0011;	M26 底孔加工子程序
	G1 G91 Z-0.5 F500;	
	X3;	
	G3 I-3 J0;	
	G1 X3;	
	G3 I-6 J0;	
	G1 X3;	
	G3 I-9 J0;	
	G1 X-9;	
	M99;	

第 20 工序铣加工只需将 M20 底孔铣削至 ϕ18,再用 M10×1 螺纹铣刀按第 15 工序第 N7 段铣削 M20 螺纹即可,程序比较简单,此处不再赘述。

第 25 工序铣加工只需将 M4 底孔钻至 ϕ3.3,再用刚性攻丝即可,程序参考第 15 工序第 N1 段、第 N3 段及第 N4 段即可。

任务三　上机训练

5.7　设备与用具

设备：AVL650e 立式加工中心。

刀具：ϕ2 中心钻、ϕ4.5 麻花钻、ϕ6.8 麻花钻、M8 机用丝锥、ϕ4 立铣刀、ϕ6 立铣刀、M10×1 螺纹刀。

夹具：125mm 精密平口钳。

工具：什锦锉刀。

量具：0～150mm 游标卡尺、0.02mm 杠杆百分表(配安装杆)、寻边器等。

毛坯：75mm×45mm×15mm。

辅助用品：钩扳子、橡胶锤、毛刷等。

5.8　开机检查

检查机床外观各部位(如防护罩、脚踏板等部位)是否存在异常；检查机床润滑油、冷却液是否充足；检查刀架、夹具、导轨护板上是否有异物；检查机床面板各旋钮状态是否正常；在开机后，检查机床是否存在报警等。可参考表 5-5 对机床状态进行点检。

表 5-5　机床开机准备卡片

检查项目		检查结果	异常描述
机械部分	主轴部分		
	进给部分		
	换刀机构		
	夹具系统		
电器部分	主电源		
	冷却风扇		
数控系统	电气元件		
	控制部分		
	驱动部分		
辅助部分	冷却系统		
	压缩空气		
	润滑系统		

5.9　加工前准备

(1) 在机床启动后，各坐标轴按操作说明书返回机床原点。

(2) 安装工件，工件伸出平口钳顶面须大于 3mm。

(3) 用寻边器找正工件左下角，并在 G54 坐标系中设定为 X0、Y0。

(4) 安装刀具。

(5) 对刀。在对刀时,可以以精密平口钳固定钳口顶面为基准。

(6) 程序的编辑。因程序比较长,建议先在微机上进行程序的编辑,在确认检查无误后,再使用存储卡将程序导入数控机床中。

5.10　零件加工

在程序自动运行前,应当在数控机床上进行一次试运行,以确保程序的正确性,防止加工事故的发生。

在图形校验过程验证无问题后,即可进行零件加工。在零件加工前,应详细了解机床的安全操作要求,穿戴好劳动保护服装和用具。在进行零件加工时,应熟悉加工中心各操作按键的功能和位置,了解紧急状况的处置方法。

注意:如已进行图形校验操作,则在操作完成后,必须执行机床回归机械零点操作,然后再进行其他的相关操作。如果未回零点而进行了相关操作,则会造成坐标偏移,甚至出现撞机、程序乱跑等异常现象,从而导致危险。

在操作面板上选择"记忆"模式(MEM),即程序的自动执行模式,调入需要进行加工的数控加工程序,按循环启动功能按键进行自动加工。

数控加工程序的首次自动运行应在调试状态下进行。机床进给倍率旋钮先调低至0%,按下操作面板上 单節執行 功能按键,通过单击显示屏底部"程序检查"按钮切换到运行状态检查窗口,如图 5-9 所示。

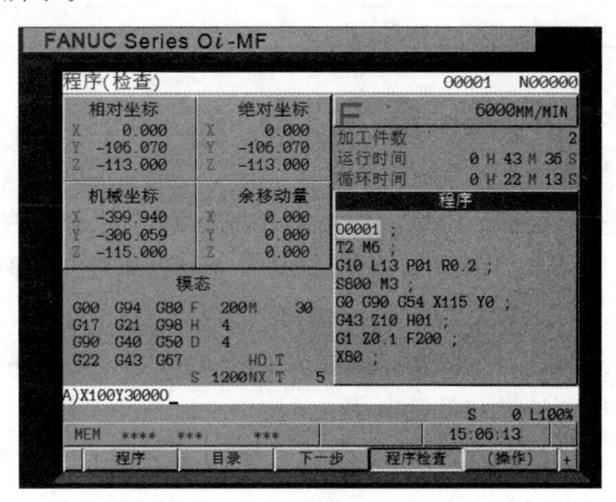

图 5-9　程序自动运行状态检查

在此状态下,每按一次循环启动功能按键程序都只自动执行光标所在的一行。在按循环启动功能按键前,应观察刀具与工件间的距离是否安全;在按下循环启动功能按键后,通过进给倍率旋钮控制机床的运动速度,同时对照显示屏"剩余移动量"栏显示的剩余移动量,观察刀具与工件之间的实际距离。若实际距离与剩余移动量相差过大,则应果断停机检查,以免发生撞机事故。在程序调试过程中,还应密切注意显示屏的"模态"状态显示,确保主轴转速、进给速度、工件坐标系号、补偿状态及补偿号等无异常发生。

5.11 零件检测

在零件加工完成后，应当认真清理工件，并按照质量管理的相关要求，对加工完成的零件进行相关检验，保证生产质量。机械加工零件“三级”检验卡片如表5-6所示。

表5-6 机械加工零件“三级”检验卡片

零部件图号		零部件名称		工　序　号	
材料		送检日期		工序名称	
检验项目	自检结果	互检结果	专业检验	备注	
检验结论	□合格　□不合格　□返修　□让步接收 检验签章： 年　月　日				
不符合项描述					

项 目 总 结

通过缸体安装座数控铣削加工，需要掌握数控铣削程序的基本格式和基本切削加工指令的使用方法，能够应用基本切削加工指令、刀具半径补偿指令、子程序，以及坐标变换指令等进行零件切削加工。掌握刚性攻丝指令G84的使用方法。

掌握立式加工中心的基本操作方法，包括开关机、刀具安装、工件找正、对刀、程序编辑、图形校验、数控加工程序调试及自动运行等。

通过任务训练，养成良好的职业素养，培养正确的加工中心安全操作规范，养成基本的机械加工质量意识。

课 后 习 题

一、选择题

1. 用FANUC系统指令“G92 X(U)Z(W)F;”加工双线螺纹，则该指令中的“F”是指(　　)。

A. 螺纹螺距　　B. 螺纹导程

C. 每分钟进给量　　D. 每转进给量

2. 在铣削矩形工件两侧垂直面时，选用机用平口钳装夹工件，若铣出的平面与基准面之间的夹角小于 90°，则应在固定钳口(　　)，垫入铜片或纸片。

A. 中部　　B. 下部

C. 上部　　D. 以上答案都可以

3. 机用虎钳主要用于装夹(　　)。

A. 矩形工件　　B. 轴类零件　　C. 套类零件　　D. 盘形工件

4. 加工中心的自动换刀装置由驱动机构、(　　)组成。

A. 刀库和机械手　　B. 刀库和控制系统

C. 机械手和控制系统　　D. 控制系统

5. 小直径内螺纹的加工刀具为(　　)。

A. 螺纹车刀　　B. 螺纹铣刀　　C. 丝锥　　D. 螺纹滚压刀具

二、判断题

1. 螺纹铣削加工需要用到螺旋插补指令。螺旋线的形成是在刀具作圆弧插补运动时，与其同步地作轴向运动。(　　)

2. 在分多层切削加工螺纹时，应尽可能平均分配每层切削的背吃刀量。(　　)

3. 在加工多线螺纹时，在加工完一条螺纹后，加工第 2 条螺纹的起点应与第 1 条螺纹的起点相隔 1 个导程。(　　)

4. 组合夹具最适宜加工位置精度要求较高的工件。(　　)

5. 螺纹切削循环指令 G92 只能用于车削直螺纹，而不能用于车削锥螺纹。(　　)

三、填空题

1. 螺纹铣刀的主要类型有________、________、________。

2. 螺纹铣削是借助数控加工中心机床的________功能及________螺旋插补指令，完成螺纹铣削工作。

3. 螺纹铣削加工与传统螺纹加工方式相比，在加工精度、加工效率方面具有极大的优势，且在加工时不受________和________的限制。

4. 平口钳主要由________、________、________、________等部分组成。

四、简答题

1. 简述刚性攻丝与柔性攻丝的区别。

2. 简述刚性攻丝指令 G84 的使用方法。

3. 简述在螺纹铣削加工编程时螺纹铣刀的运动方式。

五、解释题

1. 解释 M24×1.5-6g-LH 的螺纹标注含义。

2. 解释 M20×2-6H-LH 的螺纹标注含义。

3. 解释 M24×Ph2P1-LH 的螺纹标注含义。

4. 解释 M30×1.5-LH 的螺纹标注含义。

自我学习检测评分表如表 5-7 所示。

表 5-7 自我学习检测评分表

项目	目标要求	分值	评分细则	得分	备注
学习关键知识点	（1）掌握正六面体铣削加工时夹具的正确选择 （2）熟悉正六面体铣削加工的顺序 （3）熟悉螺纹铣削加工刀具的选择及加工工艺 （4）掌握螺纹铣削加工的编程方法 （5）掌握刚性攻丝指令的使用	20	理解与掌握		
工艺准备	（1）能够正确识读零件图 （2）能够根据零件图分析、确定工艺过程 （3）能够根据工序加工工艺，编写正确的加工程序	30	理解与掌握		
上机训练	（1）会正确选择相应的设备与用具 （2）能够正确操作数控车床，并根据加工情况调整加工参数	50	（1）理解与掌握 （2）操作流程		

思政小课堂

项目六　偏心轮铣削编程加工训练

➢思维导图

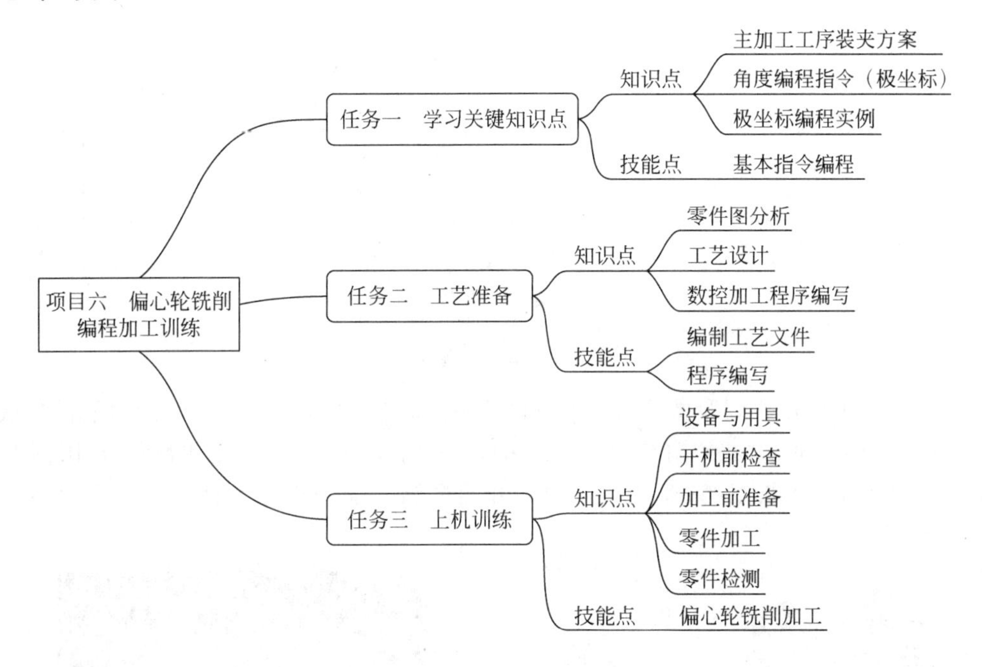

➢学习目标

知识目标

(1) 了解偏心轮的工作原理。

(2) 了解偏心轮在本机构中的用途和关键要求。

能力目标

(1) 掌握平板类异形零件的加工技巧。

(2) 掌握极坐标编程指令的使用方法。

(3) 能够独立确定加工工艺路线，并正确填写工艺文件。

(4) 能够正确操作数控加工中心，并根据加工情况调整加工参数。

(5) 能够根据零件结构特点和精度合理选用量具，并正确、规范地测量相关尺寸。

素养目标

(1) 培养学生的科学探究精神和态度。

(2) 培养学生的工程意识。

(3) 培养学生的团队合作能力。

任务一　学习关键知识点

6.1　主加工工序装夹方案

如图 6-1 所示，偏心轮毛坯为厚度 3mm 的金属板，通过线切割工序加工出零件外形，铣削加工部分只加工弧槽及 M5 孔。

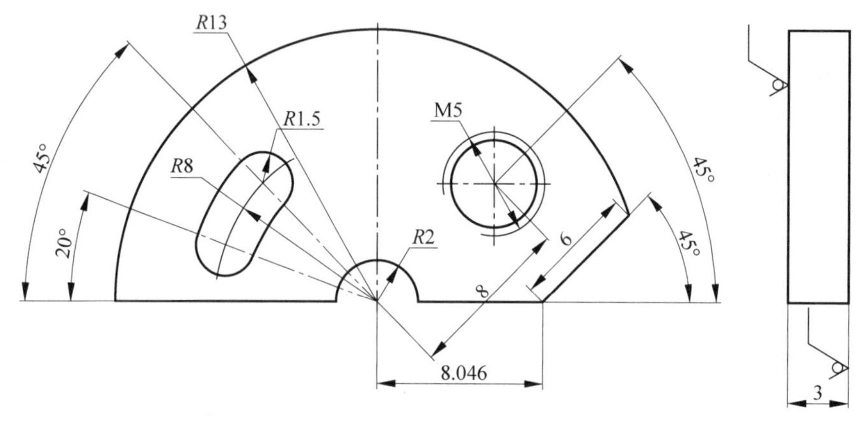

图 6-1　偏心轮零件图

铣削加工可采用带 V 形槽的精密平口钳装夹，如图 6-2 所示。零件底面加平行垫铁定位，限制 Z 向移动和 X 向、Y 向转动 3 个自由度；平行面与平口钳固定钳口贴合，限制 Y 向移动和 Z 向转动 2 个自由度；V 形块与 $R13$ 配合限制 X 向移动 1 个自由度，从而实现工件的完全定位。

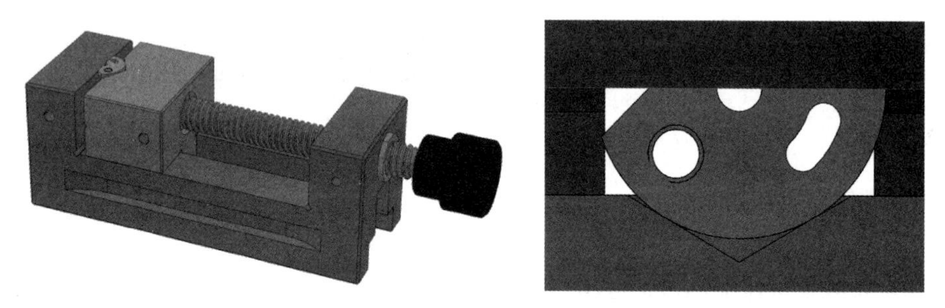

图 6-2　精密平口钳（带 V 形槽）

6.2　角度编程指令（极坐标）

极坐标属于二维坐标系统，创始人是艾萨克·牛顿，主要应用于数学领域，如图 6-3 所示，极坐标是指在平面内取一个定点 O，称为极点，引一条射线 OX，称为极轴，再选定一个长度单位和角度的正方向（通常取逆时针方向）。对于平面内任何一点 M，用 ρ 表示线段 OM 的长度（有时也用 r 表示），θ 表示从 OX 到 OM 的角度，ρ 称为点 M 的极径，θ 称为点 M 的极角，有序数对 (ρ,θ) 称为点 M 的极坐标，这样建立的坐标系称为极坐标系。在通常情况下，点 M 的极径坐标单位为 1(长度单位)，极角坐标单位为 rad(或°)。

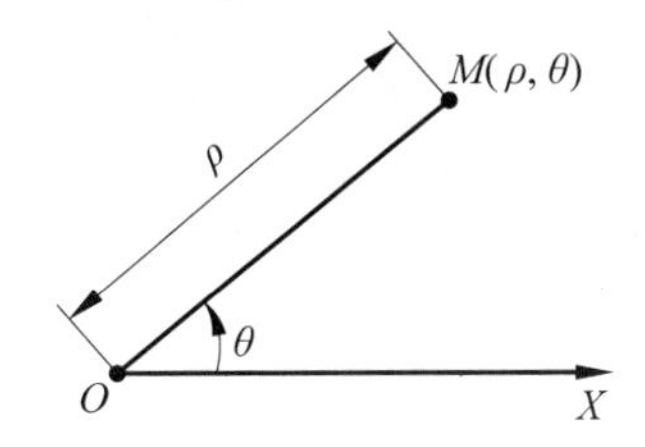

图 6-3　极坐标系

在 FANUC 0i MF 系统中，极坐标编程的基本格式为：

G◇◇ G○○ G16；	调用极坐标指令(极坐标方式)
G0 IP_；	
…	极坐标描述的加工指令
…	
G16；	取消极坐标指令(极坐标方式)

其中，G◇◇表示极坐标指令的平面选择(G17、G18、G19)；G○○表示极坐标指令的极点(中心)选择(G90 代表以工件坐标系原点为极点，G91 代表以当前点位置为极点)；IP_为极坐标指令，第 1 轴表示极径、第 2 轴表示极角。

如图 6-4 所示，极坐标应用有两种常用方式。

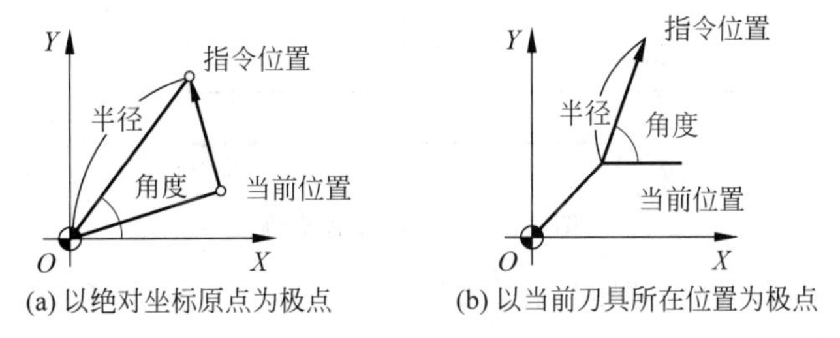

图 6-4　极坐标应用方式

1. 以绝对坐标原点为极点

在以绝对坐标原点为极点时，X 轴正方向为 0°，沿逆时针方向从 X 轴正方向到原点与指令位置所有直线的角度为极角 θ，原点到指令位置的距离为极径 ρ。其程序格式为：

G16；	调用极坐标指令
G1 Xρ Yθ F_；	向指令位置做直线运动
G15；	取消极坐标指令

在实际应用中，可以使用 G52 指令指定临时的加工坐标系原点，配合 G16 指令，以任意位置为极点进行极坐标的加工编程。

2. 以当前刀具所在位置为极点

在以当前刀具所在位置为极点时，当前位置水平向右为 0°，沿逆时针到当前位置与指令位置所在直线构成的角度为极角 θ，当前位置到指令位置的距离为极径 ρ。其程序格式为：

G91 G16；	调用极坐标指令
G1 Xρ Yθ F_；	向指令位置做直线运动
G90 G15；	取消极坐标指令

6.3 极坐标编程实例

应用极坐标指令编写图 6-5 所示的图形加工程序。

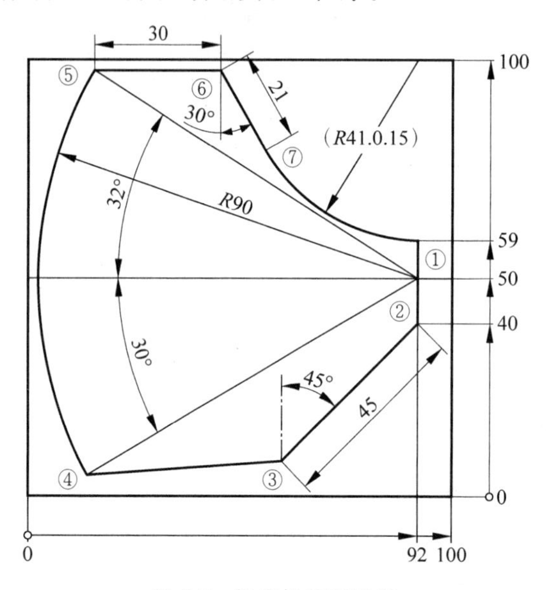

图 6-5　极坐标编程实例

编写图 6-5 所示图形程序片断如表 6-1 所示。

表 6-1　极坐标编程片断

段　号	程序语句	注　释
	…	省略程序头及下刀、切入
N1	G1 X92 Y40；	点①→点②
N2	G91 G16；	定义当前点②为极点
	G1 X45 Y225；	点②→点③
	G90 G15；	取消极坐标编程
N3	G52 X92 Y50；	临时将点(92,50)定义为加工原点
	G16；	定义临时原点为极点
	G1 X90 Y210；	点③→点④
	G2 X90 Y148 R90；	从点④到点⑤加工 *R*90 圆弧
	G15；	取消坐标编程
	G52 X0 Y0；	将加工坐标系还原到点(0,0)

续表

段　号	程序语句	注　释
N4	G1 G91 X30；	相对当前点⑤向右加工 30mm 到点⑥
N5	G91 G16；	定义点⑥为极点
	G1 X21 Y300；	以极径 21、极角 300°直线运动到点⑦
	G90 G15；	取消坐标编程
N7	G3 X92 Y59 R41.015；	沿逆时针圆弧从点⑦加工到点⑧
N8	G1 X92 Y50；	直线从点⑧运动到点①
	…	

任务二　工艺准备

6.4　零件图分析

图 6-6 所示为偏心轮的零件图，零件上、下表面为非切削加工表面，故毛坯材料可以直接选择厚度为 3mm 的 62 黄铜板，另外，为节省材料，可选择厚度为 3mm、宽度为 15mm 的条料加工；为便于在线切割机床上装夹，将条料长度控制在 250mm；在线切割外形后，铣床加工弧槽及 M5 孔，加工过程采用平口钳配合 V 形块装夹，应用极坐标指令进行数控编程。

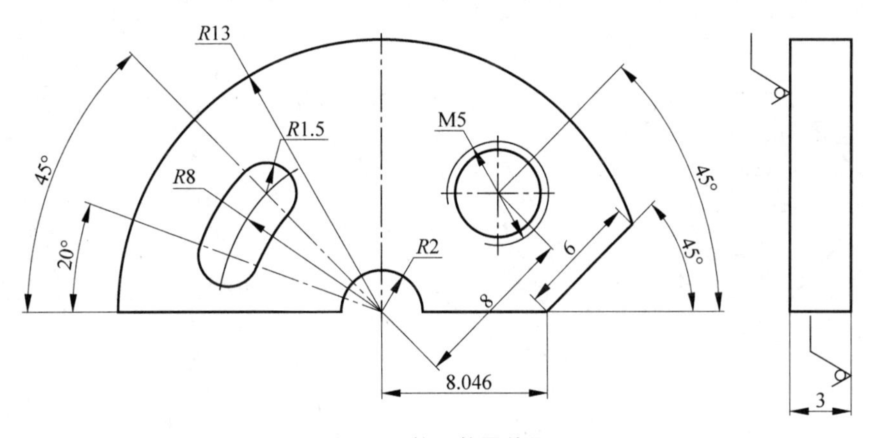

图 6-6　偏心轮零件图

注意：在铣削加工时，应当适当控制夹紧力，以防止零件表面损伤。

6.5　工艺设计

根据零件图分析，确定工艺过程，如表 6-2 所示。

表 6-2　工艺过程卡片

机械加工工艺过程卡片		产品型号	STL-01	零部件序号	STL-03		第 1 页
		产品名称	斯特林发动机	零部件名称	偏心轮		共 1 页
材料牌号	H62	毛坯规格	250mm×15mm　$t=3$	毛坯质量	0.1kg	数量	9

工序号	工序名	工序内容	工段	工艺装备	工时/min	
					准结	单件
5	备料	材料厚度 $t=3$mm，按 250mm×15mm 尺寸备料	外购	锯床		
10	线切割	采用夹板装夹，线切割加工零件外形，在加工后手动处理接丝痕迹。在批量加工条件下，可堆叠加工	线切割	线切割机床 游标卡尺	40	30
15	铣加工	使用精密平口钳配合 V 形块装夹，钻 M5 底孔至 $\phi4.2$；攻 M5 螺纹孔；铣削加工圆弧槽	铣	加工中心 游标卡尺	30	20
20	清理	清理工件，锐角倒钝	钳		15	5
25	检验	检验工件尺寸	检		15	5

本训练任务针对第 15 工序铣加工，进行工序设计，制订工序卡片，如表 6-3 所示。

表 6-3　第 15 工序铣加工工序卡片

机械加工工序卡片	产品型号	STL-01	零部件序号	STL-03	第 2 页
	产品名称	斯特林发动机	零部件名称	偏心轮	共 3 页

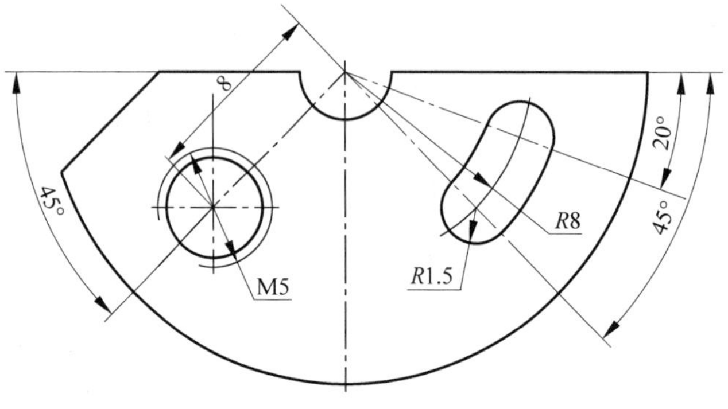

项目	内容
工序号	15
工序名	铣加工
材料	H62
设备	加工中心
设备型号	VAL650e
夹具	精密平口钳
量具	游标卡尺
准结工时	40min
单件工时	30min

工步	工步内容	刀具	S/(r/min)	F/(mm/r)	a_p/mm	a_e/mm	工步工时/min	
							机动	辅助
1	工件安装，底面贴紧专用平行垫铁，注意加工透后的避让；直面贴紧固定钳口，V 形块夹 $R13$ 处，注意夹紧力，防止夹伤；以 $R13$ 圆心为加工坐标系原点							5
2	钻中心孔	$\phi2$ 中心钻	1500	100	0.5	1	2	
3	钻 M5 螺纹底孔，$\phi4.2$ 通孔	$\phi4.2$ 麻花钻	1500	200	0.1	2.25	3	

续表

工步	工步内容	刀具	S/(r/min)	F/(mm/r)	a_p/mm	a_e/mm	工步工时/min	
							机动	辅助
4	攻 M5 螺纹	M5 机用丝锥	300	100	0.5	1	5	
5	粗铣圆弧槽	ϕ2 立铣刀	4000	500	0.3	1.5	10	
6	精铣圆弧槽	ϕ2 立铣刀	4000	500	3	0.05	2	
7	拆卸、清理工件							3
8								
9								
10								

6.6　数控加工程序编写

根据第 15 工序铣加工工序卡片，编写第 15 工序铣加工数控程序，如表 6-4 所示。

表 6-4　第 15 工序铣加工数控程序

段　号	程序语句	注　释
	O0001；	
N1	T1 M6；	调用 ϕ2 中心钻
	S1500 M3	
	G16；	调用极坐标编程指令
	G0 G90 G54 X8 Y225；	定位到 M5 孔中心位置，极径为 8mm、极角为 225°
	G43 Z10 H1；	
	G81 Z-1.5 R1 F100；	钻中心孔
	G0 G80 Z100；	
	M5；	
N2	T2 M6；	调用 ϕ4.2 麻花钻
	S1500 M3；	
	G0 G90 G54 X8 Y225；	定位到 M5 孔中心位置，极径为 8mm、极角为 225°
	G43 Z10 H2；	
	G83 Z-5 R1 Q1 F200；	钻 ϕ4.2 孔
	G0 G80 Z100；	
	M5；	
N3	T3 M6；	调用 M5 机用丝锥
	S1500 M3；	
	G0 G90 G54 X8 Y225；	定位到 M5 孔中心位置，极径为 8mm、极角为 225°
	G43 Z10 H3；	
	M29 S300；	调用刚性攻丝指令
	G84 Z-5 R1 F100；	刚性攻丝 M5
	G0 G80 Z100；	
	M5；	
N4	T4 M6；	调用 ϕ2 立铣刀
	S2500 M3；	
	G0 G90 G54 X8.45 Y-45；	定位到圆弧槽起点，极径为 8.45mm、极角为 −45°

续表

段　号	程序语句	注　释
	G43 Z10 H4；	
	G1 Z0.2 F500；	
	M98 P0110010；	调用圆弧槽粗加工子程序
	G1 X8 Y-20；	精加工侧面
	G1 G41 X6.5 Y-45 D4；	
	G3 X9.5 Y-45 R1.5；	
	X9.5 Y-20 R9.5；	
	X6.5 Y-20 R1.5；	
	G2 X6.5 Y-45 R6.5；	
	G1 G40 X8 Y-45；	
	G1 Z10；	
	G15；	取消极坐标编程指令
	G0 Z100；	
	M5；	
	M30；	
	%	
	O0010；	圆弧槽粗加工子程序
	G3 G91 X0 Y25 Z-0.5 R8.45 F500；	
	G90 X7.55 Y-20 R0.45；	
	X7.55 Y-45 R7.55；	
	X8.45 Y-45 R0.45；	
	M99；	
	%	

任务三　上机训练

6.7　设备与用具

设备：AVL650e 立式加工中心。

刀具：ϕ2 中心钻、ϕ4.2 麻花钻、M5 机用丝锥、ϕ2 立铣刀。

夹具：125mm 精密平口钳，活动钳口带 V 形块。

工具：什锦锉刀。

量具：0～150mm 游标卡尺、0.02mm 杠杆百分表（配安装杆）、寻边器等。

毛坯：第 10 工序加工后半成品。

辅助用品：钩扳子、橡胶锤、毛刷等。

6.8　开机检查

检查机床外观各部位（如防护罩、脚踏板等部位）是否存在异常；检查机床润滑油、冷却液是否充足；检查刀架、夹具、导轨护板上是否有异物；检查机床面板各旋钮状态是否正常；在开机后，检查机床是否存在报警等。可参考表 6-5 对机床状态进行点检。

表 6-5　机床开机准备卡片

检查项目		检查结果	异常描述
机械部分	主轴部分		
	进给部分		
	换刀机构		
	夹具系统		
电器部分	主电源		
	冷却风扇		
数控系统	电气元件		
	控制部分		
	驱动部分		
辅助部分	冷却系统		
	压缩空气		
	润滑系统		

6.9　加工前准备

(1) 在机床启动后，各坐标轴按操作说明书返回机床原点。

(2) 安装工件。

(3) 用寻边器分中固定钳口左、右两侧设为 $X0$，找正固定钳口贴合面设为 $Y0$，并将其设定为 G54 坐标系中的 $X0$、$Y0$。

(4) 安装刀具。

(5) 对刀。在对刀时，可以以精密平口钳固定钳口顶面为基准。

(6) 程序的编辑，因程序比较长，建议先在微机上进行程序的编辑，在确认检查无误后，再使用存储卡将程序导入数控机床中。

6.10　零件加工

在程序自动运行前，应当在数控机床上进行一次试运行，以确保程序的正确性，防止加工事故的发生。

在图形校验过程验证无问题后，即可进行零件加工。在零件加工前，应详细了解机床的安全操作要求，穿戴好劳动保护服装和用具。在进行零件加工时，应熟悉加工中心各操作按键的功能和位置，了解紧急状况的处置方法。

注意：如已进行图形校验操作，则在操作完成后，必须执行机床回归机械零点操作，然后再进行其他的相关操作。如果未回零点而进行了相关操作，则会造成坐标偏移，甚至出现撞机、程序乱跑等异常现象，从而导致危险。

在操作面板上选择“记忆”模式(MEM)，即程序的自动执行模式，调入需要进行加工的数控加工程序，按循环启动功能按键进行自动加工。

数控加工程序的首次自动运行应在调试状态下进行。机床进给倍率旋钮先调低至

0%,按下操作面板上【单段执行】功能按键,通过单击显示屏底部“程序检查”按钮切换到运行状态检查窗口,如图6-7所示。

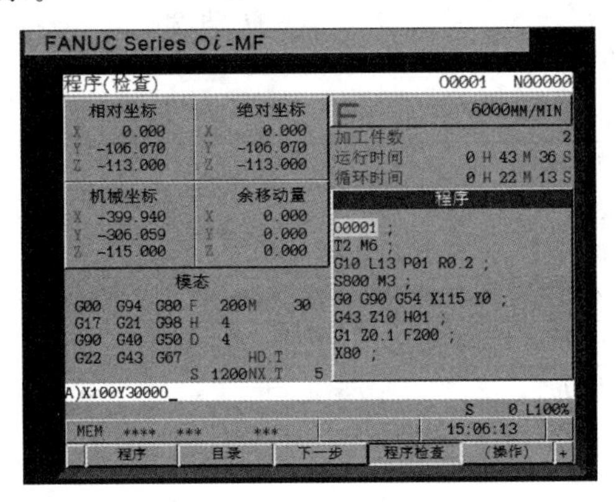

图6-7 程序自动运行状态检查

在此状态下,每按一次循环启动功能按键程序都只自动执行光标所在的一行。在按循环启动功能按键前,应观察刀具与工件间的距离是否安全;在按下循环启动功能按键后,通过进给倍率旋钮控制机床的运动速度,同时对照显示屏“剩余移动量”栏显示的剩余移动量,观察刀具与工件之间的实际距离。若实际距离与剩余移动量相差过大,则应果断停机检查,以免发生撞机事故。在程序调试过程中,还应密切注意显示屏的“模态”状态显示,确保主轴转速、进给速度、工件坐标系号、补偿状态及补偿号等无异常发生。

6.11 零件检测

在零件加工完成后,应当认真清理工件,并按照质量管理的相关要求,对加工完成的零件进行相关检验,保证生产质量。机械加工零件“三级”检验卡片如表6-6所示。

表6-6 机械加工零件“三级”检验卡片

零部件图号		零部件名称		工　序　号	
材料		送检日期		工序名称	
检验项目	自检结果	互检结果	专业检验	备注	
检验结论	□合格　□不合格　□返修　□让步接收 检验签章: 年　月　日				
不符合项描述					

项 目 总 结

通过偏心轮数控铣削加工,需要掌握数控铣削程序的基本格式和基本切削加工指令的使用方法,能够应用基本切削加工指令、刀具半径补偿指令、子程序,以及坐标变换指令等进行零件切削加工。掌握极坐标编程指令 G16 的使用方法。

掌握立式加工中心的基本操作方法,包括开关机、刀具安装、工件找正、对刀、程序编辑、图形校验、数控加工程序调试及自动运行等。

通过任务训练,养成良好的职业素养,培养正确的加工中心安全操作规范,养成基本的机械加工质量意识。

课 后 习 题

一、选择题

1. 在数控编程时,应首先设定(　　)。

A. 机床原点　　B. 机床参考点　　C. 机床坐标系　　D. 工件坐标系

2. 加工中心执行顺序控制动作和控制加工过程的中心是(　　)。

A. 基础部件　　B. 主轴部件　　C. 数控系统　　D. ATC

3. 在用轨迹法切削槽类零件时,槽两侧表面,(　　)。

A. 两面均为逆铣

B. 两面均为顺铣

C. 一面为顺铣,一面为逆铣,因此,两侧质量不同

D. 一面为顺铣,一面为逆铣,但两侧质量相同

4. 在测量凸轮坐标尺寸时,应选用(　　)。

A. 内径余弦规　　B. 万能工具显微镜　　C. 内径三角板　　D. 块规

5. 在 FANUC 系统中,程序段“G17 G16 G90 X100.0 Y30.0”中,X 指令是(　　)

A. X 轴坐标位置　　B. 极坐标原点到刀具中心距离

C. 旋转角度　　D. 时间参数

二、判断题

1. G16 G17 G90 极坐标建立 G01 X_Y_F_;
其中,X 指终点到原点的距离,也就是半径。(　　)

2. 偏心轮就是凸轮。(　　)

3. 铣削加工可采用带 V 形槽的精密平口钳装夹。零件底面加平行垫铁定位,只限制 Z 向移动一个自由度。(　　)

4. 带 V 形槽的精密平口钳是专用夹具。(　　)

5. 在角度编程中,角度是看与 Z 的负向产生的夹角,顺时针为正,逆时针为负。(　　)

三、填空题

1. 若所有坐标点的坐标值均从坐标系原点计量，则称为________坐标；若运动轨迹的终点坐标相对于线段的起点来计量，则称为________坐标。

2. 极坐标应用的两种常用方式为________、________。

3. 数控铣床主要用于加工零件的平面和曲面“轮廓”，还可以加工________的零件，如凸轮、样板、模具、螺旋槽等。同时，也可以对零件进行钻、扩、铰、锪和镗孔加工。

4. G16 指令是极坐标编程，在使用 G16 指令后，X 代表________，Y 代表________。

四、简答题

1. 简述极坐标编程指令的使用方法及注意事项。

2. 查阅资料，列举极坐标编程的其他指令方法及特点。

3. 简述极坐标的适用场合。

4. 简述异形类零件加工难点，以及异形类零件的数控加工效率低的原因。

5. 简述偏心轮的工作原理，以及偏心轮与凸轮的区别。

自我学习检测评分表如表 6-7 所示。

表 6-7 自我学习检测评分表

项 目	目标要求	分 值	评分细则	得 分	备 注
学习关键知识点	(1)了解主加工工序的装夹方案 (2) 掌握角度编程指令(极坐标)的基本格式及使用方法 (3) 能够应用极坐标指令编写相应的加工程序	20	理解与掌握		
工艺准备	(1) 能够正确识读零件图 (2) 能够根据零件图分析、确定工艺过程 (3) 能够根据工序加工工艺，编写正确的加工程序	30	理解与掌握		
上机训练	(1) 会正确选择相应的设备与用具 (2) 能够正确操作数控车床，并根据加工情况调整加工参数	50	(1) 理解与掌握 (2) 操作流程		

思政小课堂

Projects Guidance

CNC machining technology has been widely used in the manufacturing industry at home and abroad. The rapid development of the CNC machining industry also puts forward higher requirements for CNC programming and operation technicians. Guided by practical engineering projects and oriented by practical application, the scientific inquiry ability and problem-solving ability of students can be better cultivated. Stirling engine outputs power through a cycle of cooling, compression, heat absorption and expansion of the working medium in the cylinder, so it is also called a hot gas engine. In this book, the relevant parts of the Stirling engine model (Figure 0-1) are taken as the carrier, and six typical projects (Figure 0-2) are taken as the main line. Focus on learning the professional knowledge and operation skills of CNC turning and milling, such as process formulation, CNC programming and operation.

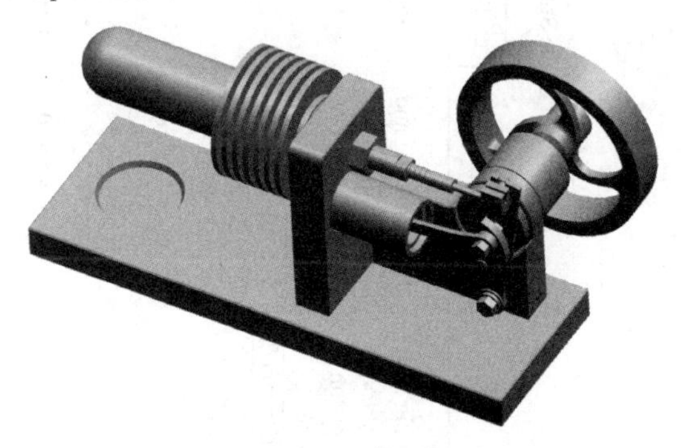

Figure 0-1 Model of the Stirling engine

This model of the Stirling engine contains 26 typical parts and several standard components. The professional knowledge and operation skills required for CNC machining of these 6 parts are studied and trained respectively to complete the machining of the parts and finally complete the assembly of the mechanism. Each project involves the learning of the specialized knowledge and skills necessary to complete the machining of parts, such as equipment, fixtures, cutters, basic instructions, and machine operations. According to the technological design of CNC machining, targeted learning and training are carried out to master the technological design, CNC programming and machine operation of CNC turning and milling.

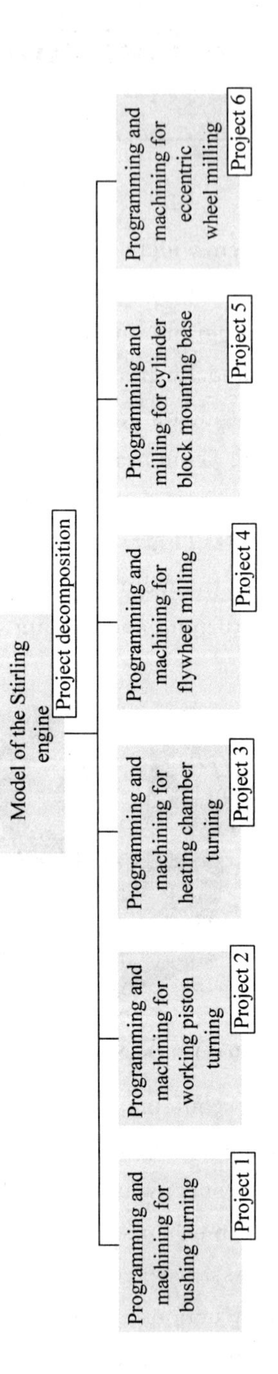
Model of the Stirling engine
Project decomposition
Programming and machining for bushing turning
Project 1
Programming and machining for working piston turning
Project 2
Programming and machining for heating chamber turning
Project 3
Programming and machining for flywheel milling
Project 4
Programming and milling for cylinder block mounting base
Project 5
Programming and machining for eccentric wheel milling
Project 6

Figure 0-2 Typical projects

Project 1 Programming and Machining Training for Bushing Turning

➢ Mind map

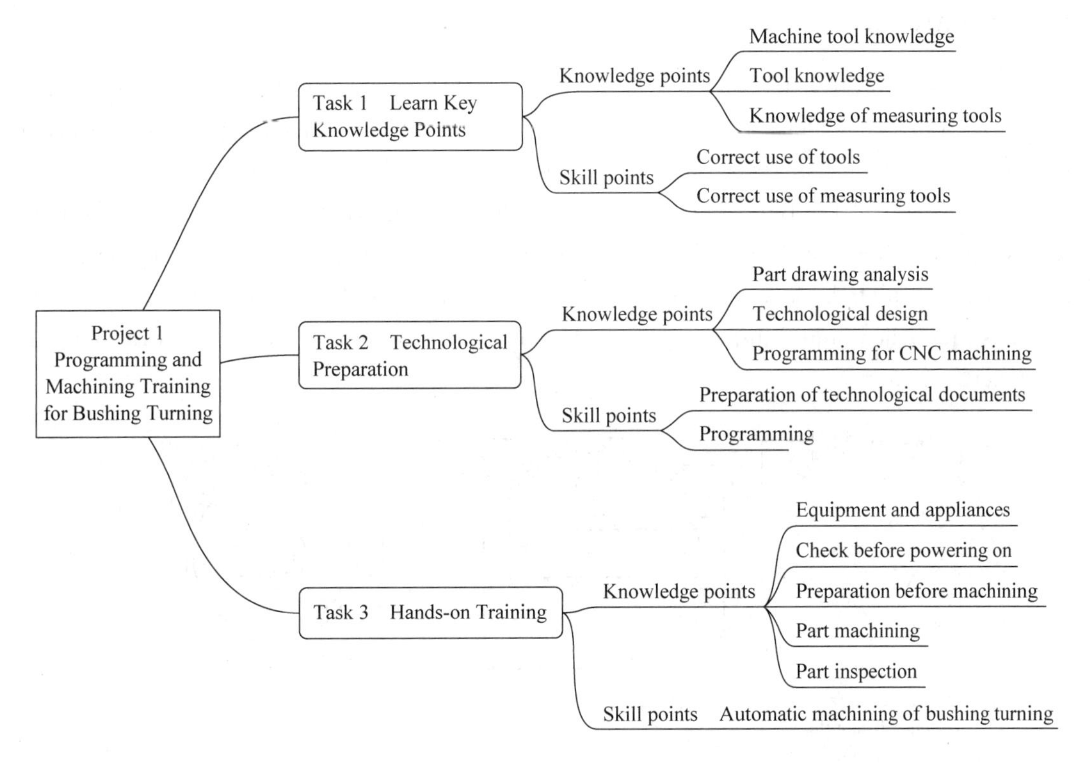

➢ Learning objectives

Knowledge objectives

(1) Be able to recognize drawings of sleeve parts.

(2) Know the use and characteristics of bushing block.

(3) Understand the meaning of each parameter of the longitudinal machining cycle instruction.

Ability objectives

(1) Master the selection and use of external turning cutters, hole-boring cutters, cut-off tools and external screw cutting tools.

(2) Be able to independently determine the process routine and fill in the technological documents correctly.

(3) Be able to operate the CNC lathe correctly and adjust the machining parameters according to the machining conditions.

(4) Be able to select measuring tools reasonably according to the accuracy and the structural characteristics of parts, and be able to measure the relevant dimensions correctly and normatively.

Literacy goals

(1) Develop students' scientific spirit and attitude.

(2) Cultivate students' engineering awareness.

(3) Develop students' teamwork skills.

➢ Task introduction

Sleeve parts refer to parts with holes, which are mainly used for support and guidance, and are widely used in production and life. The bushing part is a bearing support part, which has strict requirements for coaxiality between holes.

According to the requirements of the part drawing Figure 1-1, the processing technology and the CNC machining program are developed, and then the machining of the pin part is completed. As a typical sleeve part, it is made of 6061 aluminum alloy, and its blank is quenched and tempered, and the surface is required to be smooth without scratches.

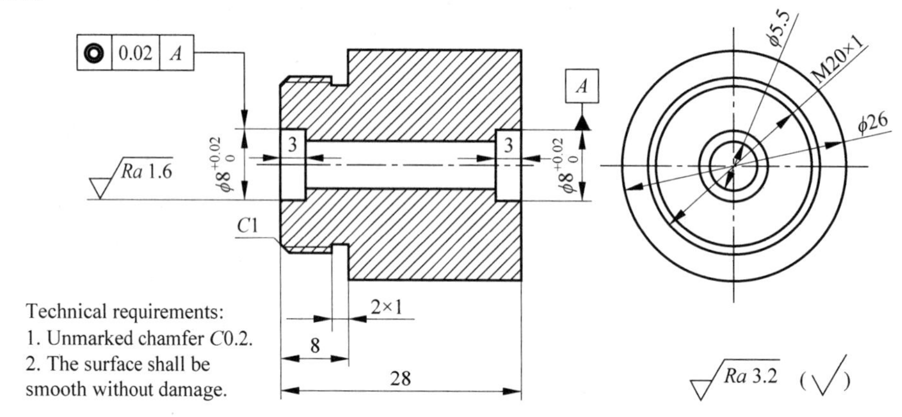

Figure 1-1 The part drawing of the bushing

Task 1 Learn Key Knowledge Points

1.1 Machine tool knowledge

1.1.1 Machine tailstock

The tailstock of CNC lathe is divided into automatic tailstock and manual tailstock according to the control mode. The center line of its sleeve coincides with the center line of the spindle and is mainly used for installing the lathe centers and drills. As shown in Figure 1-2, the manual tailstock is mainly composed of tailstock body, sleeve, handwheel, tailstock locking handle and sleeve locking handle.

Figure 1-2 Manual tailstock

1.1.2 Drill chuck

As shown in Figure 1-3, the drill chuck is mainly used for clamping tools such as center drills, straight shank drills and reamers to complete the processing of drilling center hole, drilling and reaming with tailstock of CNC lathe.

Figure 1-3 Drill chuck

1.1.3 Morse taper shank reduction sleeve

As shown in Figure 1-4, the Morse taper handle reduction sleeve is divided into 1-5 sizes according to the size specification, and has self-locking function. The reduction sleeve is used for the installation between the tailstock sleeve and the drill chuck, taper shank drill, reamer, etc. , and for the processing of drilling and reaming with the tailstock of CNC lathe. It can be used after direct installation, but the removal process needs to be completed with the help of wedges.

Figure 1-4 Morse taper shank reduction sleeve

1.2 Tool knowledge

1.2.1 Center drill

As shown in Figure 1-5, the length-diameter ratio of the center drill is small, so it has good rigidity. Its main function is to guide the twist drill to process a hole and reduce the error in processing.

Figure 1-5 Center drills

1.2.2 Drills

As shown in Figure 1-6, twist drills are the most common in hole processing, with a standard tool tip angle of 118°. A twist drill is used for drilling and is named because its chip holding groove is spiral and looks like a twist. The commonly used materials are high-speed steel and hard alloy. According to the shape of tool shank, it can be divided into straight shank drill and taper shank drill.

Figure 1-6 Twist drills

Figure 1-7 shows the hole-boring cutters, which are used for turning the inner hole. According to the installation position of the blade, they can be divided into left hand turning tools and right hand turning tools. When using, the appropriate blade shape should be selected according to the profile, and attention should be paid to whether the secondary cutting edge will crash with the workpiece.

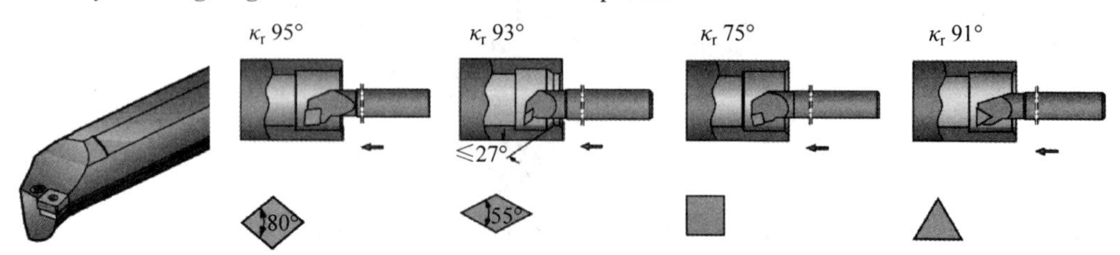

Figure 1-7 Hole-boring cutters

1.3 Knowledge of measuring tools

As shown in Figure 1-8, the inner diameter dial indicator is a measuring tool that uses the comparative measurement method to detect the diameter and shape accuracy of the inner hole and deep hole of the part. It is composed of an internal measuring lever-type measuring frame and a dial indicator. It is a measuring instrument that converts the linear displacement of the probe into the angular displacement of the pointer.

Figure 1-8 The inner diameter dial indicator

The use method is as follows.

(1) According to the size and tolerance of the measured dimension, first select a suitable micrometer (the common graduation value is 0.01mm, and the graduation value of the digital display is 0.001mm).

(2) Adjust the micrometer to the nominal size of the measured value and lock it.

(3) Hold the inner diameter dial indicator with the right hand and the micrometer with the left hand. Place the measuring head of the dial indicator in the micrometer, make the pressure gauge quantity around 0.3-0.5mm, and calibrate the indicator needle and set it to zero. Pay attention to make the measuring rod of the dial indicator perpendicular to the micrometer as far as possible.

(4) Use the adjusted internal dial indicator to measure the hole diameter of the part.

Task 2 Technological Preparation

1.4 Part drawing analysis

According to the operation requirements of the part, 45 steel is selected as the blank material of this part, and the blanking dimension is set as $\phi 30 \times 32$. Using the excircle of the blank with a diameter of $\phi 30$ as the rough reference, first process the $\phi 5.5$ through hole and the right $\phi 8$ step hole, and then process the $\phi 26$ outer circle, and finish machining to the size requirements. Then turn around, clamp the finished cylindrical surface of $\phi 26$, use a dial indicator to align to the coaxiality requirements, and ensure the length dimension of the cutting end face. Then machine the left $\phi 8$ step hole, tool retracting groove and M20 thread.

Note that when clamping the blank, attention should be paid to the extended length of the bar to avoid collision between the cutter and the chuck.

1.5 Technological design

According to the analysis of the part drawing, the technological process is designed as shown in Table 1-1.

Table 1-1 Technological process card

<table>
<tr><td colspan="2" rowspan="2">Machining process card</td><td colspan="2">Product model</td><td colspan="2">STL-00</td><td>Part number</td><td>STL-01</td><td colspan="2">Page 1</td></tr>
<tr><td colspan="2">Product name</td><td colspan="2">Stirling engine model</td><td>Part name</td><td>Bushing</td><td colspan="2">Total 1 page</td></tr>
<tr><td colspan="2">Material grade</td><td>6061</td><td>Blank size</td><td>$\phi 30 \times 32$</td><td>Blank quality</td><td>kg</td><td>Quantity</td><td colspan="2">1</td></tr>
<tr><td colspan="6">Working procedure</td><td rowspan="2">Work section</td><td rowspan="2">Technical equipment</td><td colspan="2">Man-hours/min</td></tr>
<tr><td>No.</td><td>Name</td><td colspan="4">Content</td><td>Preparation & conclusion</td><td>Single piece</td></tr>
<tr><td>5</td><td>Preparation</td><td colspan="4">Prepare the material according to the size of $\phi 30 \times 32$</td><td>Outsourcing</td><td>Sawing machine</td><td></td><td></td></tr>
<tr><td>10</td><td>Turning</td><td colspan="4">Using the excircle of $\phi 30$ as the rough reference, turn the end face</td><td>Turning</td><td>Lathe</td><td>10</td><td>5</td></tr>
<tr><td>15</td><td>Turning</td><td colspan="4">Keep the clamping position unchanged and process the $\phi 5.5$ through-hole</td><td>Turning</td><td>Lathe</td><td>10</td><td>5</td></tr>
<tr><td>20</td><td>Turning</td><td colspan="4">Keep the clamping position unchanged and process the right $\phi 8$ step hole</td><td>Turning</td><td>Lathe, Internal micrometer</td><td>15</td><td>10</td></tr>
<tr><td>25</td><td>Turning</td><td colspan="4">Keep the clamping position unchanged and process the $\phi 26$ outer circle</td><td>Turning</td><td>Lathe, Vernier caliper</td><td>10</td><td>8</td></tr>
<tr><td>30</td><td>Turning</td><td colspan="4">turn around, process the end face, and ensure the total length</td><td>Turning</td><td>Lathe, Vernier caliper</td><td>10</td><td>8</td></tr>
<tr><td>35</td><td>Turning</td><td colspan="4">Machine the left $\phi 8$ step hole</td><td>Turning</td><td>Lathe, Internal micrometer</td><td>15</td><td>10</td></tr>
<tr><td>40</td><td>Turning</td><td colspan="4">Machine the left thread major diameter and shaft shoulder</td><td>Turning</td><td>Lathe, micrometer</td><td>15</td><td>10</td></tr>
<tr><td>45</td><td>Turning</td><td colspan="4">Machine the tool retracting groove</td><td>Turning</td><td>Lathe, Vernier caliper</td><td>10</td><td>6</td></tr>
</table>

Continued

Machining process card		Product model	STL-00	Part number	STL-01	Page 1	
		Product name	Stirling engine model	Part name	Bushing	Total 1 page	
50	Turning	Machine the M20 thread		Turning	Lathe, Thread ring gauge	10	8
55	Cleaning	Clean the workpiece, debur sharp corner		Benching			5
60	Inspection	Check the workpiece dimensions		Examination			5
20	Inspection	Check the workpiece dimensions		Examination			5

In this training task, the 10th process, namely turning, is designed detailed, and the corresponding working procedure card is formulated as shown in Table 1-2.

Table 1-2 Working procedure card for turning

Machining working procedure card	Product model	STL-00	Part number	STL-01	Page 1	
	Product name	Stirling engine model	Part name	Bushing	Total 1 page	
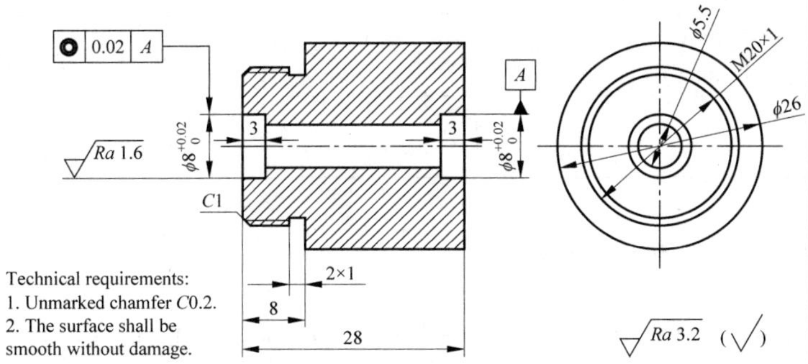 0.02 A; A; Ra 1.6; $\phi8^{+0.02}_{0}$; 3; C1; 2×1; 8; 28; $\phi5.5$; M20×1; $\phi26$; Ra 3.2 (√) Technical requirements: 1. Unmarked chamfer C0.2. 2. The surface shall be smooth without damage.				Procedure No.	01	
				Procedure name	Turning	
				Material	6061	
				Equipment	CNC lathe	
				Equipment model	CK6150e	
				Fixture	Three-jaw self-centering chuck	
				Measuring tool	Vernier caliper	
					Micrometer	
					Internal micrometer, Inner diameter dial indicator	
				Preparation & Conclusion time	100min	
				Single piece time	80min	

Continued

Work step	Content	Cutters	S/ (r/min)	F/ (mm/r)	a_p/ mm	Step hours/min	
						Mechanical	Auxiliary
1	Machine the right end face	End face turning tool	1200	0.18	1	5	5
2	Drill the center hole	A3 center drill	1600			5	5
3	Drill hole	ϕ5.5 drill	1300			5	5
4	Roughing the ϕ8 stepped hole	Hole-boring cutter	900	0.18	1	2	5
5	Finishing the ϕ8 step hole	Hole-boring cutter	1200	0.1	0.2	3	5
6	Roughing the ϕ26 cylindrical surface	Excircle turning tool	1100	0.2	1.5	2	5
7	Finishing ϕ26 cylindrical surface	Excircle turning tool	1300	0.12	0.2	5	5
8	Turn around and align, and machine the left end face	End face turning tool	1200	0.18	1	5	15
9	Roughing the left ϕ8 step hole	Hole-boring cutter	900	0.18	1	2	5
10	Finishing the left ϕ8 step hole	Hole-boring cutter	1200	0.1	0.2	3	5
11	Roughing the left thread major diameter and shaft shoulder	Excircle turning tool	1100	0.2	1.5	2	5
12	Finishing the left thread major diameter and shaft shoulder	Excircle turning tool	1300	0.12	0.2	5	5
13	Machine tool retracting groove	Cut-off tool (width: 2mm)	800	0.06		3	3
14	Machine M20 thread	External thread turning tool	600	1		5	5
15	Dismantling and cleaning workpieces						5

1.6 Programming for CNC machining

According to the technology of the working procedure, the machining program is written as shown in Table 1-3.

Table 1-3 CNC machining program for bushing turning

No.	Program statement	Annotation
	O0001;	**CNC machining program for the right side**
N1	T0202;	Call the hole-boring turning cutter

Continued

No.	Program statement	Annotation
	G97 G99 S900 M03;	Set constant rotative speed control, unit of feed mount is mm/r, spindle speed is 900r/min, forward rotation
	G0 X5 Z2 M8;	Rapidly locate to the beginning of the cycle (*X*5, *Z*2) and open the coolant. Note that the axial position of this point should be on the right side of the workpiece, and the radial position should not be greater than the diameter of the bottom hole
	G71 U1 R0. 5;	Call the longitudinal roughing cycle instruction
	G71 P10 Q20 U-0. 4 W0. 05 F0. 18;	Radial finishing allowance is 0. 4mm (unilateral 0. 2mm), the axial finishing allowance is 0. 05mm, and the feed rate is 0. 18mm/r. Note that the radial finishing allowance *U* when machining the inner hole and the wear compensation in the machine tool system are all negative values
N10	G0 G41 X8;	Call left compensation of nose radius instruction
	G1 Z-3;	
	X5. 4;	
N20	G1 G40 X5;	Cancel compensation of nose radius instruction
	G0 X100 Z150;	Quickly retract to the point(*X*100, *Z*150)
	M5;	Stop the spindle
	M9;	Coolant off
	M01;	Optional stop (the optional stop button needs to be pushed) to observe the completion of roughing
N2	T0202;	Call the hole-boring turning cutter
	G97 G99 S1200 M03;	Set constant rotative speed control instruction, unit of feed mount is mm/r, spindle speed is 1200r/min, forward rotation
	G0 X5 Z2 M8;	Rapidly locate to the beginning point of the cycle (*X*5, *Z*2) and open the coolant. Note that the axial position of this point should be on the right side of the workpiece, and the radial position should not be greater than the diameter of the bottom hole
	G70 P10 Q20 F0. 1;	Call the finishing cycle instruction
	G0 X100 Z150;	Quickly retract to the point(*X*100, *Z*150)
	M5;	Stop the spindle
	M9;	Coolant off
	M01;	Optional stop (the optional stop button needs to be pushed) to observe the completion of fine machining
N3	T0101;	Call the cylindrical turning tool
	G97 G99 S1100 M03;	Set constant rotative speed control instruction, unit of feed mount is mm/r, spindle speed is 1100r/min, forward rotation
	G0 X32 Z2 M8;	Rapidly locate to the beginning point of the cycle (*X*32, *Z*2) and open the coolant
	G71 U1. 5 R0. 5;	Call the longitudinal roughing cycle instruction
	G71 P30 Q40 U0. 4 W0. 05 F0. 2;	Radial finishing allowance is 0. 4mm (unilateral 0. 2mm), the axial finishing allowance is 0. 05mm, the feed rate is 0. 2mm/r

Continued

No.	Program statement	Annotation
N30	G0 G42 X7；	Call right compensation of nose radius instruction
	G1 Z0；	
	X26 C0.2；	Insert *C*0.2 chamfer
	Z-21；	
N40	G1 G40 X32；	
	G0 X100 Z150；	Quickly retract to the point(*X*100, *Z*150)
	M5；	Stop the spindle
	M9；	Coolant off
	M01；	Optional stop (the optional stop button needs to be pushed) to observe the completion of roughing
N4	T101；	Call the cylindrical turning tool
	G97 G99 S1300 M03；	Set constant rotative speed control instruction, unit of feed mount is mm/r, spindle speed is 1300r/min, forward rotation
	G0 X32 Z2 M8；	Rapidly locate to the beginning point of the cycle (*X*5, *Z*2) and open the coolant. Note that the axial position of this point should be on the right side of the workpiece, and the radial position should not be greater than the diameter of the bottom hole
	G70 P30 Q40 F0.12；	Call the finishing cycle instruction
	G0 X100 Z150；	Quickly retract to the point(*X*100, *Z*150)
	M5；	Stop the spindle
	M9；	Coolant off
	M30；	End of the program
	O0002；	**CNC machining program for the left side**
N1	T0202；	Call the hole-boring turning cutter
	G97 G99 S900 M03；	Set constant rotative speed control instruction, unit of feed mount is mm/r, spindle speed is 900r/min, forward rotation
	G0 X5 Z2 M8；	Rapidly locate to the beginning point of the cycle (*X*5, *Z*2) and open the coolant. Note that the axial position of this point should be on the right side of the workpiece, and the radial position should not be greater than the diameter of the bottom hole
	G71 U1 R0.5；	Call the longitudinal roughing cycle instruction
	G71 P10 Q20 U-0.4 W0.05 F0.18；	Radial finishing allowance is 0.4mm (unilateral 0.2mm), the axial finishing allowance is 0.05mm, the feed rate is 0.18mm/r. Note that the radial finishing allowance *U* when machining the inner hole and the wear compensation in the machine tool system are all negative values
N10	G0 G41 X8；	Call left compensation of nose radius instruction
	G1 Z-3；	
	X5.4；	
N20	G1 G40 X5；	Cancel left compensation of nose radius instruction
	G0 X100 Z150；	Quickly retract to the point(*X*100, *Z*150)

Continued

No.	Program statement	Annotation
	M5;	Stop the spindle
	M9;	Coolant off
	M01;	Optional stop (the optional stop button needs to be pushed) to observe the completion of roughing
N2	T0202;	Call the hole-boring turning cutter
	G97 G99 S1200 M03;	Set constant rotative speed control instruction, unit of feed mount is mm/r, spindle speed is 1200r/min, forward rotation
	G0 X5 Z2 M8;	Rapidly locate to the beginning point of the cycle (*X*5, *Z*2) and open the coolant. Note that the axial position of this point should be on the right side of the workpiece, and the radial position should not be greater than the diameter of the bottom hole
	G70 P10 Q20 F0.1;	Call the finishing cycle instruction
	G0 X100 Z150;	Quickly retract to the point(*X*100, *Z*150)
	M5;	Stop the spindle
	M9;	Coolant off
	M01;	Optional stop (the optional stop button needs to be pushed) to observe the completion of finishing
N3	T0101;	Call the cylindrical turning tool
	G97 G99 S1100 M03;	Set constant rotative speed control instruction, unit of feed mount is mm/r, spindle speed is 1100r/min, forward rotation
	G0 X32 Z2 M8;	Rapidly locate to the beginning point of the cycle (*X*32, *Z*2) and open the coolant
	G71 U1.5 R0.5;	Call the longitudinal roughing cycle instruction
	G71 P30 Q40 U0.4 W0.05 F0.2;	Radial finishing allowance is 0.4mm (unilateral 0.2mm), the axial finishing allowance is 0.05mm, the feed rate is 0.2mm/r
N30	G0 G42 X7;	Call right compensation of nose radius
	G1 Z0;	
	X19.9 C1;	Insert *C*1 chamfer
	Z-8;	
	X27;	
N40	G1 G40 X30;	Cancel right compensation of nose radius
	G0 X100 Z150;	Quickly retract to the point(*X*100, *Z*150)
	M5;	Stop the spindle
	M9;	Coolant off
	M01;	Optional stop (the optional stop button needs to be pushed) to observe the completion of roughing
N4	T101;	Call the cylindrical turning tool
	G97 G99 S1300 M03;	Set constant rotative speed control instruction, unit of feed mount is mm/r, spindle speed is 1300r/min, forward rotation
	G0 X32 Z2 M8;	Rapidly locate to the beginning point of the cycle (*X*32, *Z*2) and open the coolant. Note that the axial position of this point should be on the right side of the workpiece, and the radial position should not be greater than the diameter of the bottom hole

Continued

No.	Program statement	Annotation
	G70 P30 Q40 F0.12;	Call the finishing cycle instruction
	G0 X100 Z150;	Quickly retract to the point($X100$, $Z150$)
	M5;	Stop the spindle
	M9;	Coolant off
	M01;	Optional stop (the optional stop button needs to be pushed) to observe the completion of finishing
N5	T0303;	Call the grooving cutter (cutter width: 2mm)
	G97 G99 S800 M03;	Set constant rotative speed control instruction, unit of feed mount is mm/r, spindle speed is 800r/min, forward rotation
	G0 X28 Z2 M8;	
	G0 Z-8;	
	G1 X18 F0.06;	
	G1 X28 F0.3;	
	G0 X100 Z150;	Quickly retract to the point($X100$, $Z150$)
	M5;	Stop the spindle
	M9;	Coolant off
	M01;	Optional stop (the optional stop button needs to be pushed) to observe the completion of roughing
N6	T0404;	Call the external thread turning tool
	G97 G99 S600 M03;	Set constant rotative speed control instruction, unit of feed mount is mm/r, spindle speed is 600r/min, forward rotation
	G0 X22 Z2 M8;	Locate to the starting point of the cycle ($X22$, $Z2$) for thread processing
	G92 X19.3 Z-6.2 F1;	
	X18.9;	
	X18.7;	
	X18.7;	
	G0 X100 Z150;	Quickly retract to the point($X100$, $Z150$)
	M5;	Stop the spindle
	M9;	Coolant off
	M30;	The end of the program

Task 3 Hands-on Training

1.7 Equipment and appliances

Equipment: CK6150e CNC lathe.

Cutters: center drill, drill bit, cylindrical turning tool, hole-boring tool, grooving tool and thread turning tool (tool width: 2mm).

Fixture: the self-centering three-jaw self-centering chuck.

Tools: chuck wrenches, tool holder wrenches, etc.

Gauges: the 0-150mm vernier caliper, internal micrometer or internal diameter dial indicator, M20×1 thread ring gauge.

Blank: $\phi30\times32$.

Auxiliary appliances: cutter shim, brushes and so on.

1.8 Check before powering on

Refer to Table 1-4 to check the machine status.

Table 1-4 Preparing card for machine start-up

Check item		Test result	Abnormal description
Mechanical part	Spindle		
	Feed part		
	Tool holder		
	Three-jaw self-centering chuck		
Electrical part	Main power supply		
	Cooling fan		
CNC system	Electrical components		
	Controlling part		
	Driving section		
Auxiliary part	Cooling system		
	Compressed air system		
	Lubricating system		

1.9 Preparation before machining

Before machining, the tools required for this task should be prepared and installed correctly. The origin of the workpiece is set according to the process requirements, and the CNC machining program is entered and the graphic verification is performed.

1.10 Part machining

After the graphic verification process has verified that there is no problem, the parts machining can be carried out. Before the parts are processed, you should understand the safety operation requirements of the machine tool in detail, and wear labor protection clothing and utensils. When processing parts, you should be familiar with the functions and positions of the operation buttons of the CNC lathe, and understand the methods of dealing with emergency situations. During the machining process, especially before cutting, the actual distance between the cutter and the workpiece should be observed, and at the same time, the remaining movement amount displayed in the "Remaining

Movement Amount" column on the display screen should be compared. When the difference between the actual distance and the remaining movement amount is too large, the vehicle should be stopped and checked to avoid collision. If there is any abnormality, the machine tool movement should be stopped in time.

1.11 Part inspection

After the parts are processed, the workpiece should be carefully cleaned, and in accordance with the relevant requirements of quality management, the processed parts should be subject to relevant inspections to ensure the production quality. The "three-level" inspection cards for machined parts is shown in Table 1-5.

Table 1-5 "Three-level" inspection cards for machined parts

<table>
<tr><th>Part drawing number</th><th></th><th>Part name</th><th></th><th>Working step number</th><th></th></tr>
<tr><td>Material</td><td></td><td>Inspection date</td><td></td><td>Working step name</td><td></td></tr>
<tr><td>Inspection items</td><td>Self-inspection result</td><td>Mutual inspection result</td><td colspan="2">Professional inspection</td><td>Remark</td></tr>
<tr><td></td><td></td><td></td><td colspan="2"></td><td></td></tr>
<tr><td></td><td></td><td></td><td colspan="2"></td><td></td></tr>
<tr><td></td><td></td><td></td><td colspan="2"></td><td></td></tr>
<tr><td></td><td></td><td></td><td colspan="2"></td><td></td></tr>
<tr><td></td><td></td><td></td><td colspan="2"></td><td></td></tr>
<tr><td></td><td></td><td></td><td colspan="2"></td><td></td></tr>
<tr><td></td><td></td><td></td><td colspan="2"></td><td></td></tr>
<tr><td>Conclusion</td><td colspan="5">□ Qualified □ Unqualified □ Repair □ Concession to receive
Inspection signature:
Date:</td></tr>
<tr><td>Non-conforming item description</td><td colspan="5"></td></tr>
</table>

Project Summary

As a typical machining part of CNC lathe, shaft sleeve is widely used in various equipment. According to equipment conditions and precision requirements, there will be some differences in the processing technology. Programmers and operators need to formulate the processing technology reasonably according to the processing conditions to improve the processing accuracy of the parts and the production efficiency.

Exercises After Class

1. Choice questions

(1) When machining parts with many and complicated surfaces, the working

procedure is usually divided ( ).

A. according to the tool used

B. according to the processing site

C. according to the installation times

D. according to rough or finish machining

(2) The vernier caliper has 50 scale lines, which is the same width as the main ruler's 49 scale lines, so the minimum reading of this vernier caliper is ( ).

A. 0.1mm B. 2cm C. 0.02mm D. 0.4mm

(3) When measuring the hole diameter with an inner diameter dial indicator, the accurate measurement result should be obtained until ( ) is found in the axial direction.

A. the minimum value B. the average value

C. the maximum value D. the limiting value

(4) After the machine tool is powered on, check first whether ( ) is normal.

A. processing route

B. each switch button and key

C. voltage, oil pressure and processing route

D. workpiece accuracy

(5) For workpieces with high position accuracy requirements, it is not appropriate to use ( ).

A. special fixture B. universal fixture

C. combined fixture D. three-jaw self centering chuck

(6) ( ) instruction is the intermittent longitudinal cutting cycle instruction.

A. G74 B. G71 C. G72 D. G73

2. True or false

(1) The direction of the tool away from the workpiece in a certain axis direction is the negative direction of the coordinate axis. ( )

(2) A program segment should contain all functional words. ( )

(3) In CNC turning processing, the feed speed directly affects the surface roughness value and turning efficiency. ( )

(4) According to the control trajectory of the machine tool movement, the CNC lathe belongs to the point-to-point control. ( )

(5) The sleeve parts usually only play a supporting role. ( )

3. Fill in the blanks

(1) The CNC machine tool adopts ________ coordinate system.

(2) CNC programming is generally divided into ________ and ________.

(3) The main processing method for the cylindrical surface are ________ and ________.

(4) The processing action of CNC lathe is mainly divided into two parts: the motion of ________ and the motion of ________.

(5) The main contents of CNC programming are ________, determining the processing procedure, ________, preparing part machining program sheet; ________, ________ and try cutting of the first workpiece.

(6) ________ is a type of part that is often encountered in machining, and its application range is very wide.

4. Short answer questions

(1) What are the characteristics of CNC turning programming?

(2) What are the types of drills?

(3) What are the uses of Morse reduction sleeve?

(4) Briefly describe the use method of inner diameter dial indicator.

5. Comprehensive programming

According to the part drawing(Figure 1-9), determine the processing technology, write the program, and automatically process.

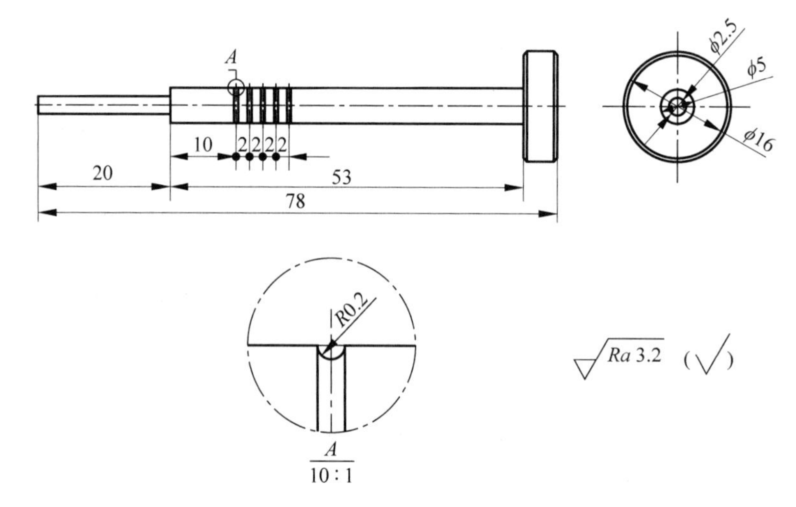

Figure 1-9 Part drawing for question 5

Self-learning test score sheet is shown in Table 1-6.

Table 1-6 Self-learning test score sheet

Task	Task requirements	Score	Scoring rules	Score	Remark
Learn key knowledge points	(1) Understand the basic structure of the machine tool (2) Master the basic knowledge of center drill and drill bit (3) Master the characteristics and use method of the inner diameter dial indicator	20	Understand and master		

Continued

Task	Task requirements	Score	Scoring rules	Score	Remark
Technological preparation	(1) Be able to correctly read the part drawings (2) Be able to analyze and determine the process according to the part drawing (3) Be able to write the correct program according to the process	30	Understand and master		
Hands-on training	(1) The corresponding equipment and utensils will be selected correctly (2) Be able to correctly operate the CNC lathe and adjust the processing parameters according to the machining situation	50	(1) Understand and master (2) Operation process		

Ideological and Political Classroom

Project 2　Programming and Machining Training for Working Piston Turning

➢ Mind map

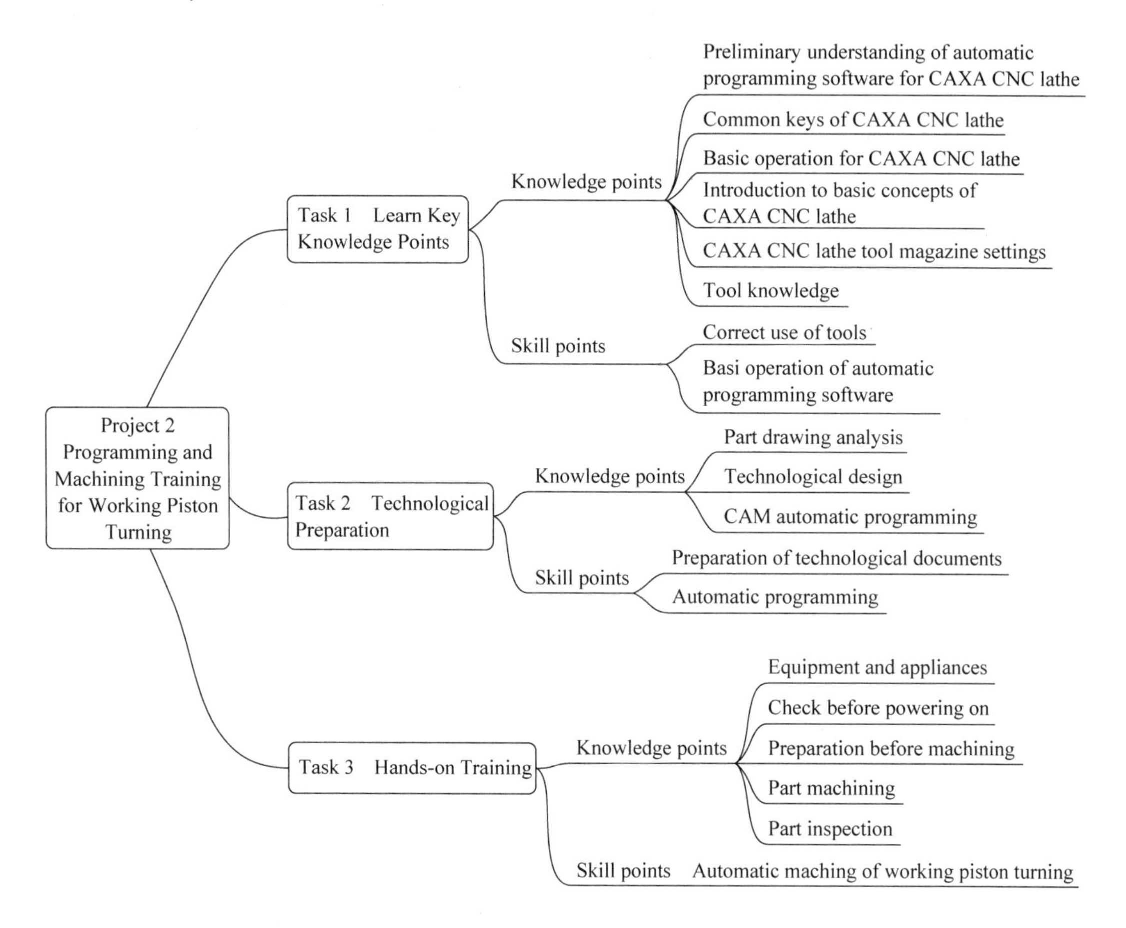

➢ Learning objectives

Knowledge objectives

(1) Understand the processing characteristics of circular groove.

(2) Understand the application characteristics of CAM software of CAXA CNC lathe.

(3) Understand the basic principle of automatic programming.

Ability objectives

(1) Be able to independently determine the process routine and fill in the technological

documents correctly.

(2) Be able to reasonably select the processing method of "CAXA CNC lathe", reasonably set the processing parameters, and generate the CNC processing program.

(3) Be able to select measuring tools reasonably according to the accuracy and the structural characteristics of parts, and be able to measure the relevant dimensions correctly and normatively.

Literacy goals

(1) Develop students' scientific spirit and attitude.

(2) Cultivate students' engineering awareness.

(3) Develop students' teamwork skills.

➢ Task introduction

The piston cooperates with the cylinder block and is driven by the connecting rod to complete the three auxiliary stroke functions of intake, compression and exhaust. And convert linear motion into circular motion to realize power output. As the main part of engine, piston is widely used in the field of engine.

According to the requirements of the part drawing Figure 2-1, the processing technology is formulated, the CNC machining program is developed, and the processing of handle part is completed. The blank material of this part is yellow brass, which is required surface smoothing.

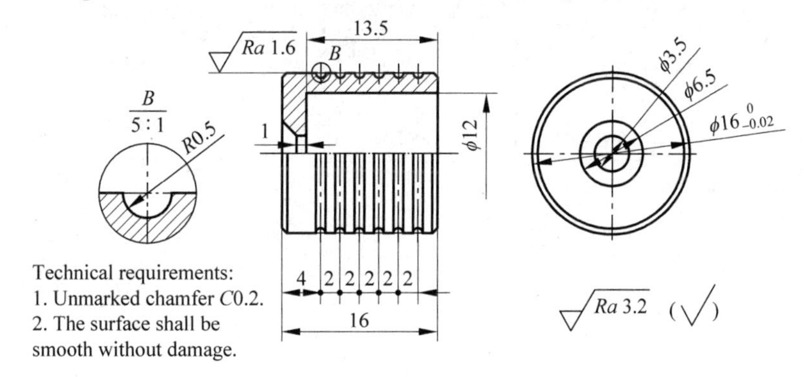

Figure 2-1 The part drawing of working piston

Task 1 Learn Key Knowledge Points

2.1 Preliminary understanding of automatic programming software for CAXA CNC lathe

CAXA CNC lathe has powerful drawing function of CAD software and perfect external data interface. It can draw any complex graphics and exchange data with other systems through DXF, IGES and other data interfaces. CAXA CNC lathe has the characteristics of powerful function, simple track generation and general post-processing function. The software provides a powerful and concise path generation method, which can generate a variety of complex graphics according to the processing requirements. The general post-processing module enables CAXA CNC lathe to meet the code format of various machine tools, output G code, and verify and process the generated code. CAXA CNC lathe provides a good solution for your two-dimensional drawing and CNC lathe processing. Combining CAXA CNC lathe with CAXA professional design software will fully meet your CAD/CAM requirements. CNC machine tool automatically processes the parts according to the processing program prepared in advance. The programmer compiles the process route, process parameters, tool motion path, displacement, cutting parameters and auxiliary functions of the parts into a processing program sheet according to the command code and program format specified by the CNC machine tool. Then record the contents of this program sheet on the control medium, and then input them into the numerical control device, so as to control the machine tool to realize the processing of parts.

Figure 2-2 shows the software interface of CAXA CNC lathe 2020, including menu bar, toolbar, title bar, management tree bar status bar and plot area.

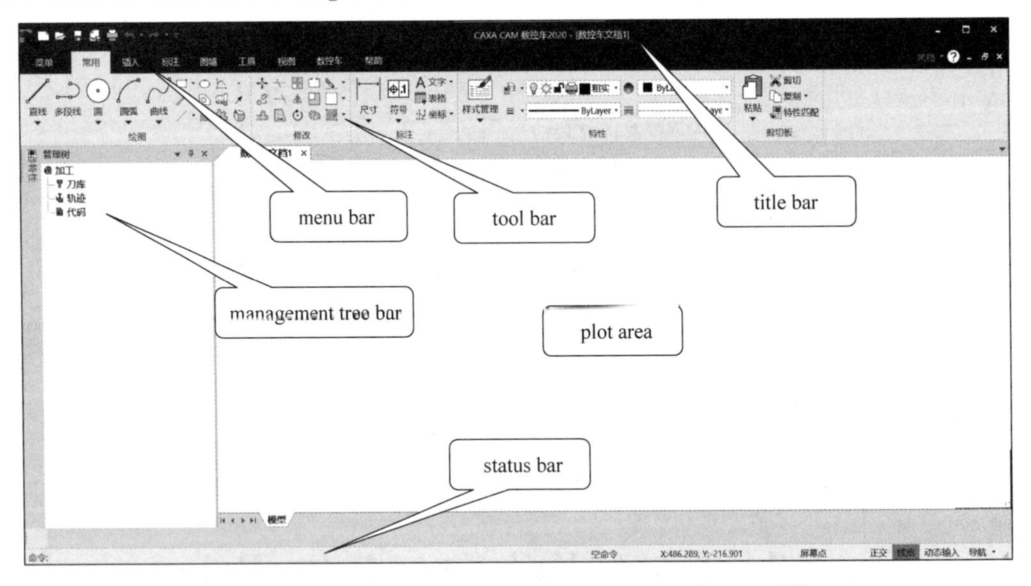

Figure 2-2 The software interface of CAXA CNC lathe 2020

2.2 Common keys of CAXA CNC lathe

2.2.1 Basic key and shortcut key operation

The left mouse button can be used to activate the menu to determine location points, pick elements, etc. For example, to run the line drawing function, first move the mouse cursor to the "Line" icon and then press the left mouse button to activate the line drawing function. At this moment, the prompt for the next operation appears in the command prompt area: "Enter the starting point". Move the mouse cursor to the drawing area, press the left mouse button, enter a location point, and then enter a second location point according to the prompt, and then a line is generated.

The right mouse button is used to confirm the pick, end operation, and end commands. For example, after picking the elements to be deleted, press the right mouse button to end picking, and the picked elements will be deleted. Another example, in the function of generating spline curve, after inputting a series of points in sequence, press the right mouse button to end the operation of inputting points, and the spline curve is generated.

Enter key and numeric key. In CAXA CNC lathe, when the system requires to enter a point, enter key and numeric key can activate a coordinate input bar in which coordinate values can be entered. If the coordinate value starts with @, it represents a relative coordinate relative to the previous input point. In some cases, you can also enter a string.

The spacebar will pop up the Point Tools menu. For example, when the system requires you to enter a point, press the spacebar to pop up the Point Tools menu.

Hotkey. CAXA CNC lathe provides users with hotkey operation. For a skilled CAXA CNC lathe user, hotkeys will greatly improve work efficiency. Users can also customize the desired hotkeys. See Table 2-1 for several function hotkeys set in CAXA CNC lathe.

Table 2-1 Common shortcut keys and definition methods of CAXA CNC lathe

Shortcut key	Definition	Shortcut key	Definition	Shortcut key	Definition
F1	Help files	F3	Show all	F5	Coordinate system switch
F6	Snap mode switch	F7	3D view navigation switch	F8	Orthogonal mode switch
F9	Interface switch	Delete	Delete	Ctrl+P	Print

Other shortcut keys are consistent with common software, such as Ctrl+C to copy.

2.2.2 Immediate menu

Immediate menu is a unique interactive mode provided by CAXA CNC lathe. The interactive mode of the immediate menu greatly improves the interactive process. The traditional interaction mode is a completely sequential step-by-step question and answer mode, and users need to input one by one according to the interaction mode route set by the system. The system provides the immediate menu when using the immediate menu interaction mode: during the interaction process, if necessary, you can modify the default values provided in the immediate menu at any time, breaking the complete sequence of interaction process.

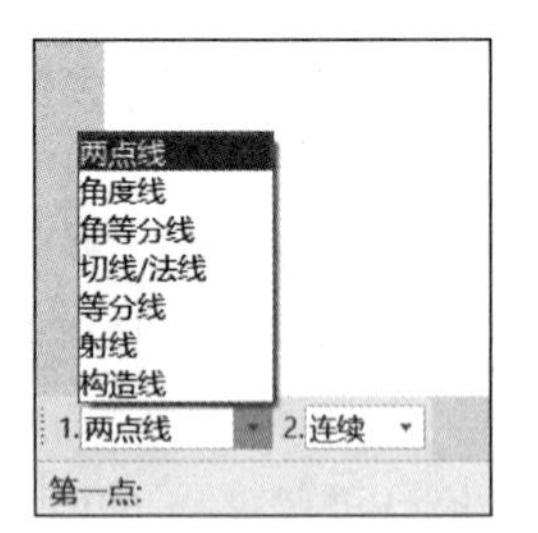

Figure 2-3 Example of the line immediate menu

Another main function of the immediate menu is to control the options of the function. Thanks to this mechanism of the immediate menu, the function can be closely organized. For example, the immediate menu option in Figure 2-3 is provided in the Line function. There are two-point line, angle line, angle equipartition line, curve tangent/normal line, equipartition line, ray and tectonic line in the way of generating straight line.

2.2.3 Input of points

In the process of interaction, it is often encountered to input precise positioning points. In this case, the system provides the Point Tools menu. You can use the Point Tools menu to precisely locate a point. The representation of the Point Tools is shown in Figure 2-4.

Use the spacebar to activate the Point Tools menu. For example, for generating a line, after the system prompts Enter start point, press the spacebar to pop up the Point Tools menu. Select a point positioning method according to the required method. You can also use the hotkey to switch to the desired point state. Hotkey is the first letter of each point in the menu. For example, if you need to locate the center of a circle for generating a line, after the system prompts Enter start point, press C to switch the point state to the center point state. The following is the specific meaning of various point states.

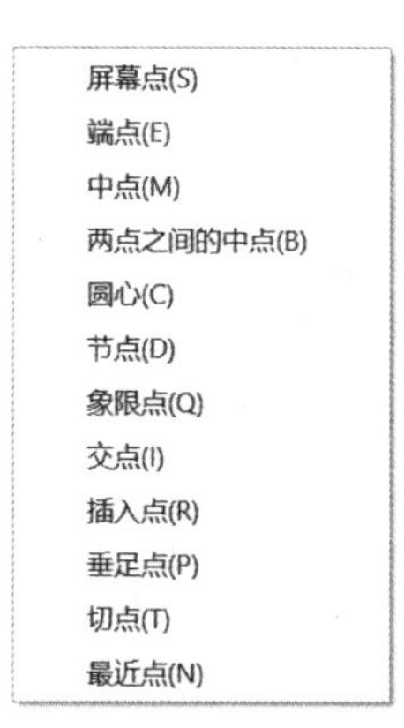

Figure 2-4 Point Tools menu

(1) Screen point (S): the point on the current plane taken by the mouse on the screen.

(2) End point (E): the start and end point of the curve, whichever is closer to the pick point.

(3) Midpoint (M): the arc length bisection point of the curve.

(4) Intersection point (I): the intersection of two curves, whichever is closer to the pick point.

(5) Center point (C): the center of a circle or arc.

(6) Quadrant point (Q): coordinate quadrant point of a given point.

(7) Perpendicular point (P): used for making vertical line.

(8) Tangent point (T): tangent and tangent arc.

(9) Nearest point (N): the point on the curve closest to the input point.

All kinds of points can be entered quadrant points, which can be inputted with one of rectangular coordinate system, polar coordinate system and spherical coordinate system.

The system provides immediate menu switch and input value. In the default point state, the system automatically determines the endpoint, midpoint, intersection, screen point according to the mouse position. When entering the system, the system's point status is default point. The user can choose whether to lock the status of the Point Tools or not, which can be done in the System parameter setting function (the user can select the corresponding options according to needs and habits, see the later introduction for details). When the Point Tools status is locked, once the Point Tools status is specified, it will not change until it is reassigned. However, the incremental point is the exception. It will be restored to the non-relative point state after use. When you choose not to lock the tool point status, the Point Tools will be restored to the default point status after being used once.

2.3 Basic operation of CAXA CNC lathe

2.3.1 Concept of objects

In the electronic drawing board, various curves, text, blocks and other drawing element entities drawn in the drawing area are called primitive objects, or objects for short. A solid that can be picked individually is an object. In the electronic drawing board, objects such as blocks can also contain several sub-objects. In addition to editing environment parameters, also contains the process of generating and editing objects during the process of drawing in the electronic drawing board.

2.3.2 Pick objects

In the electronic drawing board, if you want to operate on the generated objects, you must pick the objects. The methods of picking objects can be divided into point selection, box selection and all selection. The selected objects are highlighted. The specific effect of highlighting can be set in the system options.

1. Point selection

Point selection means to move the cursor to the line or entity within the object and click the left button. The entity will be directly selected.

2. Box selection

Box selection refers to selecting two diagonal points in the drawing area to form a selection box to pick objects. Box selection can not only select a single object, but also select multiple objects at one time. Box selection can be divided into two forms: positive selection and reverse selection.

Positive selection means that the first corner is on the left and the second corner is on the right during the selection process. That is, the abscissa of the first point is smaller than the second point. When the selection is positive, the color of the selection box is blue and the box line is solid, and the object will be selected only when all its points are in the selection box.

Reverse selection means that the first corner is on the right and the second corner is on the left during the selection process. That is, the abscissa of the first point is greater than the second point. When the selection is reversed, the color of the selection box is green and the box line is dashed, and the object will be selected as long as one point of the object is in the selection box.

3. All selection

All selection can pick all the objects that can be selected in the drawing area at one time. Select All has the shortcut key Ctrl+A. It should be noted that the pick filter settings will also affect the entities that can be selected by selecting all.

In addition, when an object has been selected, you can still use the above method to add a pick directly on the basis of the existing selection.

2.3.3 Deselection

After using the general command to end the operation, the selected objects will also be automatically deselected. If you want to cancel all current selections manually, you can click ESC on the keyboard. You can also call the Deselect All instruction by right-click menu in the drawing area to cancel all selections. If you want to cancel the selection status of one or several objects in the current selection set, you can press the Shift key on the keyboard to select the objects to be removed.

2.3.4 Command operation

In the electronic drawing board, commands must be called no matter what operation is performed. There are three methods to call commands: left clicking the main menu or icon, keyboard commands and shortcut keys. Left clicking the main menu or icon means finding the option or icon of the command in the main menu, toolbar or function area, and left click to call.

2.3.5 Command status

The command status of the electronic drawing board can be divided into three types, namely "empty command status", picking entity status and execution command status.

In Empty Command Status, you can directly enter the arithmetic formula to obtain the calculation result, it is shown as follows.

Enter the 8−2 command, and the input area will display 8−2=6.

Enter the 4 * 3 command, and the input area will display 4 * 3=12.

Enter the 2^3 command, and the input area will display 2^3=8.

2.4 Basic operation for CAXA CNC lathe

The CAXA CNC lathe provides integrated tool magazine management as shown in Figure 2-5. The tool magazine management function refers to the management of various tool types such as contour turning tools, grooving tools, thread tools, drilling tools, etc., which is convenient for users to obtain tool information from the tool magazine and maintain the tool magazine.

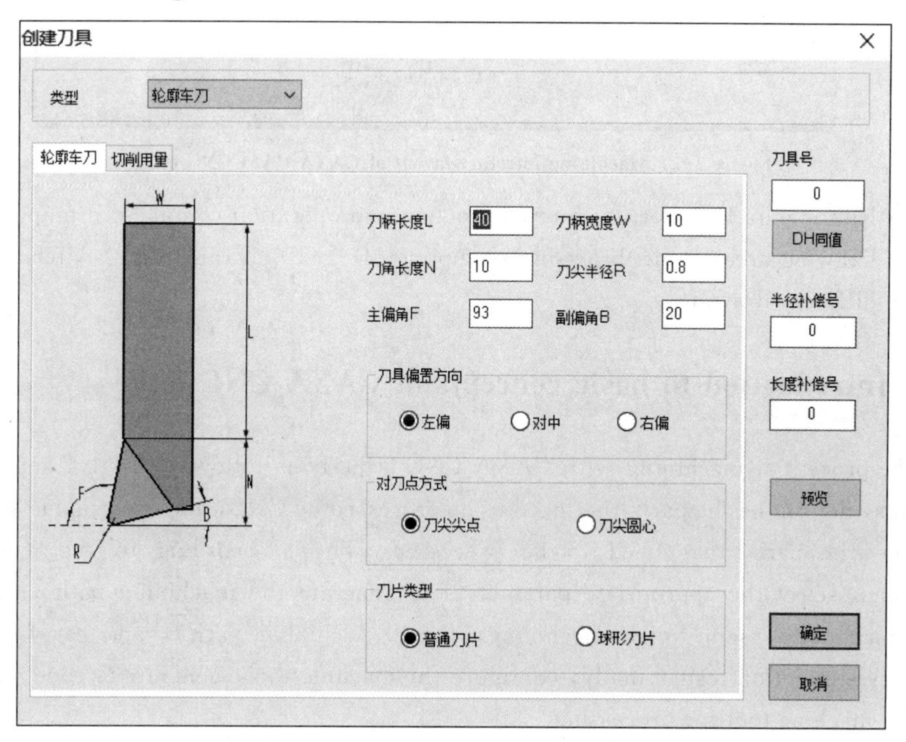

Figure 2-5 Create Tool dialog box of CAXA CNC lathe

In the 2020 version, it is convenient to create a tool directly through the Create Tool command or by using Create Tool in the right-click menu. Moreover, the cutting quantity of this tool are associated with the geometric parameters, and its cutting elements are also called while calling this tool, which simplifies the call time of the tool on the cutting quantity.

2.5 Introduction to basic operations of CAXA CNC lathe

The layout of the machining interface is shown in Figure 2-6.

(1) Tabs: All function commands can be found in the tab area.

(2) Management tree: All tools, machining paths and G code information will be recorded and displayed on the management tree.

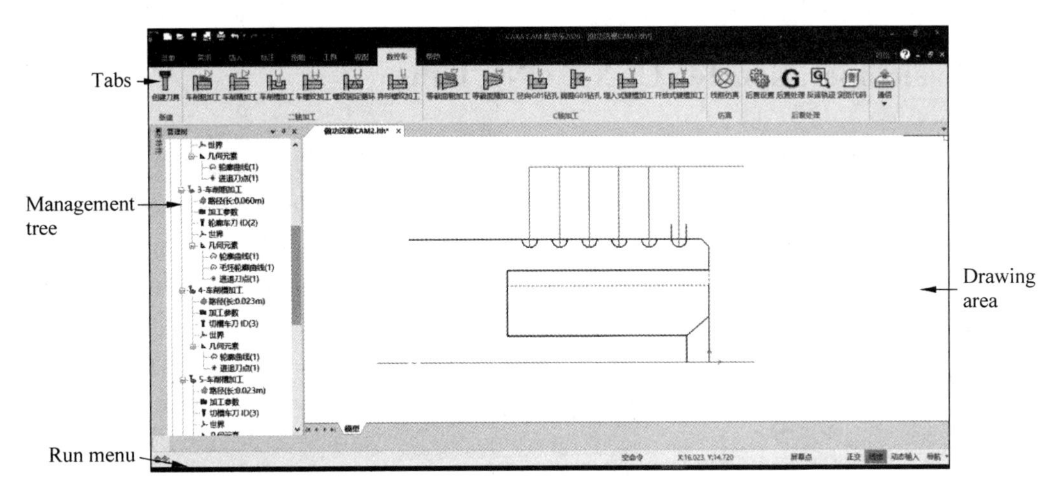

Figure 2-6　Machining interface layout of CAXA CAM CNC lathe

(3) Run menu: Run options for tab function and operation command prompt.

(4) Drawing area: multi-browsing is supported, and you can freely switch and edit between different drawings.

2.6　Introduction to basic concepts of CAXA CNC lathe

The process of machining with CAXA CNC lathe is as follows. Firstly, analyze the drawings, determine the parts that need to be processed by CNC, and use graphic software to shape the parts that need to be processed. Then, according to the machining conditions, select the appropriate parameters to generate the machining path (including rough machining, semi-finish machining, finish machining path), and carry out the trajectory simulation test. Finally, configure the machine tool, generate G code and send it to the machine tool for processing.

2.6.1　Two-axis machining

In CAXA CNC lathe, the *Z* axis of the machine coordinate system is the *X* axis of the absolute coordinate system, and the plane graphics are uniformly projected onto the *XOY* plane of the absolute coordinate system.

2.6.2　Contour

The contour is a collection of a series of end-to-end curves, as shown in Figure 2-7.

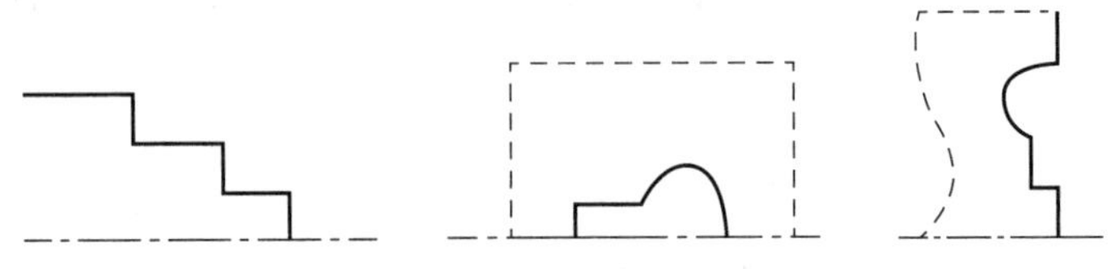

Figure 2-7　Contours

2.6.3 Blank contour

For rough turning, it is necessary to appoint the blank of the processed body. The blank contour is a collection of a series of end-to-end curves, as shown in Figure 2-8.

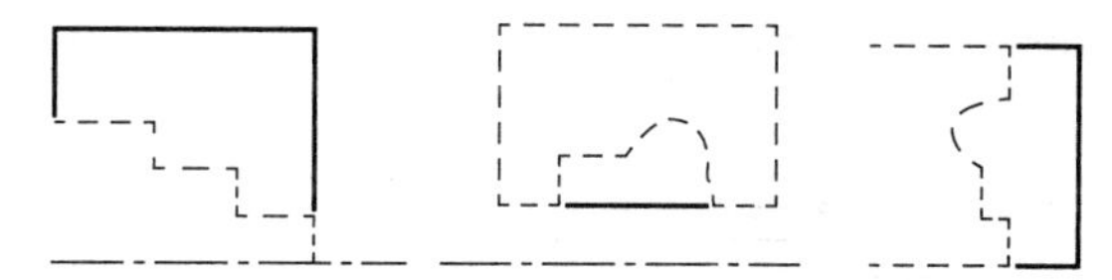

Figure 2-8 Blank contour

When interactively specifying the graphics to be processed in CNC programming, users are often required to specify the contour of the blank to define the surface to be processed or the blank itself to be processed. If the blank contour is used to define the machined surface, the specified contour is required to be closed. If the blank contour itself is processed, the blank contour can also not be closed.

2.6.4 Speed parameters of CNC lathe

The speed parameters of CNC lathe, including spindle speed, approaching speed, feed rate and tool withdrawal speed, are shown in Figure 2-9.

(1) The spindle speed is the angular velocity of the machine tool spindle rotation during cutting.

(2) The feed rate is the linear velocity of the tool during normal cutting.

(3) The approaching speed is the linear speed of the tool from the feed point to the front of the workpiece, also known as the feeding speed.

(4) The tool withdrawal speed is the linear speed of the tool when the tool leaves the workpiece and returns to the tool withdrawal position.

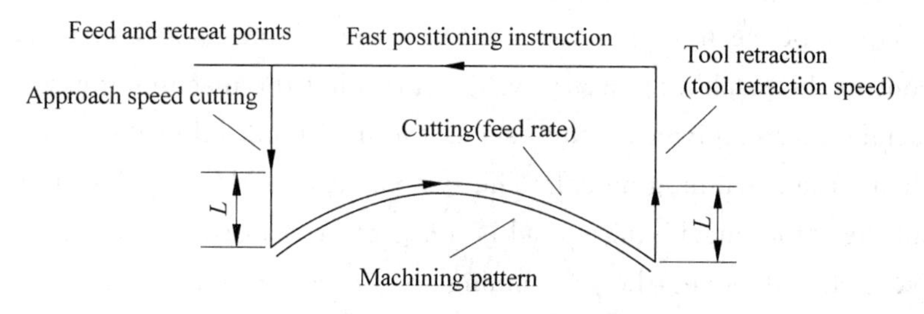

Figure 2-9 The speed parameters of CNC lathe

The setting of these speed parameters generally depends on the user's experience. In principle, they are related to the machine itself, the material of the workpiece, the tool material, the machining accuracy and the surface rough requirements of the workpiece.

Speed parameters are closely related to processing efficiency.

2.6.5 Tool path and cutter-location points

The tool path is the path of the cutter generated by the system according to the given process requirements when cutting a given machining pattern, as shown in Figure 2-10. The system displays the tool path graphically. The tool path consists of a series of ordered cutter-location points and straight lines (linear interpolation) or arcs (arc interpolation) connecting these cutter-location points.

The tool path of this system is displayed according to the position of the tool tip.

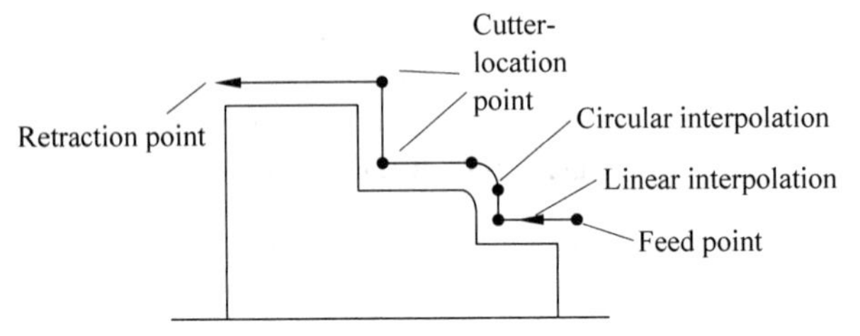

Figure 2-10 Tool path and cutter-location points

2.6.6 Machining allowance

Turning is a process of removing surplus, that is, removing surplus materials gradually from the blank to get the required parts. This process often consists of rough machining and finishing, and semi-finishing is also required when necessary. That is, it needs to go through multiple processes. In the previous process, it is often necessary to leave a certain margin for the next process.

The actual machining model is the result of the specified machining model equidistant according to the given machining allowance. As shown in Figure 2-11.

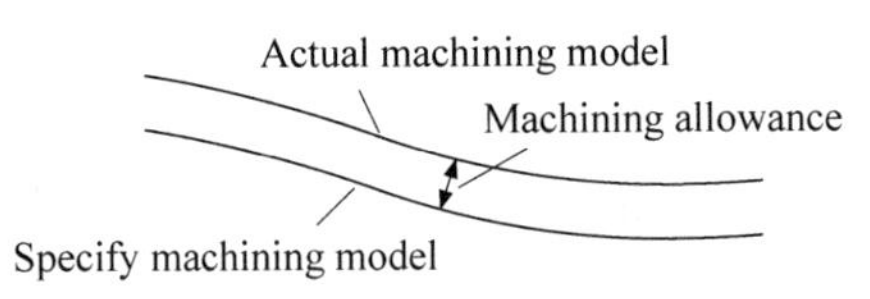

Figure 2-11 Machining allowance

2.6.7 Machining error

The deviation between tool path and actual machining model is machining error. Users can control the machining accuracy by controlling the machining error.

The machining error given by the user is the maximum allowable deviation between the tool path and the machining model. The system ensures that the deviation between the tool path and the actual machining model is not greater than the machining error.

The user should give the machining error according to the actual process requirements. For example, the machining error can be relatively large during rough machining, otherwise the processing efficiency will be affected unnecessarily. When finishing, the machining error shall be given according to the surface requirements.

In two-axis machining, there is no machining error for straight line and arc machining. Machining error refers to the error when the spline is approached by a broken line segment when machining. As shown in Figure 2-12.

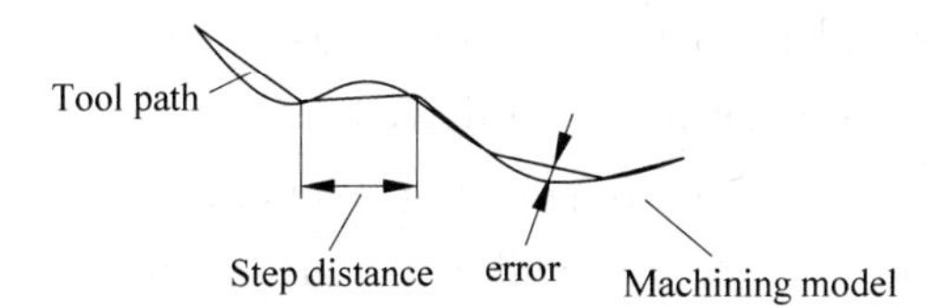

Figure 2-12 Machining error

2.6.8 Machining interference

When cutting the machined surface, the tool cuts the part that should not be cut, which is called interference or over-cutting.

In CAXA CNC lathe system, interference is divided into the following two situations:

(1) The over-cutting phenomenon occurs when there is part of the machined surface that cannot be cut by the tool.

(2) There is over-cutting between the tool and the unprocessed surface during cutting.

2.7 CAXA CNC lathe tool magazine settings

Before processing with CNC lathe software, the tool, CNC system and machine tool need to be set up, which will directly affect the machining trajectory and the generated G code. This chapter will introduce these settings in detail.

This function defines and determines the relevant data of the tool, so that the user can obtain the tool information from the tool magazine and maintain the tool magazine. The tool magazine management function includes the management of four tool types: contour turning tools, grooving tools, thread turning tools and drilling tools.

Operation method:

(1) Select menu bar→CNC lathe→Create Tool, and the system will pop up the Create Tool dialog box, where users can add new tools according to their needs. The newly created tool list will be displayed under the tool magazine node of the management tree on the left side of the drawing area.

(2) Double-click the tool node under the tool magazine node to open the tool editing dialog box to change the tool parameters.

(3) Select the Export Tools command from the menu that pops up after right-clicking the tool magazine node to save the information of all tools to a file.

(4) Select the Import Tools command from the menu that pops up after right-clicking on the tool magazine node to read all the tool information saved in the file into the document and add it to the tool magazine node.

(5) It should be pointed out that all kinds of tools in the tool magazine are only abstract descriptions of the same type of tools, and are not detailed tool magazine

conforming to national standards or other standards. Therefore, only some parameters that have an impact on path generation are listed, while other tool parameters related to specific machining processes are not listed. For example, all kinds of external contour, internal contour, and end face coarse and fine turning tools are classified as contour turning tools, which has no effect on track generation.

2.7.1 Contour turning tools

In Contour turning tools tab as shown in Figure 2-13, the specific parameters that need to be configurated are as follows.

(1) Tool No. : the serial number of the tool, which is used for the automatic tool change command of post-processing. The tool number is unique and corresponds to the tool magazine of the machine tool.

(2) Tool compensation No. : the serial number of the tool radius compensation value, whose value corresponds to the database of the machine tool.

(3) Tool shank length: the length of the clamping section of the tool.

(4) Tool shank width: the width of the clamping section of the tool.

(5) Tool angle length: the length of the cutting section of the tool.

(6) Nose radius: the radius of the arc used for cutting in the tool tip part.

(7) Main deflection angle: the included angle between the front edge of the tool and the rotation axis of the workpiece.

(8) Secondary deflection angle: the included angle between the rear edge of the tool and the rotating axis of the workpiece.

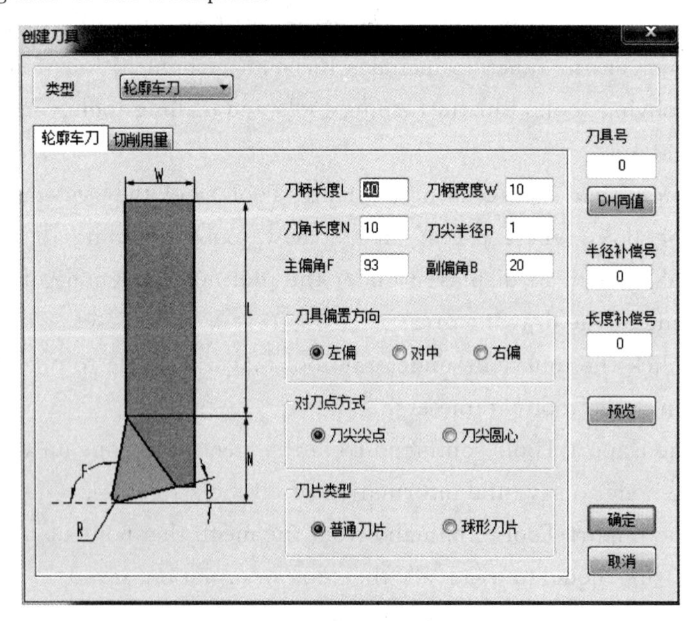

Figure 2-13 Contour turning tool tab

2.7.2 Grooving tools

In Grooving tool tab as shown in Figure 2-14, the specific parameters that need to be configurated are as follows.

(1) Tool No. : the serial number of the tool, which is used for the automatic tool change command of post-processing. The tool number is unique and corresponds to the tool magazine of the machine tool.

(2) Tool compensation No. : the serial number of the tool radius compensation value, whose value corresponds to the database of the machine tool.

(3) Tool shank length: the length of the clamping section of the tool.

(4) Tool shank width: the width of the clamping section of the tool.

(5) Tool angle length: the length of the cutting section of the tool.

(6) Nose radius: the radius of the arc used for cutting in the tool tip part.

(7) Main deflection angle: the included angle between the front edge of the tool and the rotation axis of the workpiece.

(8) Secondary deflection angle: the included angle between the rear edge of the tool and the rotating axis of the workpiece.

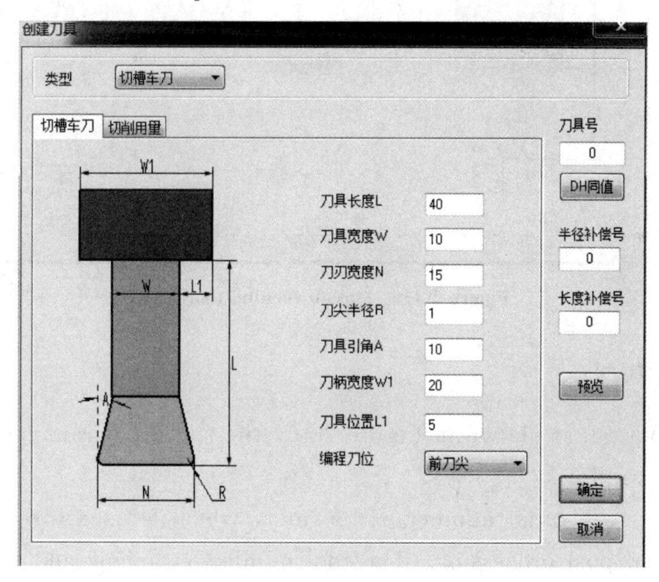

Figure 2-14 Grooving tool tab

2.7.3 Thread turning tools

In Thread turning tool tab as shown in Figure 2-15, the specific parameters that need to be configurated are as follows.

(1) Tool No. : the serial number of the tool, which is used for the automatic tool change command of post-processing. The tool number is unique and corresponds to the tool magazine of the machine tool.

(2) Tool compensation No.: the serial number of the tool radius compensation value, whose value corresponds to the database of the machine tool.

(3) Tool shank length: the length of the clamping section of the tool.

(4) Tool shank width: the width of the clamping section of the tool.

(5) Blade length: the length of the top of the main cutting edge of the tool.

(6) Tool nose width: the width of thread root.

(7) Tool angle: the angle between the two sides of the cutting section of the tool and the direction perpendicular to the cutting direction, which determines the tooth shape angle of the turned thread.

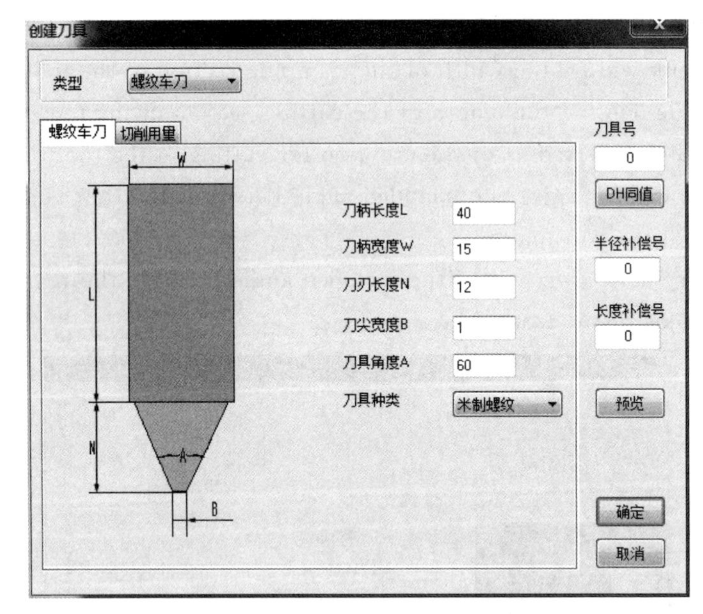

Figure 2-15 Thread turning tool tab

2.7.4 Drilling tools

In Drilling tool tab as shown in Figure 2-16, the specific parameters that need to be configurated are as follows.

(1) Tool No.: the serial number of the tool, which is used for the automatic tool change command of post-processing. The tool number is unique and corresponds to the tool magazine of the machine tool.

(2) Tool compensation No.: the serial number of the tool compensation value, whose value corresponds to the database of the machine tool.

(3) Diameter: the diameter of the tool.

(4) Nose angle: the angle of the nose of the drill.

(5) Blade length: the length of the cutter arbor that can be used for cutting.

(6) Cutter arbor length: the distance between the tool nose and the tool shank. The cutter arbor length should be greater than the effective blade length.

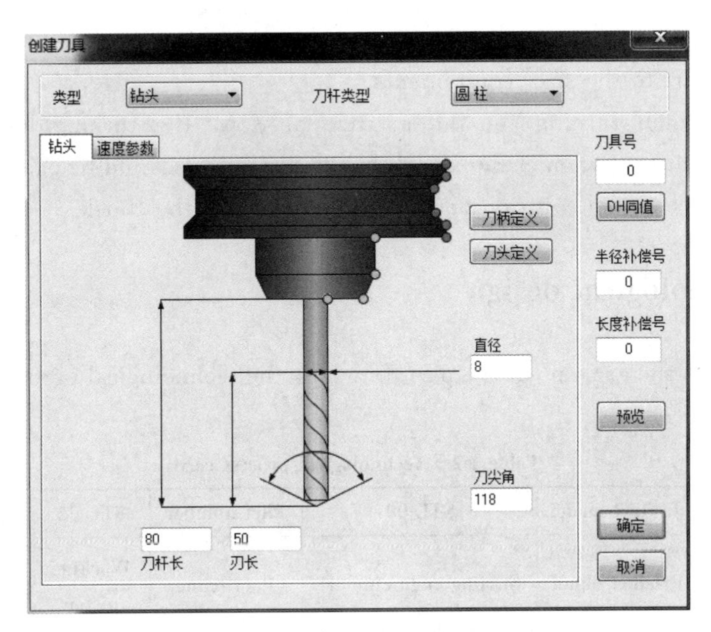

Figure 2-16 Drilling tool tab

2.8 Tool knowledge

As shown in Figure 2-17, this tool is an arc grooving tool, which belongs to the forming tool, and is used for turning arc grooves.

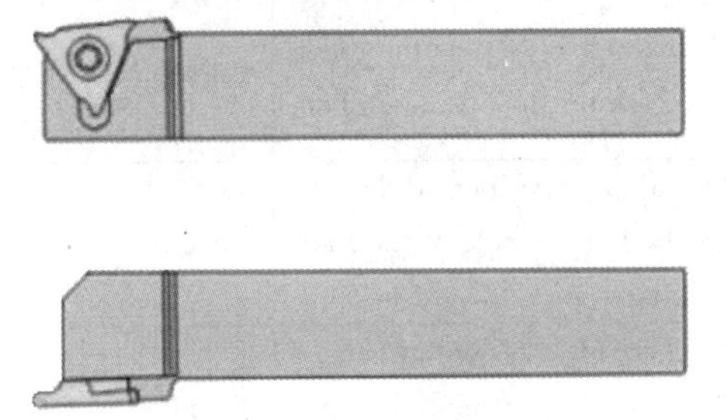

Figure 2-17 Arc grooving tool

Task 2 Technological Preparation

2.9 Part drawing analysis

According to the operation requirements of the part, 45 steel is selected as the blank material of this part, and the blanking dimension is set as $\phi 20 \times 40$. Using the excircle of the blank with a diameter of $\phi 20$ as the rough reference, rough and finish machining the right outer circle and internal control part to the required size, then cut it off, leave allowance in the length direction, and finally turn around to clamp the outer circle of $\phi 16$

(pay attention to protection during clamping to prevent surface clamping), and process the left end of the part to the size requirements.

Note that when turning the outer circle of $\phi 16$, the turning length should be sufficient. In addition, when clamping the blank, attention should be paid to the extended length of the bar to avoid collision between the cutter and the chuck.

2.10 Technological design

According to the analysis of the part drawing, the technological process is designed as shown in Table 2-2.

Table 2-2 Technological process card

<table>
<tr><td colspan="2" rowspan="2">Machining process card</td><td colspan="2">Product model</td><td colspan="2">STL-00</td><td>Part number</td><td>STL-02</td><td colspan="2">Page 1</td></tr>
<tr><td colspan="2">Product name</td><td colspan="2">Stirling engine model</td><td>Part name</td><td>Working piston</td><td colspan="2">Total 1 page</td></tr>
<tr><td colspan="2">Material grade</td><td>H65</td><td>Blank size</td><td>$\phi 20 \times 50$</td><td>Blank quality</td><td>kg</td><td>Quantity</td><td colspan="2">1</td></tr>
<tr><td colspan="6">Working procedure</td><td rowspan="2">Work section</td><td rowspan="2">Technical equipment</td><td colspan="2">Man-hours/min</td></tr>
<tr><td>No.</td><td>Name</td><td colspan="4">Content</td><td>Preparation & conclusion</td><td>Single piece</td></tr>
<tr><td>5</td><td>Preparation</td><td colspan="4">Prepare the material according to the size of $\phi 20 \times 50$</td><td>Outsourcing</td><td>Sawing machine</td><td></td><td></td></tr>
<tr><td>10</td><td>Turning</td><td colspan="4">Using the excircle of $\phi 20$ as the rough reference, process the $\phi 6.5$ through-hole</td><td>Turning</td><td>Lathe, Vernier caliper</td><td>15</td><td>10</td></tr>
<tr><td>15</td><td>Turning</td><td colspan="4">Keep the clamping position unchanged and process the bottom hole of the right inner cavity</td><td>Turning</td><td>Lathe, Vernier caliper</td><td>10</td><td>5</td></tr>
<tr><td>20</td><td>Turning</td><td colspan="4">Keep the clamping position unchanged and rough and finish machining the right $\phi 16$ outer circle</td><td>Turning</td><td>Lathe, micrometer</td><td>15</td><td>10</td></tr>
<tr><td>25</td><td>Turning</td><td colspan="4">Keep the clamping position unchanged and rough and finish machining the right $\phi 12$ inner hole</td><td>Turning</td><td>Lathe, Inside micrometer, Inside diameter dial indicator</td><td>20</td><td>15</td></tr>
<tr><td>30</td><td>Turning</td><td colspan="4">Machining the circular groove on the cylindrical surface</td><td>Turning</td><td>Lathe</td><td>15</td><td>10</td></tr>
</table>

Continued

Working procedure			Work section	Technical equipment	Man-hours/min	
No.	Name	Content			Preparation & conclusion	Single piece
35	Turning	Cut off and ensure the total length	Turning	Lathe, Vernier caliper	10	5
40	Turning	Turn around and machine the left taper hole	Turning	Lathe, Vernier caliper	15	10
45	Cleaning	Clean the workpiece, debur sharp corner	Benching			
50	Inspection	Check the workpiece dimensions	Examination			

This training task is aimed at the turning process of the 10th, 15th and 20th working procedure, and the corresponding working procedure card is formulated as shown in Table 2-3.

Table 2-3 Working procedure card for turning

Machining working procedure card	**Product model**	**STL-00**	**Part number**	**STL-02**	**Page 1**
	Product name	**Stirling engine model**	**Part name**	**Working piston**	**Total 1 page**

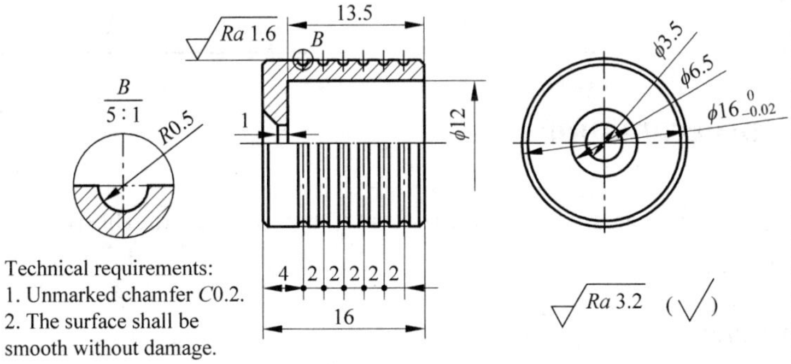
B
5:1
R0.5
Ra 1.6
B
13.5
1
ϕ12
4 2 2 2 2 2
16
ϕ3.5
ϕ6.5
$\phi16_{-0.02}^{0}$
Ra 3.2 (√)
Technical requirements:
1. Unmarked chamfer C0.2.
2. The surface shall be smooth without damage.

Procedure No.	15
Procedure name	Turning
Material	6061
Equipment	CNC lathe
Equipment model	CK6150e
Fixture	three-jaw self-centering chuck
Measuring tool	Vernier caliper
	micrometer
Preparation & Conclusion time	100min
Single-piece time	65min

Continued

Work step	Content	Cutters	S/ (r/min)	F/ (mm/r)	a_p/ mm	Step hours/min	
						Mechanical	Auxiliary
1	Workpiece installation						5
2	Using the excircle of ϕ20 as the rough reference, drill the center hole	A3 center drill	2000			5	5
3	Process the ϕ6.5 through-hole	ϕ6.5 drill	1200			5	5
4	Drill the bottom hole of the right inner cavity	Cylindrical rough turning tool, Hole-boring cutter	1000			5	5
5	Roughing the right ϕ16 outer circle	Excircle turning tool	1000	0.2	1.5	5	5
6	Finishing the right ϕ16 outer circle	Excircle turning tool	1300	0.12	0.3	5	5
7	Roughing the right ϕ12 inner hole	Hole-boring cutter	1200	0.15	1	10	5
8	Finishing the right ϕ12 inner hole	Hole-boring cutter	1400	0.1	0.2	5	5
9	Machining the circular groove on the cylindrical surface	Circular groove cutter	700	0.07		10	5
10	Process the left chamfer and cut off to ensure the total length	Cut-off tool	1000	0.06		5	5
11	Turn around and roughing the left taper hole	Chamfer drill head	1100			10	5
12	Cutting and chamfering						
13	Dismantling and cleaning workpieces						

2.11 CAM automatic programming

Model with CAXA CNC lathe, select the appropriate machining method, and generate a CNC machining program.

2.11.1 Contour programming

CAXA CNC lathe has powerful functions, with simple track generation and general post-processing module.

1. CAD modeling

As shown in Figure 2-18, first draw the outer contour of the part and the blank, that is, define the processing area. The model origin should be consistent with the programming origin. The current version of the CAD software does not support customized machining origin.

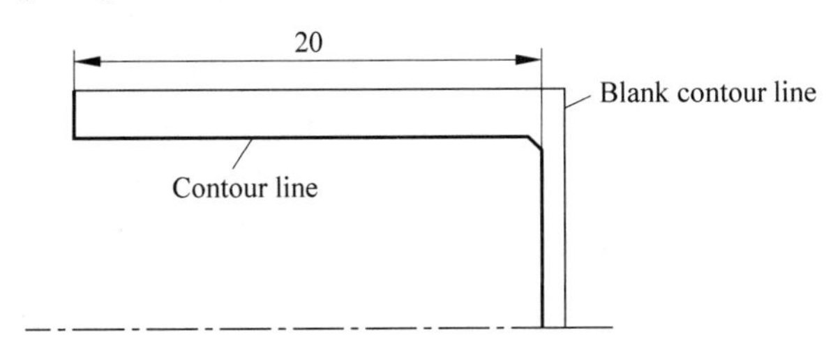

Figure 2-18 The contour of the part and the blank

2. CAM automatic programming

As shown in Figure 2-19, select the Rough turning button.

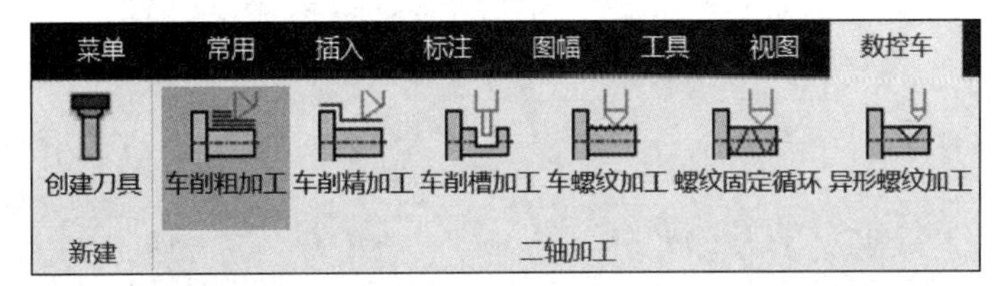

Figure 2-19 Two-axis machining method option group

Set processing parameters according to Figure 2-20.

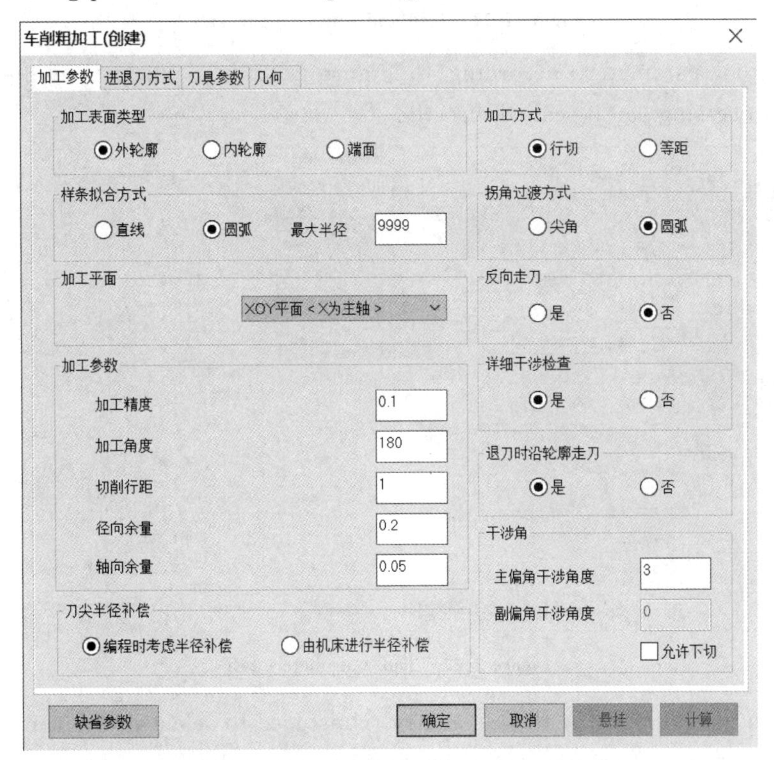

Figure 2-20 Processing parameters tab for rough turning

Set the parameters of feed and retract mode according to Figure 2-21.

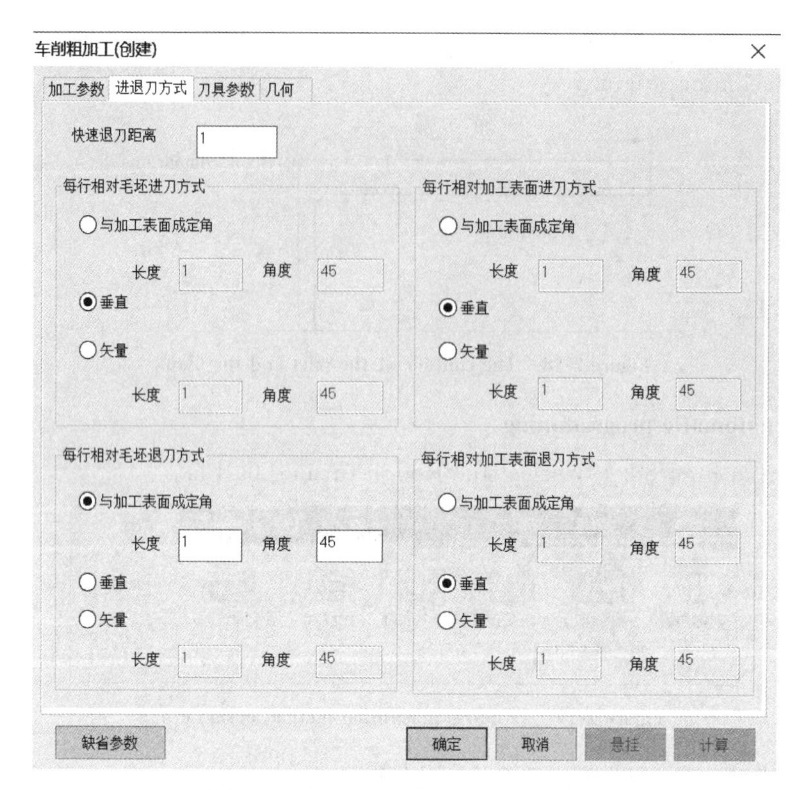

Figure 2-21　feed and retract mode Tab

Set the tool parameters according to Figure 2-22, and click Stock-in so that the subsequent processing can be called directly.

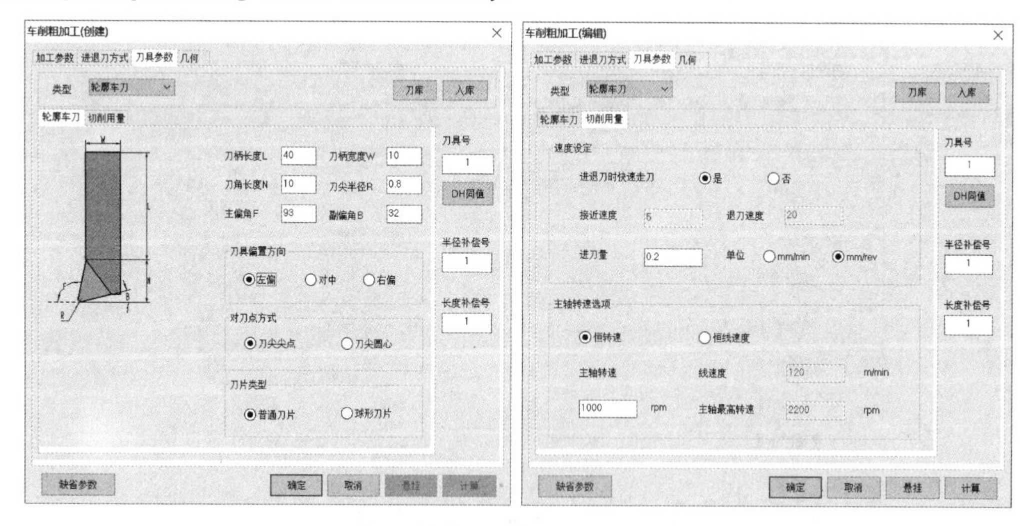

Figure 2-22　Tool parameters tab

As shown in Figure 2-23, the Geometry tab is used to select profile curves and blank profile curves. When selecting a curve, it should be noted that the selected curve is not the actual workpiece contour and blank contour. The difference of the selected curve will directly affect the generation effect of the tool path.

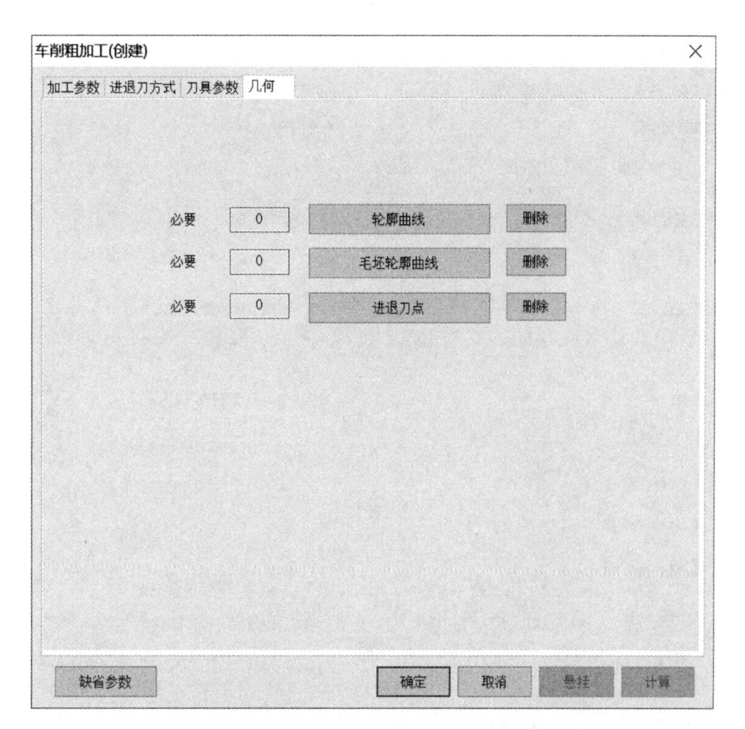

Figure 2-23 Geometry tab of rough turning

Left click the contour curve button, select the contour curve and blank contour curve respectively in the way of single pick, and then right-click to end the selection. The selection of feed and retract points should pay attention to whether the tool will interfere with the workpiece. The tool feed and retract points during contour processing should normally be outside the blank. The generated tool path is shown in Figure 2-24.

Figure 2-24 The outer contour tool path

Set the finishing parameters according to Figure 2-25.

The geometric elements do not need to select the blank contour, just select the contour line, and then generate the tool path, as shown in Figure 2-26.

2.11.2 Programming of the right $\phi 12$ inner hole

To process the rightside $\phi 12$ hole, it is necessary to pre-drill a $\phi 10$ bottom hole before boring.

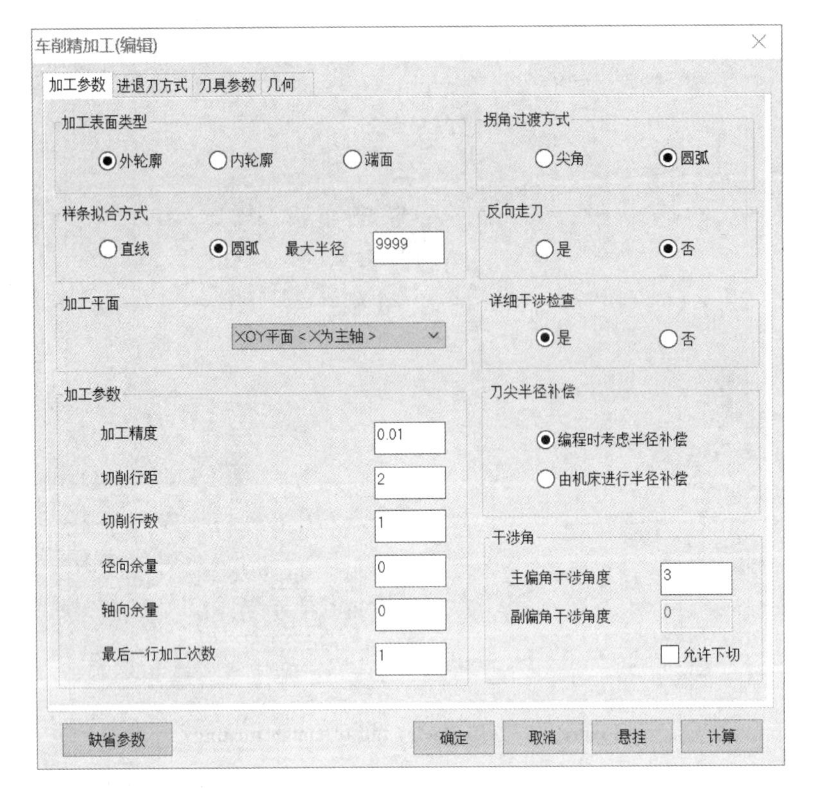

Figure 2-25 Parameters tab for finish turning

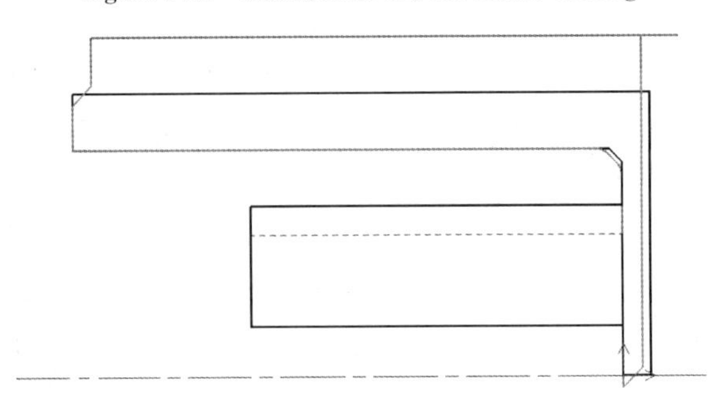

Figure 2-26 Tool path for finish turning

1. CAD modeling

As shown in Figure 2-27, first draw the contour of the part and the blank, that is, define the processing area. The model origin should be consistent with the programming origin. The current version of the CAD software does not support customized machining origin.

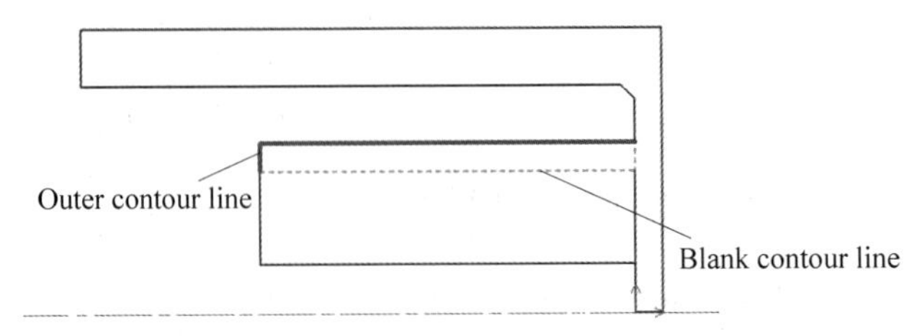

Figure 2-27 The contour of the part and the blank

2. CAM automatic programming

When setting the inner contour, the selected contour line should be broken at the intersection, otherwise the selection will fail. In addition, it is necessary to pay attention to the selection of feed and retreat points. This point should be slightly smaller than the diameter of the bottom hole, otherwise there may be interference between the tool and the workpiece.

Set the finishing parameters according to Figure 2-28. In the geometry, tab don't need to select the blank contour, just select the contour line, and then generate the tool path, as shown in Figure 2-29.

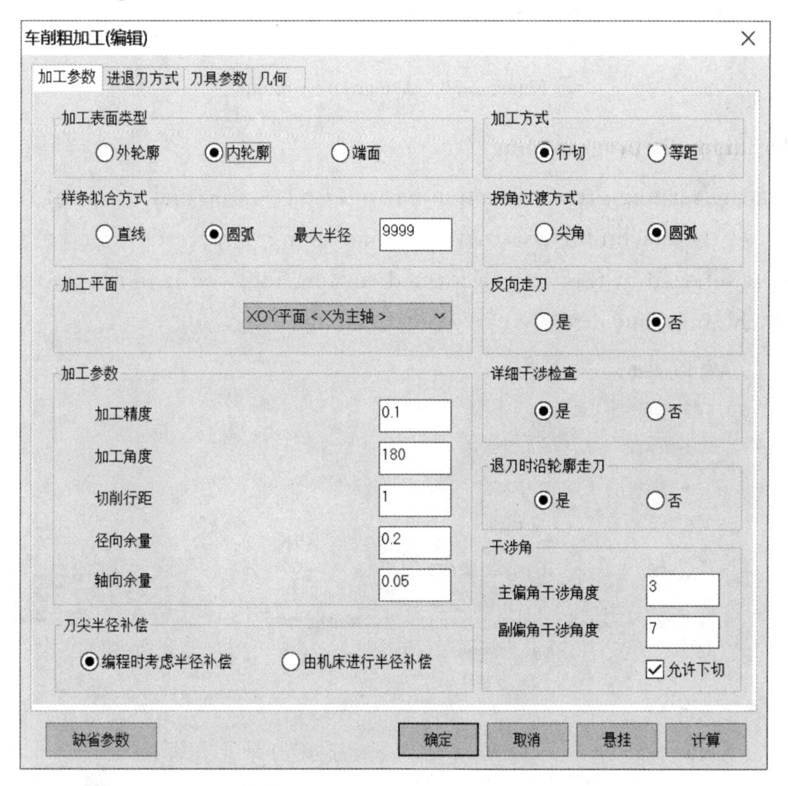

Figure 2-28 Machining parameters for the rightside $\phi 12$ inner hole

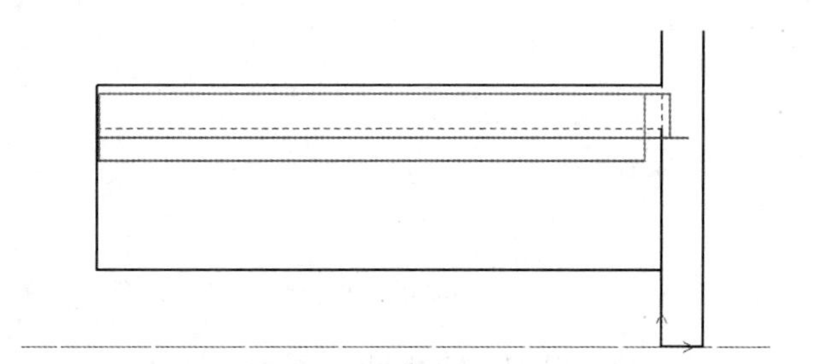

Figure 2-29 The tool path for rightside $\phi 12$ inner hole machining

The finish machining method is basically the same as the rough machining method, but it is not necessary to select the blank contour curve.

2.11.3 Programming for arc groove

1. CAD modeling

As shown in Figure 2-30, first draw the outer contour of the part and the blank, that is, define the processing area. The arc groove needs to make auxiliary contour lines before automatic programming. This step is not required for machining common rectangular grooves, and the contour can be directly selected.

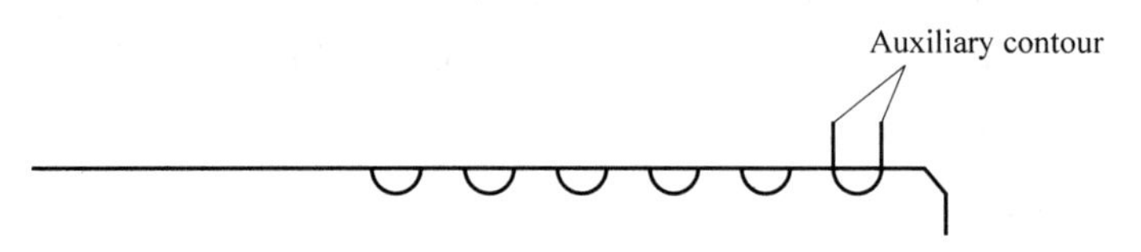

Figure 2-30 Arc groove contour

2. CAM automatic programming

Left click the turning groove function button, and pop up dialog box of turning groove process, and set the machining parameters in machining parameter tab according to Figure 2-31. The tool selected in this step is formed turning tool, so it is not necessary to carry out rough machining, but directly carry out finishing.

Figure 2-31 Machining parameters tab

Set the tool parameters according to Figure 2-32. The default tool shape is an ordinary grooving cutter. We can set the tool as an arc grooving cutter by modifying the nose radius.

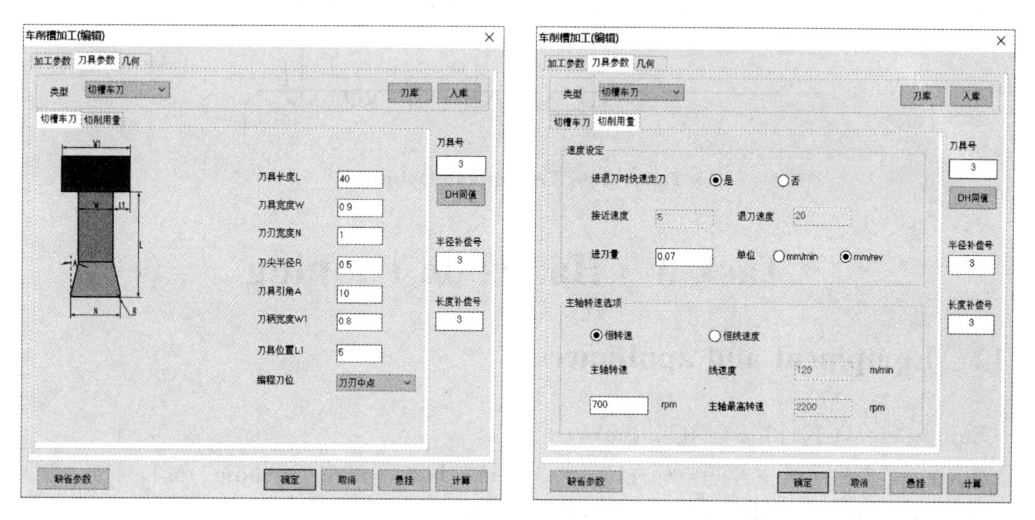

Figure 2-32 Tool parameters tab for turning groove

As shown in Figure 2-33, the first groove and the auxiliary line are selected as the contour curve, and the generated tool path is shown in Figure 2-34.

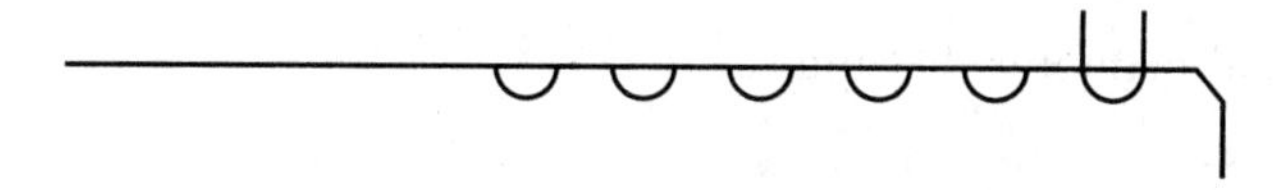

Figure 2-33 Groove contour curve

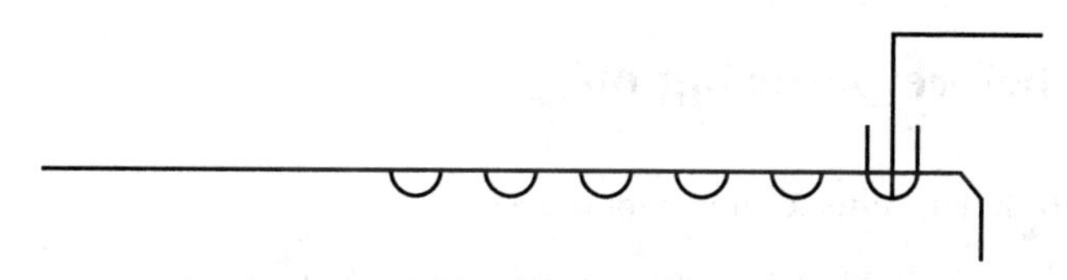

Figure 2-34 The tool path for grooving

The remaining grooves can be processed by copying. As shown in Figure 2-35, select the content that needs to be copied in the left management tree, right-click, left click the Copy with base point function, and then complete the tool path replication according to the prompt. The result is shown in Figure 2-36.

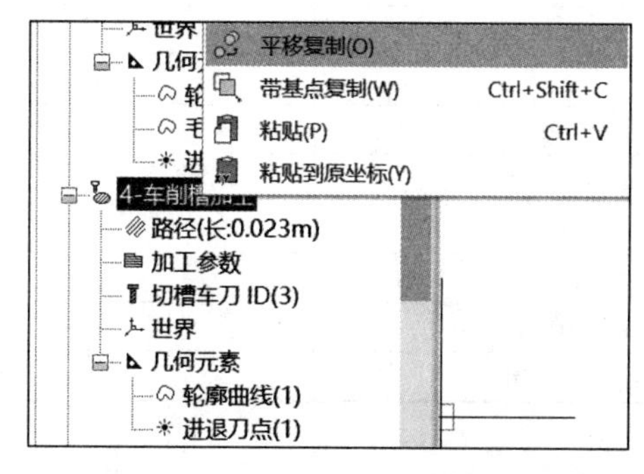

Figure 2-35 Shift copy of tool path

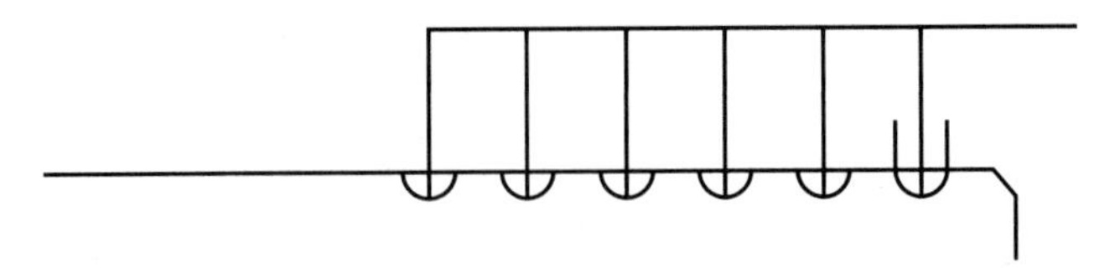

Figure 2-36 Arc groove tool path

Task 3 Hands-on training

2.12 Equipment and appliances

Equipment: CK6150e CNC lathe.

Cutters: the outer circle turning tool, the hole-boring turning tool, the circular groove cutter ($R0.5$), the cut-off tool (width: 3mm).

Fixture: the three-jaw self-centering chuck.

Tools: chuck wrenches, tool holder wrenches, etc.

Gauges: the 0-150mm vernier caliper, the 0-25mm outside micrometer, internal micrometer, internal diameter dial indicator.

Blank: $\phi 20 \times 50$.

Auxiliary appliances: cutter shim, brushes and so on.

2.13 Check before powering on

Refer to Table 2-4 to check the machine status.

Table 2-4 Preparing card for machine start-up

Check item		Test result	Abnormal description
Mechanical part	Spindle		
	Feed part		
	Tool holder		
	three-jaw self-centering chuck		
Electrical part	Main power supply		
	Cooling fan		
CNC system	Electrical components		
	Controlling part		
	Driving section		
Auxiliary part	Cooling system		
	Compressed air system		
	Lubricating system		

2.14 Preparation before machining

Before machining, the tools required for this task should be prepared and installed correctly. The origin of the workpiece is set according to the process requirements, and the CNC machining program is entered and the graphic verification is performed.

2.15 Part machining

After the graphic verification, process has verified that there is no problem, the parts machining can be carried out. Before the parts are processed, you should understand the safety operation requirements of the machine tool in detail, and wear labor protection clothing and utensils. When processing parts, you should be familiar with the functions and positions of the operation buttons of the CNC lathe, and understand the methods of dealing with emergency situations. During the machining process, especially before cutting, the actual distance between the cutter and the workpiece should be observed, and at the same time the remaining movement amount displayed in the "Remaining Movement Amount" column on the display screen should be compared. When the difference between the actual distance and the remaining movement amount is too large, the vehicle should be stopped and checked to avoid collision. If there is any abnormality, the machine tool movement should be stopped in time.

2.16 Part inspection

After the parts are processed, the workpiece should be carefully cleaned, and in accordance with the relevant requirements of quality management, the processed parts should be subject to relevant inspections to ensure the production quality. The "three-level" inspection cards for machined parts is shown in Table 2-5.

Table 2-5 "three-level" inspection cards for machined parts

<table>
<tr><th colspan="2">Part drawing number</th><th></th><th colspan="2">Part name</th><th colspan="2"></th><th>Working step number</th><th></th></tr>
<tr><td colspan="2">Material</td><td></td><td colspan="2">Inspection date</td><td colspan="2"></td><td>Working step name</td><td></td></tr>
<tr><td>Inspection items</td><td colspan="3">Self-inspection result</td><td colspan="2">Mutual inspection result</td><td colspan="2">Professional inspection</td><td>Remark</td></tr>
<tr><td></td><td colspan="3"></td><td colspan="2"></td><td colspan="2"></td><td></td></tr>
<tr><td></td><td colspan="3"></td><td colspan="2"></td><td colspan="2"></td><td></td></tr>
<tr><td></td><td colspan="3"></td><td colspan="2"></td><td colspan="2"></td><td></td></tr>
<tr><td></td><td colspan="3"></td><td colspan="2"></td><td colspan="2"></td><td></td></tr>
<tr><td></td><td colspan="3"></td><td colspan="2"></td><td colspan="2"></td><td></td></tr>
<tr><td></td><td colspan="3"></td><td colspan="2"></td><td colspan="2"></td><td></td></tr>
<tr><td></td><td colspan="3"></td><td colspan="2"></td><td colspan="2"></td><td></td></tr>
<tr><td>Conclusion</td><td colspan="8">☐ Qualified ☐ Unqualified ☐ Repair ☐ Concession to receive
Inspection signature:
Date:</td></tr>
<tr><td>Non-conforming item description</td><td colspan="8"></td></tr>
</table>

Project Summary

As a typical machining part of CNC lathes, the working piston is widely used in various equipment. According to equipment conditions and precision requirements, there will be some differences in the processing technology. Programmers and operators need to formulate the processing technology reasonably according to the processing conditions to improve the processing accuracy of the parts and the production efficiency.

Exercises After Class

1. Choice questions

(1) The function of the code G is ( ).

A. tool function B. preparation function

C. coordinate address D. feed rate

(2) The ***I*** and ***K*** values in arc programming refer to the vector values of ( ).

A. the starting point to the center of the circle

B. the end point to the center of the circle

C. the center of the circle to the starting point

D. the center of the circle to the end point

(3) In the counterclockwise arc interpolation instruction G03 X_Y_R_, the value after R represents ( ) of the arc.

A. radius

B. coordinate value of the end point

C. coordinate value of the starting point

D. central angle

(4) In the following description, ( ) is wrong.

A. Each CNC program stored in the system must have a program number

B. The program segment is composed of one or more instructions, indicating all actions of the CNC machine tool

C. In most systems, the program segment number is only used as the target position indication of "jump" or "program retrieval"

D. The program notes of FANUC system are enclosed with "()"

(5) Which of the following does not belong to the processing characteristics of CNC machine tools. ( )

A. high processing accuracy B. strong adaptability

C. high production efficiency D. suitable for mass production

(6) When measuring the thickness of the convex shoulder of the workpiece, ( ) should be selected.

A. sine gauge B. outer diameter micrometer

C. triangle plate D. block gauge

2. True or false

(1) The CNC lathe is programmed with diameter dimensions. ( )

(2) The arc radius of the blade is generally 2-3 times of the amount of feed. ( )

(3) When the arc interpolation is programmed with R cede, the R takes a negative value if the central angle corresponding to the arc is equal to 180°. ( )

(4) When programming an arc with fillet transition instruction, you cannot describe an integer circle. ( )

3. Fill in the blanks

(1) CNC programming is generally divided into ________ and ________.

(2) The process of tool nose radius compensation is divided into ________ , ________ and ________.

(3) There are two kinds of tool compensation for CNC lathes: ________ and ________.

(4) The process of automatically converting the part design and processing information entered into the computer into the instructions (or information) that the NC device can read and execute is ________.

4. Short answer questions

(1) Briefly describe the function of arc turning tool.

(2) When machining arc groove parts, how to choose the tool entry point?

(3) What are the main functions of two-dimensional machining of CAXA CNC lathe?

(4) What are the precautions for boring processing?

(5) What are the commonly used CNC machining process files?

5. Analysis questions

According to the part drawing(Figure 2-37),determine the processing technology and fill in the process sheet as shown in Table 2-6.

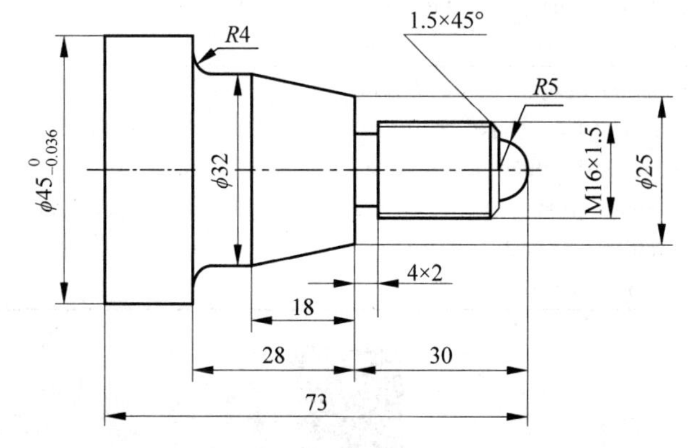

Figure 2-37 part drawing for question 5

Table 2-6 Process sheet

No.	Steps and contents	Tool No.	Tool type	Description of cutting amount

Self-learning test score sheet is shown in Table 2-7.

Table 2-7 Self-learning test score sheet

Task	Task requirements	Score	Scoring rules	Score	Remark
Learn key knowledge points	(1) Understand the basic components of automatic programming software of CAXA CNC lathe (2) Familiar with the common keys of automatic programming software of CAXA CNC lathe (3) Master the basic operation of automatic programming software of CAXA CNC lathe (4) Master basic concepts of CAXA CNC machining (5) Master the settings of the CAXA CNC tool magazine	20	Understand and master		
Technological preparation	(1) Be able to correctly read the shaft part drawings (2) Can be analyzed according to the part drawing, determine the process (3) Be able to write the correct processing program according to the processing process	30	Understand and master		
Hands-on training	(1) The corresponding equipment and utensils will be selected correctly (2) Can correctly operate the CNC lathe and adjust the processing parameters according to the machining situation	50	(1) Understand and master (2) Operation process		

Ideological and Political Classroom

Project 3 Programming and Machining Training for Heating Chamber Turning

➢ Mind map

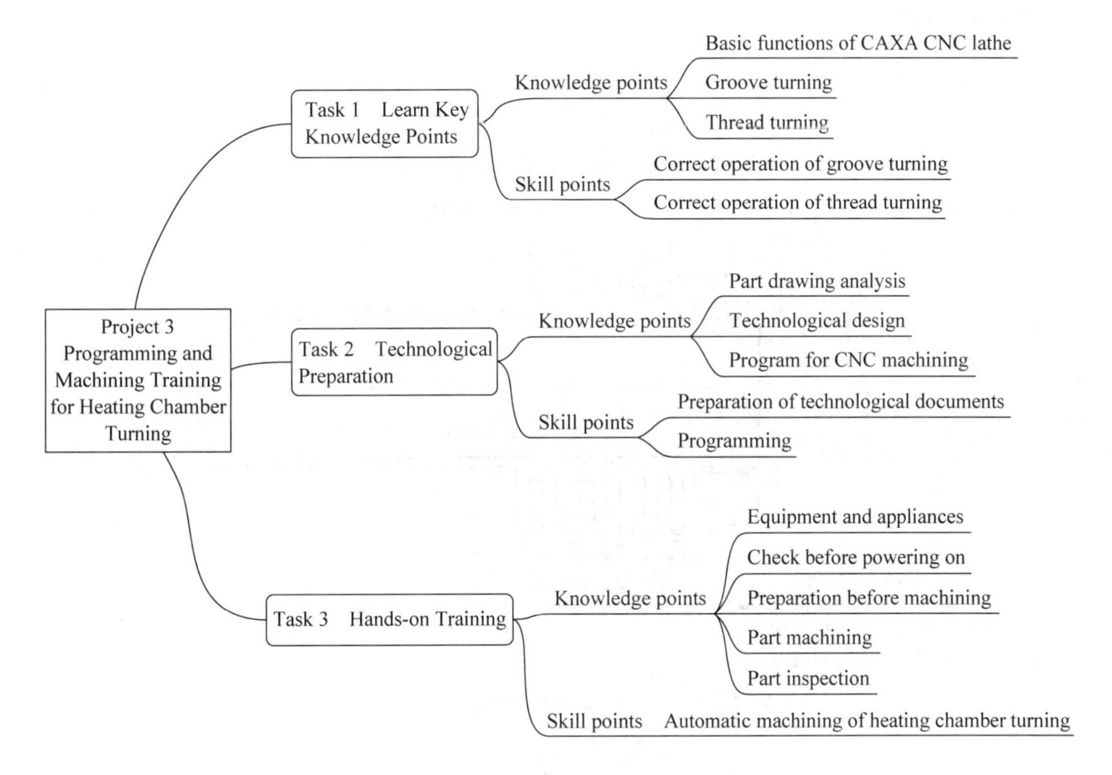

➢ Learning objectives

Knowledge objectives

(1) Understand the processing characteristics of heating chamber parts.

(2) Understand the meaning of each parameter of the hole processing cycle instruction.

Ability objectives

(1) Be able to independently determine the process routine and fill in the technological documents correctly.

(2) Be able to operate the CNC lathe correctly and adjust the machining parameters according to the machining conditions.

(3) Be able to select measuring tools reasonably according to the accuracy and the structural characteristics of parts, and be able to measure the relevant dimensions correctly and normatively.

Literacy goals

(1) Develop students' scientific spirit and attitude.

(2) Cultivate students' engineering awareness.

(3) Develop students' teamwork skills.

➢ Task introduction

According to the requirements of part drawing shown in Figure 3-1, develop the processing technology, compile the CNC processing program, and complete the processing of heating chamber parts. The blank material of the part is 45 steel, quenched and tempered, and the surface is required to be smooth without damage.

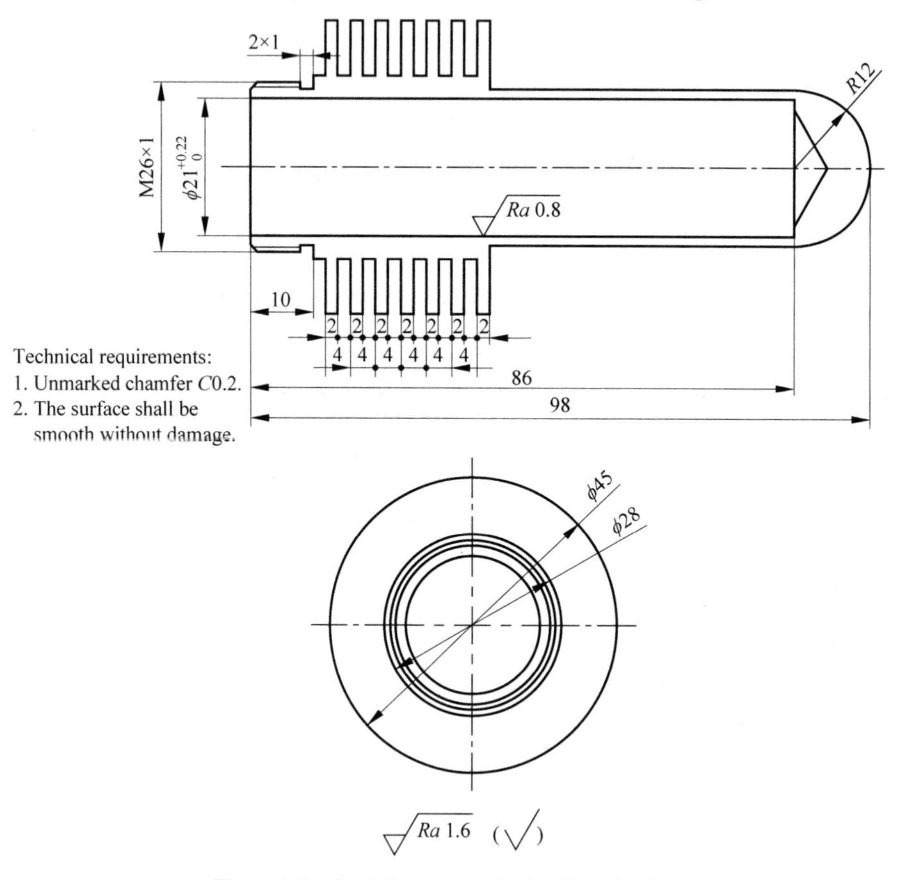

Figure 3-1 Part drawing of the heating chamber

Task 1 Learn Key Knowledge Points

3.1 Basic functions of CAXA CNC lathe

3.1.1 Turning grooves machining(create)dialog box

Turning grooves machining is used to cut grooves on the outer contour surface, inner contour surface and end face of the workpiece.

During groove cutting, the machined contour shall be determined. The machined contour is the workpiece surface contour after machining, and it cannot be closed or self-intersected.

Operation steps:

(1) Select menu bar→CNC lathe→Turning grooves machining, and the system will pop up the Turning grooves machining (create) dialog box, the machining parameter tabis as shown in Figure 3-1. In the machining parameter tab, firstly determine whether the outer contour surface, inner contour surface or end face to be processed, and then determine other processing parameters according to the processing requirements.

(2) After determining the parameters, the contour to be processed can be picked up, and the contour picking tool provided by the system can be used.

(3) After selecting the contour, determine the feed and retract point of the tool. Specifies a point for the tool location before and after machining. Right click to ignore the input of this point.

After completing the above steps, the grooving path can be generated. Select menu bar→CNC lathe→Create Tool, and pick up the newly generated tool path, and then generate the machining instructions.

3.1.2 Machining parameter tab

The machining parameters mainly define various process conditions and processing methods in grooving.

The meaning of each machining parameter is explained as follows.

1. Machining contour type option group

(1) Outer contour: The outer contour is grooved, or the outer contour is machined with a grooving knife.

(2) Inner contour: Groove the inner contour, or machine the inner contour with a grooving knife.

(3) End face: Groove the end face, or use a grooving knife to machine the end face.

2. Processing technology type option group

(1) Roughing: Only rough machining the groove.

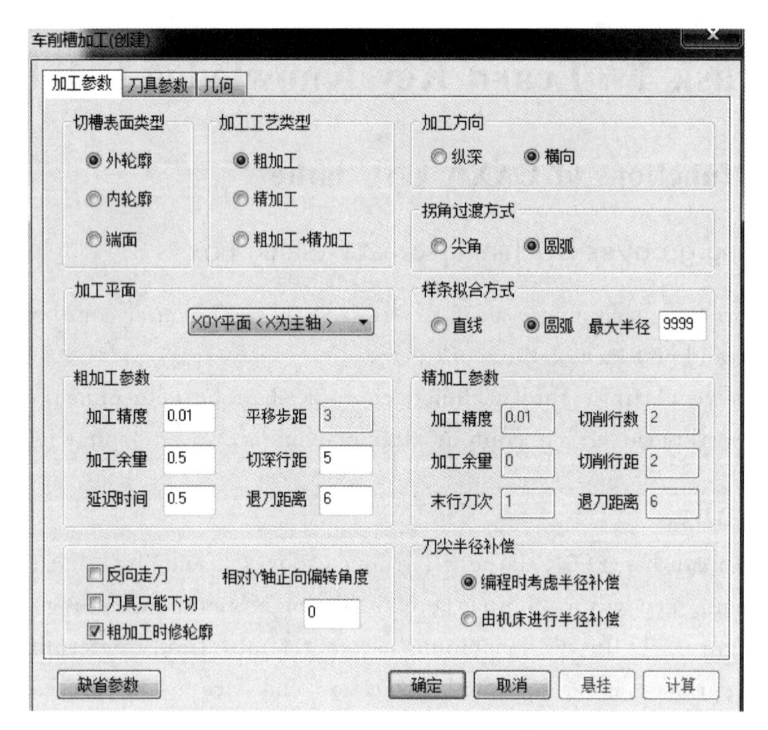

Figure 3-2 "Machining parameter" tab

(2) Finishing: Only finish machining the groove.

(3) Roughing and Finishing: Roughing the groove is followed by finishing.

3. Corner transition mode option group

(1) Round corner: When the cutting process encounters a corner, the tool transits from one side of the contour to the other side in the form of an arc.

(2) Sharp corner: When the cutting process encounters a corner, the tool transits from one side of the contour to the other side in a sharp way.

4. Roughing parameters option group

(1) Delay time: the time that the tool stays at the bottom of the groove during rough grooving.

(2) Cutting depth translation amount: the cutting amount of each longitudinal cutting of the tool (X direction of the machine tool) during rough grooving.

(3) Horizontal translation amount: the horizontal translation amount (Z direction of the machine tool) before the next cutting after the tool cuts to the specified cutting depth translation amount during rough grooving.

(4) Tool retraction distance: the distance from the tool to the outside of the groove before the next line of cutting in the rough grooving.

(5) Machining allowance: the reserved amount of the unprocessed part of the machined surface during roughing.

5. Finishing parameters option group

(1) Cutting line spacing: the distance between finishing lines.

(2) Number of cutting lines: the number of machining lines of the tool path during finishing, excluding the repetition number of the last line.

(3) Tool retraction distance: the tool retraction distance before cutting the next line after finishing one line.

(4) Machining allowance: the reserved amount of the unprocessed part of the machined surface during finishing the groove.

(5) Cutting times of the last line: when finishing the groove, in order to improve the surface quality of the processing, the last line is often turned several times under the same feed rate. This defines the number of multiple cuts for last line.

3.1.3 Grooving turning tools tab

Left click on the "Tool Parameters" tag to enter Grooving turning tools tab as shown in Figure 3-3 and Figure 3-4. This tab is used to set parameters of the grooving turning tool. For specific parameters, please refer to the instructions in Section 2. 7. 2.

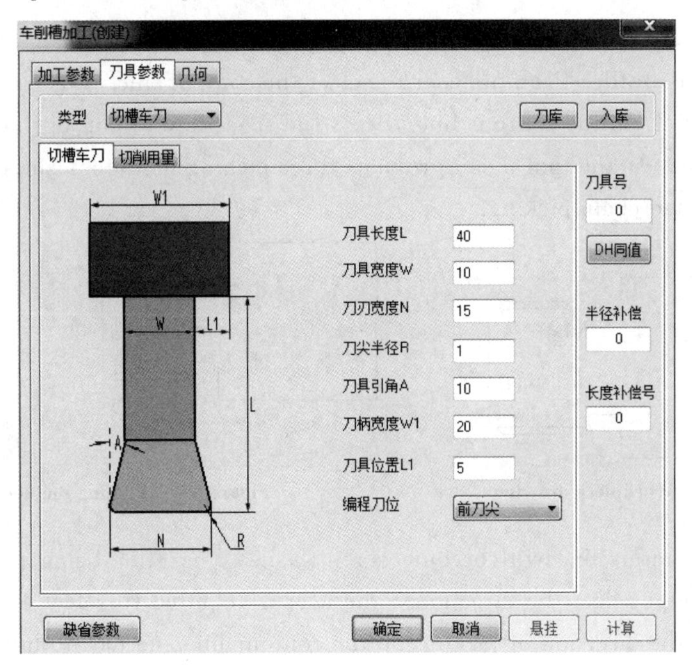

Figure 3-3 Grooving turning tools tab

3.1.4 Machining example of turning a groove

(1) Determine the contour to be machined. As shown in Figure 3-5, the groove part of the thread tool retraction is the contour to be machined.

(2) Fill in the parameter table. After filling in the parameters in the Grooving turning

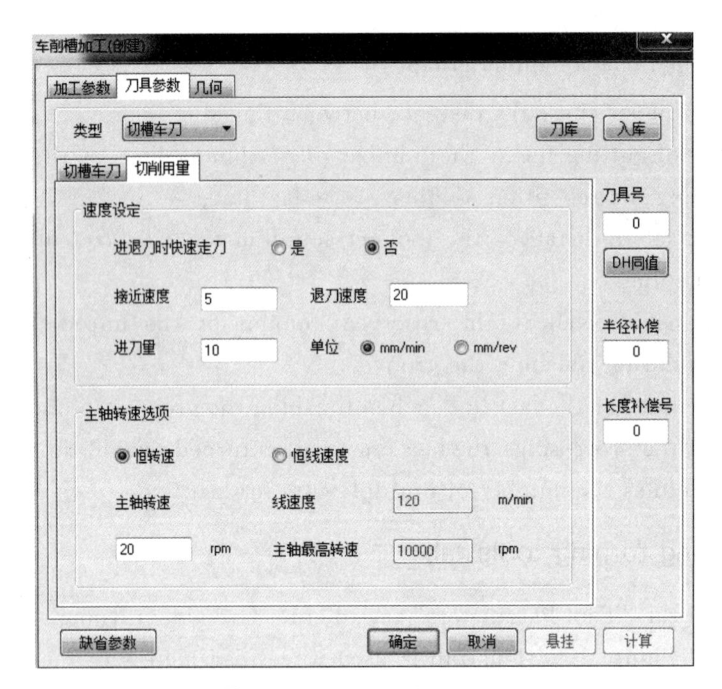

图 3-4　Cutting parameter tab

tools tab, left click the Confirm button.

(3) Pick the profile. Prompts user to select the contour line. The curve picking menu can be used to pick up the contour line. Press the space key to pop up the tool menu. As shown in Figure 3-6, the tool menu provides three picking methods: single picking, chain picking and limited chain picking.

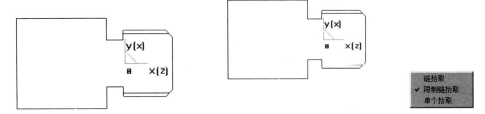

Figure 3-5　Workpiece drawing　　　　Figure 3-6　Picking methods

The first contour line will become a red dashed line after being picked. Then the system will prompt select the direction. The user is required to select a direction, which only represents the direction of picking up the contour line and has nothing to do with the machining direction of the tool, as shown in Figure 3-7.

After selecting the direction, if the chain picking method is used, the system automatically picks up the contour lines connected from end to end. If the single picking method is used, the system prompts you to continue picking the contour lines. If the limited chain picking method is used, the system continues to prompt you to select the restriction line. The selected end segment is the left part of the groove, and the groove part turns dotted line, as shown in Figure 3-8.

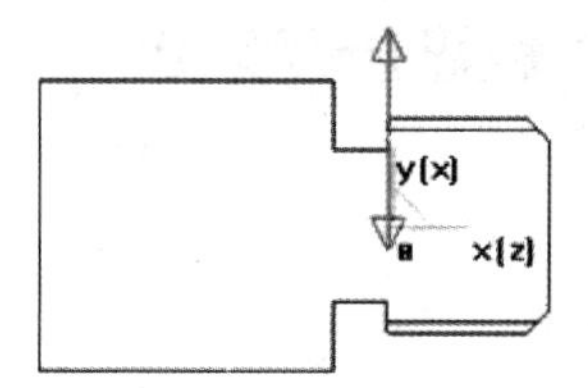

Figure 3-7 Select the contour line direction

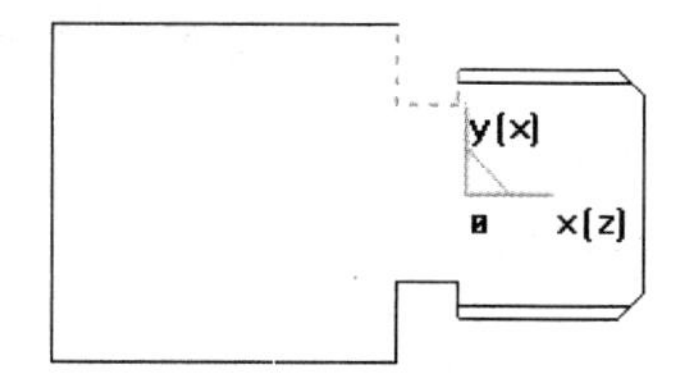

Figure 3-8 Chain picking method

(4) Determine the feed and retraction point. Specifies a point as the location of the tool before and after machining. Right click to ignore the input of this point.

(5) Generate tool path. After determining the feed and retraction point, the system generates a tool path, as shown in Figure 3-9.

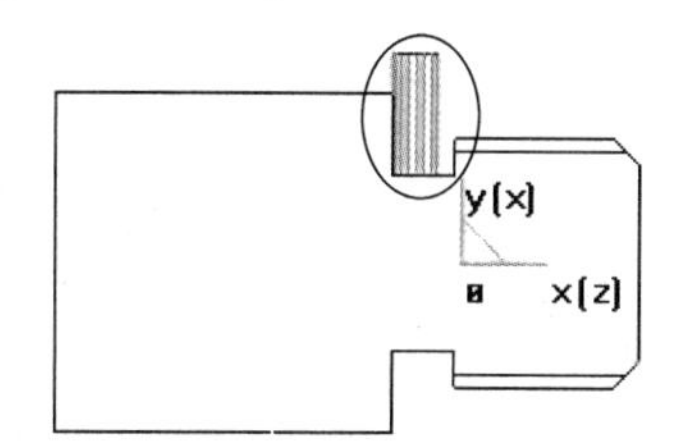

Figure 3-9 Generated tool path

Note: The contour to be processed cannot be closed or self-intersecting. The generated tool path is closely related to the parameters such as the angle radius and the blade width of the grooving turning tool. you can draw only the upper part of the tool retraction groove according to actual needs.

3.1.5 Thread turning (create) dialog box

This function is a non-fixed cycle way to process threads, which can control various process conditions and processing methods in thread processing more flexibly.

Operation steps are as follows.

(1) Select CNC lathe→Thread turning, and then pop up the Thread turning dialog box, as shown in Figure 3-10. The user can determine the processing parameters in the dialog box.

(2) Pick the starting point and end point of the thread, as well as the tool feed and retraction points.

(3) After filling in the parameters, left click the Confirm button to generate the thread turning tool path.

(4) Select CNC lathe→Post processing, and pick the tool path just generated, and then the threading instructions can be generated.

3.1.6 Thread parameters tab

Left click the Thread Parameters label in the dialog box of thread turning (create) to enter the thread parameters tab. Thread Parameters tab mainly contains parameters related to the nature of the thread, such as thread type, thread pitch, number of thread heads, etc. The starting point and end point coordinates are derived from the pickup results of the previous step and can be modified by the user.

The meaning of each thread parameter is described as follows.

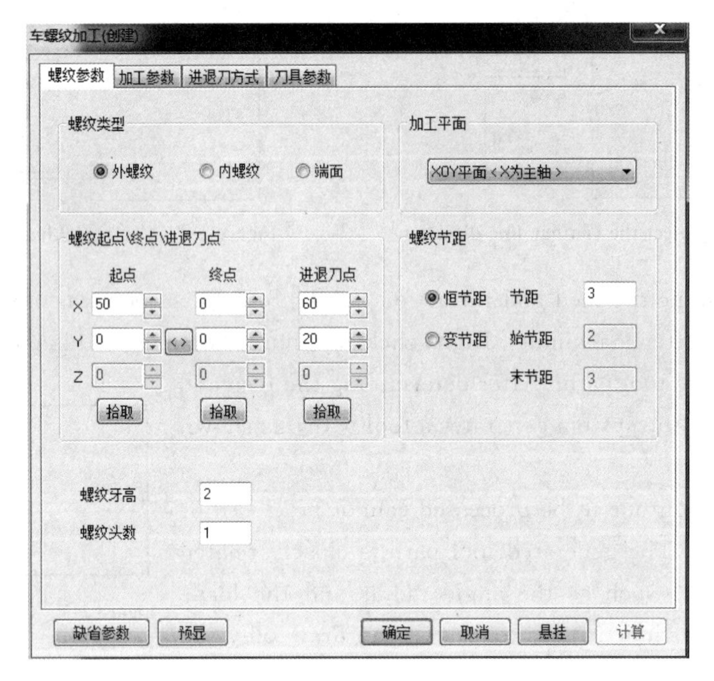

Figure 3-10　Threading parameters tab

(1) Starting point coordinate: The coordinate of the starting point of the threading, in mm.

(2) End point coordinate: The coordinate of the end point of the threading, in mm.

(3) Coordinates of the feeding and retraction points: the coordinates of the feeding and retraction points of the threading, in mm.

(4) Thread tooth height: the height of the thread tooth.

(5) Number of thread heads: The number of teeth between the starting point and the end point of the thread.

(6) Thread pitch. There are five kinds of it.

① Constant pitch: The distance between the corresponding points on two adjacent thread profiles is a constant value.

② Pitch: constant pitch value.

③ Variable pitch: the distance between the corresponding points on the two adjacent thread profiles is a variable value.

④ Start pitch: The pitch of the thread at the starting end.

⑤ Final pitch: The pitch of the thread at the termination end.

3.1.7 Machining parameters tab

The Machining Parameters tab is used to set the process conditions and machining methods during thread machining, as shown in Figure 3-11.

The meaning of each thread machining parameter is explained as follows.

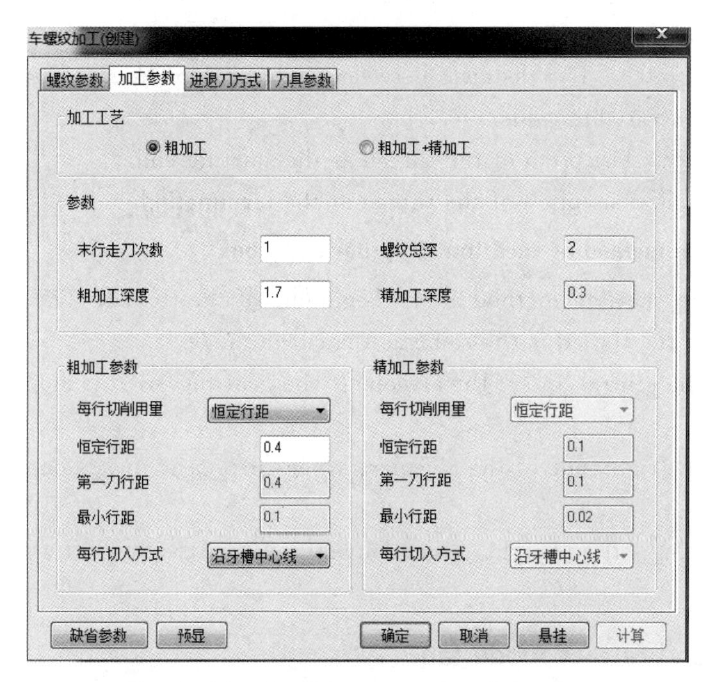

Figure 3-11　Machining parameters of threading

1. Machining technology option group

(1) Rough machining. It refers to the direct use of rough cutting to process threads.

(2) Rough and finish machining. It refers to that after rough cutting according to the specified roughing depth, the remaining allowance (finish machining depth) is cut by fine cutting method (such as using smaller line spacing).

2. Parameters option group

(1) In order to improve the quality of machined surface, the last line often needs repeated turning with the same feedrate, and it is defined as the number of multi-cutting of the last line.

(2) Total thread depth: the total cutting depth for roughing and finishing of the thread.

(3) Roughing depth: the cutting depth of thread for roughing.

(4) Finishing depth: the cutting depth of thread for finishing.

3. Cutting dosage of each line drop-down list box

Cutting dosage of each line contains the following detailed.

(1) Constant line spacing. The line spacing when the machining is carried out along a constant line spacing.

(2) Constant cutting area. In order to ensure the constant cutting area each time, the cutting depth of each time will be gradually reduced until it is equal to the minimum line spacing. The user needs to specify the first tool line spacing and the minimum line spacing. The cutting depth is specified as follows: the depth of the nth cutting is $\sqrt{n}$ times

of the depth of the first cutting.

(3) Variable pitch. The distance between the corresponding points on two adjacent thread profiles is a variable value.

(4) Start pitch. The pitch of the thread at the starting end.

(5) Final pitch. The pitch of the thread at the termination end.

4. Pitching-in method of each line drop-down list box

It refers to the cutting method at the beginning of the thread. The exit mode of the tool at the end of the thread is the same as the cut-in mode.

(1) Along the central line of the alveolar: when cutting in, it is along the central line of the alveolar.

(2) Along the right side of the alveolar: when cutting in, it is along the right side of the alveolar.

(3) Alternating left and right: when cutting in, it alternates left and right along the alveolar.

3.1.8 Feed and retract mode tab

Left click the Feed/Retract Mode tag to enter Feed/Retract mode tab, which is used to set the parameters of feed and retract mode as shown in Figure 3-12.

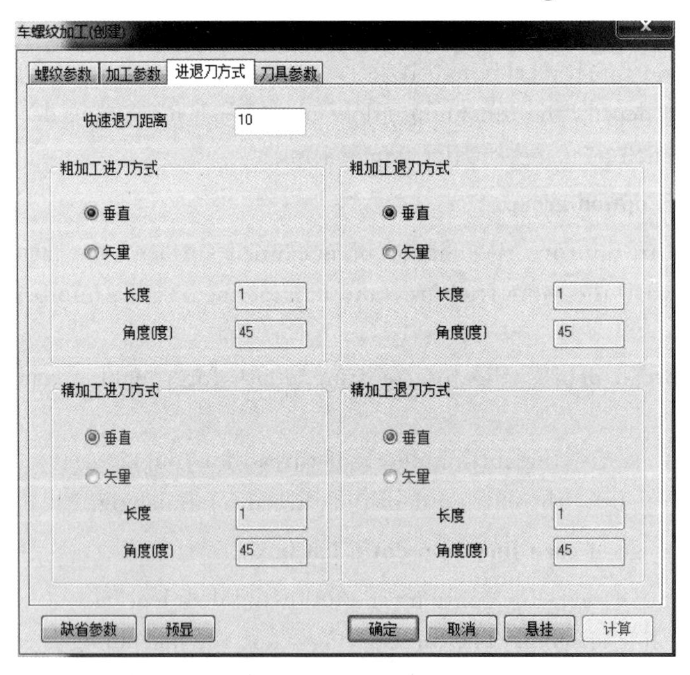

Figure 3-12　Thread turning Feed and Retract mode tab

1. Feed mode

(1) Vertical. It refers to the tool directly feeds to the starting point of each cutting line.

(2) Vector. It refers to adding a feed segment with a certain angle to the positive

direction of the system X axis (machine tool Z axis) after every cutting line. The tool feeds to the starting point of the feed segment, and then feeds along the feed segment to the cutting line.

(3) Length. It defines the length of the vector (feed segment).

(4) Angle. It defines the included angle between the vector (feed segment) and the positive direction of the system X axis.

2. Retract mode

(1) Vertical. It refers to the tool directly retracts to the starting point of each cutting line.

(2) Vector. It refers to adding a retraction segment with a certain angle to the positive direction of the system X axis (machine tool Z axis) after every cutting line. The tool retracts along the segment first, and then retracts vertically from the end point of the segment.

① Length. It defines the length of the vector (retraction segment).

② Angle. It defines the included angle between the vector (retraction segment) and the positive direction of the system X axis.

3. Tool retraction distance

It refers to the distance (relative value) that the tool retracts at the maximum feed speed allowed by the machine tool.

3.1.9 Cutting dosage tab

For the description of cutting dosage, as shown in Figure 3-13, please refer to the instructions in Section 2. 11. 1.

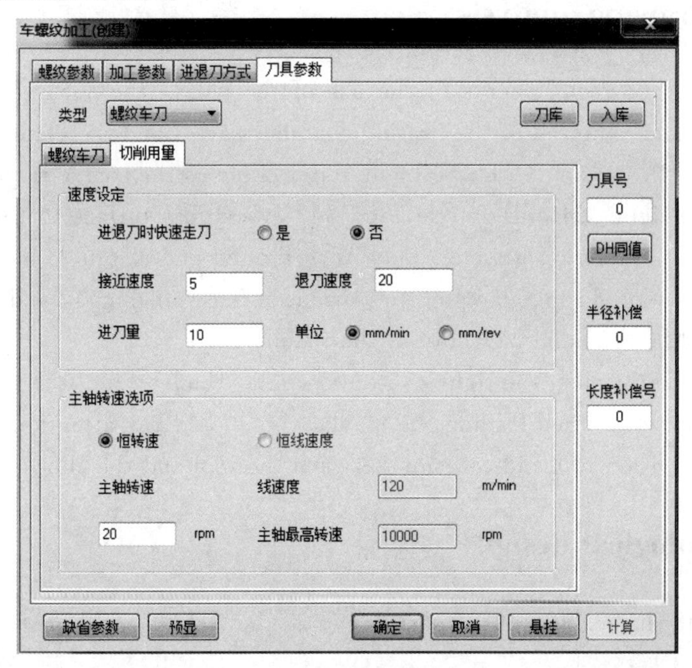

Figure 3-13 Parameters setting for cutting dosage

3.1.10 Threading tool tab

Left click the Threading Tool tag to enter Threading tool tab as shown in Figure 3-14. This tab is used to set the parameters of threading tool used in processing. For specific parameter description, please refer to the description in Section 2.7.3.

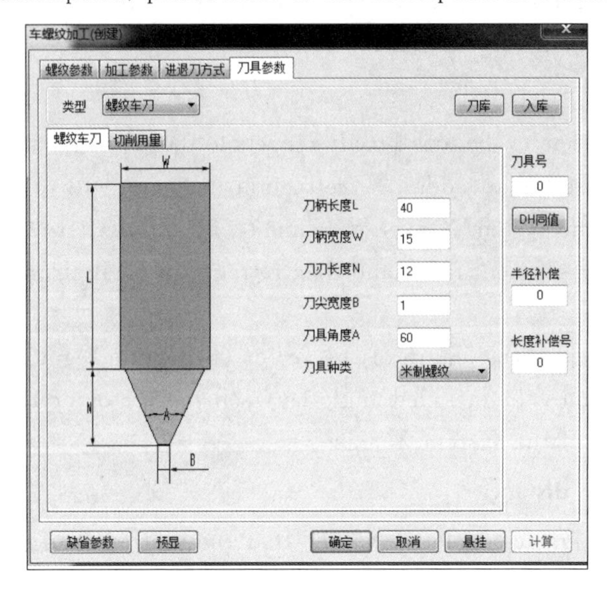

Figure 3-14 Thread turning tool tab

Task 2 Technological Preparation

3.2 Part drawing analysis

According to the requirements of the use of the parts, the 6061 aluminum alloy is selected as the blank material of the heating chamber parts, and the blank blanking size is set to $\phi 50 \times 102$. Processing with $\phi 50$ blank outer circle as the coarse reference, roughing and finishing the right $\phi 24$ outer circle, and $\phi 45$ cylindrical surface to the required size, cutting processing outer ring groove. Then turn around and clamp $\phi 24$ outer circle (pay attention to protection during clamping to prevent surface pinching), and process the left end of the part $\phi 26$, $\phi 28$ and $\phi 21$ bore to the required size.

Note that when turning the right $\phi 45$ outer circle, the length of the turning should be sufficient. In addition, when loading the blank, you should pay attention to the length of the bar stock extension to avoid collision between the tool and the chuck.

3.3 Technological design

According to the analysis of the part drawing, the technological process is designed as shown in Table 3-1.

Table 3-1 Technological process card

Machining process card	**Product model**		**STL-00**		**Part number**	**STL-03**	**Page 1**
	Product name		**Stirling engine model**		**Part name**	**Heating chamber**	**Total 1 page**
Material grade	6061	Blank size	ϕ50×102	Blank quality	kg	Quantity	1
Working procedure					Work section	Technical equipment	Man-hours/min	
No.	Name	Content					Preparation & conclusion	Single piece

This training task is to carry out process design for the heating chamber part and develop process cards, as shown in Table 3-2.

Table 3-2 Working procedure card for heating chamber turning

Machining working procedure card	**Product model**	**STL-00**	**Part number**	**STL-03**	**Page 1**
	Product name	**Stirling engine model**	**Part name**	**Heating chamber**	**Total 1 page**
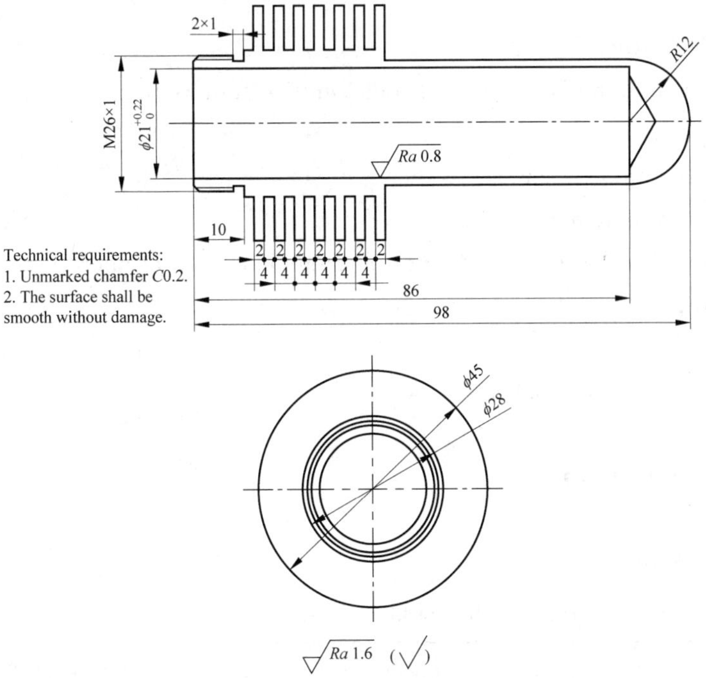				Procedure No.	15
				Procedure name	Turning
				Material	6061
				Equipment	CNC lathe
				Equipment model	CK6150e
				Fixture	three-jaw self-centering chuck
				Measuring tool	Vernier caliper
					micrometer
				Preparation & Conclusion time	
				Single-piece time	

2×1
R12
M26×1
$\phi21^{+0.22}_{0}$
Ra 0.8
10
2 2 2 2 2 2 2
4 4 4 4 4 4
86
98
Technical requirements:
1. Unmarked chamfer C0.2.
2. The surface shall be smooth without damage.
$\phi45$
$\phi28$
Ra 1.6 (√)

Continued

Work steps	Content	Cutters	S/ (r/min)	F/ (mm/r)	a_p/ mm	Step hours/min	
						Mechanical	Auxiliary

3.4 Programming for CNC machining

According to the processing technology, the CAXA CNC lathe is used to create the contour model for the right and left ends, generate the tool path and the CNC machining program.

Task 3 Hands-on Training

3.5 Equipment and appliances

Equipment: CK6150e CNC lathe.

Cutters: the outer circle turning tool, the cut-off tool (width: 2mm), the hole-boring turning tool, thread turning tool.

Fixture: the three-jaw self-centering chuck.

Tools: chuck wrenches, tool holder wrenches, etc.

Gauges: the 0-150mm vernier caliper, the 0-25mm micrometer, internal micrometer, internal diameter dial indicator.

Blank: $\phi 20 \times 50$.

Auxiliary appliances: cutter shim, brushes and so on.

3.6 Check before powering on

Refer to Table 3-3 to check the machine status.

Table 3-3 Preparing card for machine start-up

Check item		Test result	Abnormal description
Mechanical part	Spindle		
	Feed part		
	Tool holder		
	Three-jaw self-centering chuck		

Continued

Check item		Test result	Abnormal description
Electrical part	Main power supply		
	Cooling fan		
CNC system	Electrical components		
	Controlling part		
	Driving section		
Auxiliary part	Cooling system		
	Compressed air system		
	Lubricating system		

3.7 Preparation before machining

Before machining, the tools required for this task should be prepared and installed correctly. The origin of the workpiece is set according to the process requirements, and the CNC machining program is entered and the graphic verification is performed.

3.8 Part machining

After the graphic verification process has verified that there is no problem, the parts machining can be carried out. Before the parts are processed, you should understand the safety operation requirements of the machine tool in detail, and wear labor protection clothing and utensils. When processing parts, you should be familiar with the functions and positions of the operation buttons of the CNC lathe, and understand the methods of dealing with emergency situations. During the machining process, especially before cutting, the actual distance between the cutter and the workpiece should be observed, and the remaining movement amount displayed in the "Remaining Movement Amount" column on the display screen should be compared. When the difference between the actual distance and the remaining movement amount is too large, the vehicle should be stopped and checked to avoid collision. If there is any abnormality, the machine tool movement should be stopped in time.

3.9 Part inspection

After the parts are processed, the workpiece should be carefully cleaned, and in accordance with the relevant requirements of quality management, the processed parts should be subject to relevant inspections to ensure the production quality. "three-level" inspection cards for machined parts is shown in Table 3-4.

Table 3-4 "Three-level" inspection cards for machined parts

Part drawing number		Part name		Working step number	
Material		Inspection date		Working step name	
Inspection items	Self-inspection result	Mutual inspection result	Professional inspection	Remark	
Conclusion	☐ Qualified ☐ Unqualified ☐ Repair ☐ Concession to receive Inspection signature: Date:				
Non-conforming item description					

Project Summary

As a typical machining part of CNC lathes, the heating chamber is widely used in various equipment. According to equipment conditions and precision requirements, there will be some differences in the processing technology. Programmers and operators need to formulate the processing technology reasonably according to the processing conditions to improve the processing accuracy of the parts and the production efficiency.

Exercises After Class

1. Choice questions

(1) For the fixed cycle command, ( ).

A. it can complete a certain processing with only one instruction

B. it can only cycle once

C. it cannot replace with other instructions

D. it can only cycle twice

(2) When machining hole parts, the method of drilling→flat-bottom drill reaming→chamfering→fine boring is applicable to ( ).

A. step holes

B. blind holes with small aperture

C. blind holes with large aperture

D. flat bottom holes with large aperture

(3) In the CNC system, the instruction ( ) is non-modal in the machining process.

A. G01　　B. G04　　C. G17　　D. G81

(4) To make the spindle rotates clockwise at 800r/min, the instruction (　　) should be used.

A. S800 M03;　　B. S800 M04;　　C. S800 M05;　　D. S800 M06;

(5) Empty running the CNC machine tool is mainly used to check the (　　).

A. correctness of programming

B. correctness of tool path

C. operation stability of machine tool

D. correctness of machining accuracy

(6) When measuring the inner diameter of the hole, (　　) should be selected.

A. sine gauge　　B. internal micrometer

C. triangular plate　　D. block gauge

2. True or false

(1) G code of Group 00 in FANUC system are all non-modal instructions. (　　)

(2) G04 Command is modal code. (　　)

(3) Non-modal code is only valid in the program segment of the code. (　　)

(4) The G73 instruction of CNC lathe in FANUC system cannot contain macro program processing instructions. (　　)

(5) The IT value of the inner hole surface of the drilling workpiece is 5. 9. (　　)

3. Fill in the blanks

(1) For FANUC system CNC lathe, the single fixed cycle of inner and outer circle cutting is specified with the instruction ________, while the end cutting cycle is specified with the instruction ________.

(2) For FANUC system CNC lathe, the radial grooving fixed cycle is realized by the instruction ________, while the end grooving fixed cycle is realized by the instruction ________.

(3) The hole processing cycle instruction is ________. Once a hole processing cycle instruction is valid, the hole processing cycle instruction is used for hole processing in all subsequent positions until the hole processing cycle is cancelled with ________ instruction.

(4) If the width of the groove is less than the depth, use ________. If the width is greater than the depth, use ________. ________ can be used when processing slender workpiece.

4. Short answer questions

(1) Briefly describe the methods and precautions for groove processing.

(2) Complete the manual programming of the heating chamber.

(3) What are modal codes and non-modal codes? Please give examples respectively.

Self-learning test score sheet is shown in Table 3-5.

Table 3-5 Self-learning test score sheet

Task	Task requirements	Score	Scoring rules	Score	Remark
Learn key knowledge points	(1) Master the operation steps and parameters setting for turning a groove (2) Master the operation steps and parameters setting for thread turning (3) Master the setting method of feed/retract tool, cutting dosage and thread turning tool	20	Understand and master		
Technological preparation	(1) Be able to correctly read the shaft part drawings (2) Can be analyzed according to the part drawing, determine the process (3) Be able to write the correct processing program according to the processing process	30	Understand and master		
Hands-on training	(1) The corresponding equipment and utensils will be selected correctly (2) Can correctly operate the CNC lathe and adjust the processing parameters according to the machining situation	50	(1) Understand and master (2) Operation process		

Ideological and Political Classroom

Project 4 Programming and Machining Training for Flywheel Milling

➢ Mind map

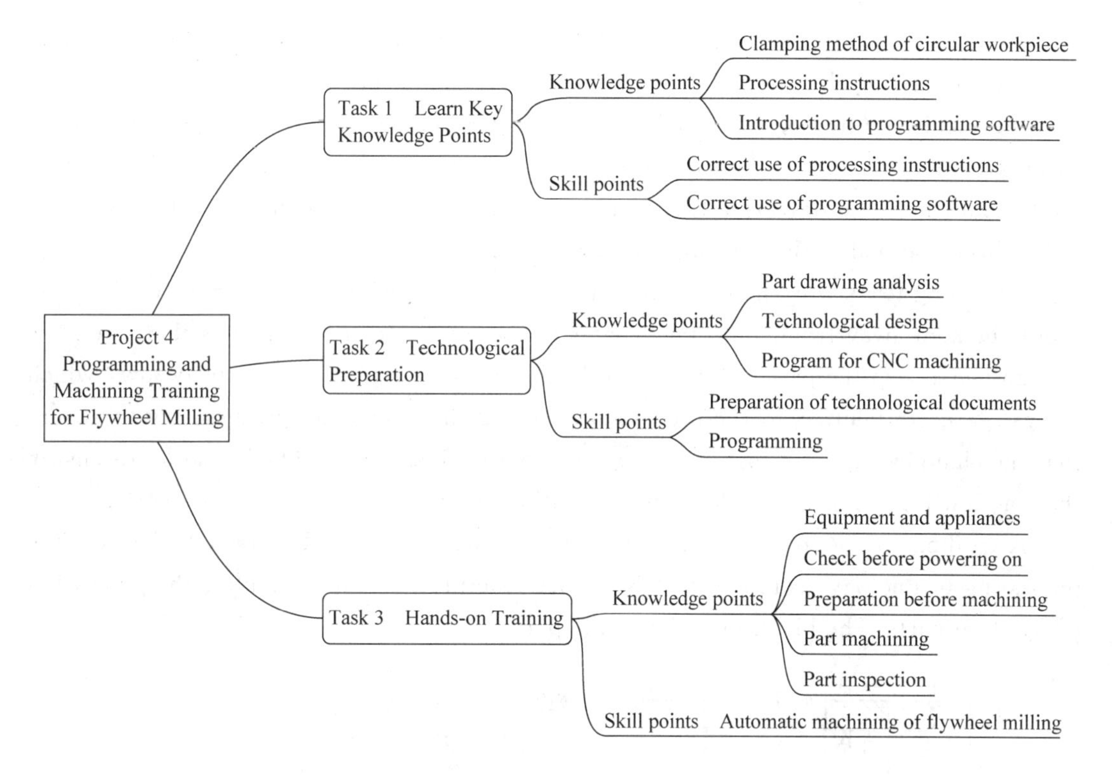

➢ Learning objectives

Knowledge objectives

(1) Have the ability to read the drawings of disc parts.

(2) Understand the purpose and characteristics of flywheel.

(3) Understand the drawing and programming methods of CAXA software.

Ability objectives

(1) Master the milling method of circular workpiece.

(2) Be able to independently determine the process routine and fill in the technological documents correctly.

(3) Be able to operate the CNC lathe correctly and adjust the machining parameters

according to the machining conditions.

(4) Be able to select measuring tools reasonably according to the accuracy and the structural characteristics of parts, and be able to measure the relevant dimensions correctly and normatively.

Literacy goals

(1) Develop students' scientific spirit and attitude.

(2) Cultivate students' engineering awareness.

(3) Develop students' teamwork skills.

➢ Task introduction

Flywheel is an important component of the Stirling engine. Stirling engine outputs power through a cycle of cooling, compression, heat absorption and expansion of the working medium (hydrogen or helium) in the cylinder. Only the expansion process outputs positive work. In order to ensure the normal cycle of the machine's working cycle, the flywheel needs to provide sufficient rotational inertia. At the same time, the existence of flywheel also makes the Stirling engine rotate more smoothly.

According to its working characteristics, during the design process, most of the mass should be kept away from the rotary center of the workpiece as far as possible to ensure enough inertia, so the part web is relatively thin and the rigidity is relatively low, which brings certain difficulties to the processing; During the machining process, try to ensure that the clamping and cutting process are concentric inside and outside, so as to ensure that the Stirling engine is stable during operation and reduce energy loss and noise.

According to the requirements of part drawing shown in Figure 4-1, develop the processing technology, compile the CNC processing program, and complete the processing of flywheel parts. The blank material of the part is 45 steel.

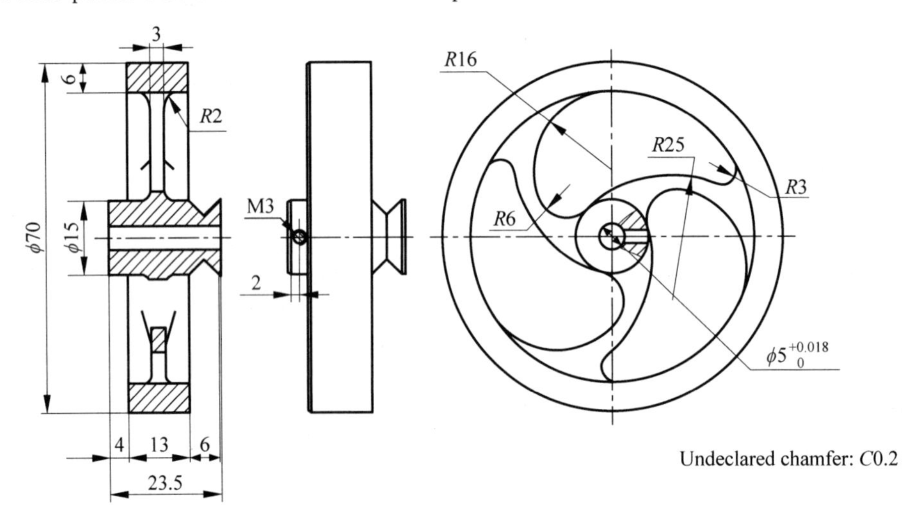

Figure 4-1 Part drawing of the flywheel

Task 1 Learn Key Knowledge Points

4.1 Clamping method of circular workpiece

In this task, we introduce the method of processing circular workpiece on the milling machine tool. Generally, the circular workpiece can be clamped by the three-jaw self-centering chuck and the V-shaped block.

1. Three-jaw self-centering chuck

The three-jaw self-centering chuck is composed of chuck body, movable jaw and jaw drive mechanism. Under the guide part of the three-jaw self-centering , there are threads that mesh with the plane threads on the back of the dish-shaped bevel gear. When a wrench is used to turn the bevel gear through the square hole, the dish-shaped gear rotates, and the plane thread simultaneously drives the three jaws to approach or exit towards the center to clamp workpieces with different diameters. Replace the three jaws with three counter jaws to clamp the workpiece with larger diameter. The self-centering accuracy of three-jaw self-centering chuck is 0.05-0.15mm. The accuracy of the workpiece processed by the three-jaw self-centering chuck is affected by the manufacturing accuracy of the chuck and the wear after use. In order to ensure the positioning accuracy, soft three-jaw is usually used to mill the same shape as the mounting surface of the clamped part (as shown in Figure 4-2), which greatly improves the alignment accuracy and reduces the damage of the clamping to the workpiece.

When installing the flywheel parts, the rigidity of the parts should be considered comprehensively. Because the processed spoke is thin and hollow, the spoke should be placed in the clamping position of the three jaws as far as possible, and the clamping force should be controlled reasonably to reduce the deformation of the workpiece.

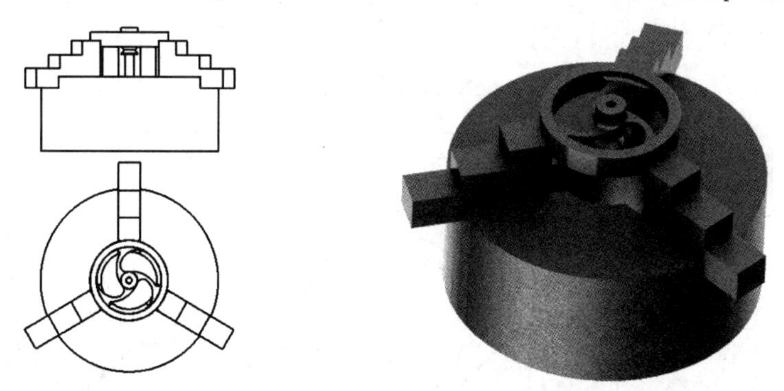

Figure 4-2 Schematic diagram of three-jaw self-centering chuck clamping

2. V-shaped block

V-shaped block is mainly used to clamp round workpiece such as shaft, sleeve, disc, etc., so as to locate the center of round workpiece. Generally, the V-shaped block is a pair of two pieces, and the plane and V-shaped groove of the two pieces are ground in one installation. The error of parallelism and perpendicularity between the surfaces of the precision V block is within 0.01mm. The center line of the V-shaped groove must be in the symmetry plane of the V-shaped frame and parallel to the bottom surface. The concentricity and parallelism error are also within 0.01mm. The half-angle error of the V-shaped groove is within ±(0.5°-1°). Precision V-shaped block can also be used for marking. V-shaped block with clamping bow can firmly clamp the cylindrical workpiece on the V-shaped block and turn it to various positions for marking.

In the milling process, the V-shaped block is used with precision flat-nose pliers. Some flat-nose pliers are delivered with V-shaped blocks (as shown in Figure 4-3).

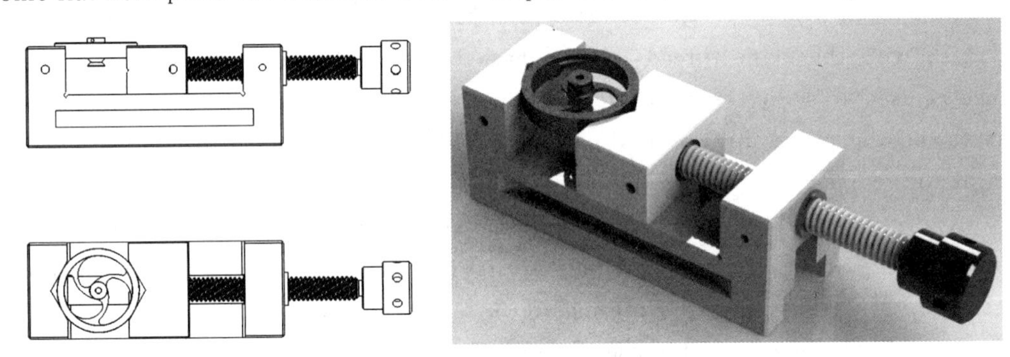

Figure 4-3 Schematic diagram of V-shaped block clamping of flat-nose plier

Generally, the flat-nose plier includes a fixed jaw and a movable jaw, so there is no self-centering function. The V-shaped block that really works is the one on the side of the fixed jaw, so the tolerance of the workpiece's outer circle has a great impact on the centering error during installation and positioning. Figure 4-4 is the schematic diagram of positioning error calculation during V-block positioning. As shown in Figure 4-4, we need to consider the offset of the positioning center caused by the machining error δ_d of the outer circle of the flywheel.

$$\overline{O_1O_2} = \frac{d}{2\sin\frac{\alpha}{2}} - \frac{d-\delta_d}{2\sin\frac{\alpha}{2}} = \frac{\delta_d}{2\sin\frac{\alpha}{2}} \tag{4-1}$$

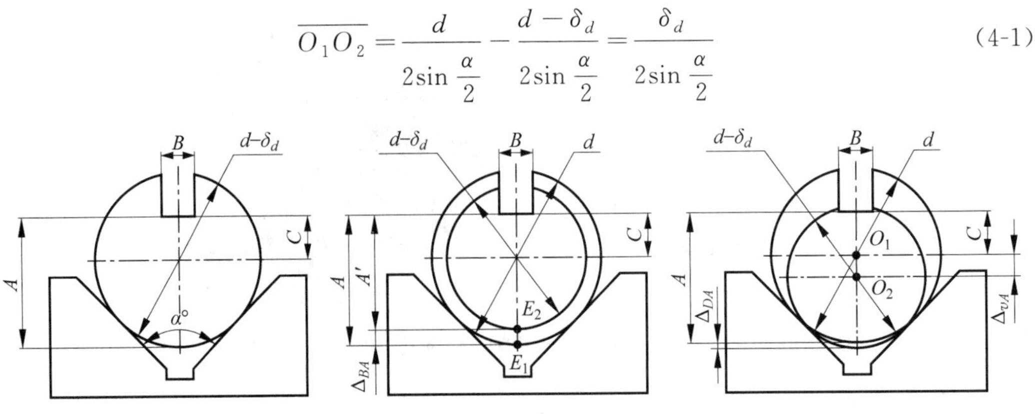

Figure 4-4 V-shaped block positioning error

When V-shaped block clamps round workpiece, the force bearing part is "line" contact. When three-jaw self-centering chuck is clamped with soft jaws, it is "face" contact. Therefore, it is better to use three-jaw self-centering chuck with soft jaws for installation. If conditions permit, it is recommended to use three-jaw self-centering chuck for workpiece clamping during training.

4.2 Processing instructions

1. Sub-program instructions (M98, M99)

In a processing program, if some of the processing contents are identical or similar, as shown in Figure 4-5, in order to simplify the program, these repeated program segments can be listed separately and written into sub-programs in a certain format. If the main program needs a sub-program during execution, call the sub-program by calling instructions, and then return to the main program after execution, and the main program continues to execute the following program segments.

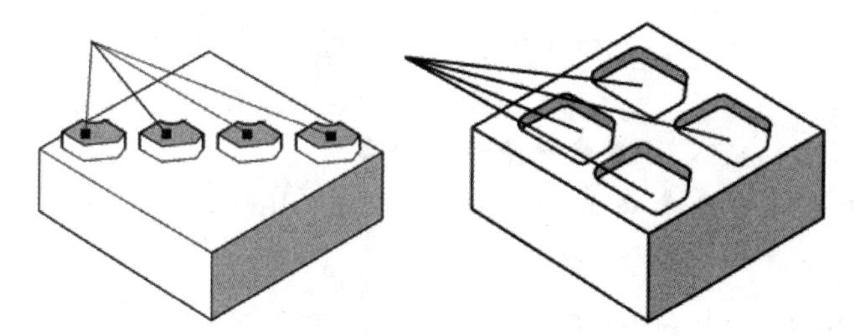

Figure 4-5 Application of sub-programs

The main functions of the sub-programs are:

(1) Several parts on the workpiece have the same contour shape. In this case, you can just write a sub-program to process the contour shape, and then use the main program to call the sub-program several times to complete the processing of the workpiece.

(2) The tool path with the same trajectory appears repeatedly in the process. If the tool path with the same trajectory appears in a certain processing area or on all levels of this area, it is convenient to use sub-program to write the processing program. In the program, the increment value is often used to determine the cutting depth.

(3) When processing more complex parts, it often includes many independent processes, and sometimes the processes need to be adjusted appropriately. In order to optimize the processing program, each independent process is compiled into a subroutine, which forms a modular program structure to facilitate the adjustment of the processing sequence. There are only instructions such as tool change and calling sub-programs in the main program.

The call of sub-program in FANUC 0*i* MF can be nested for 15 levels, and its program composition is shown in Table 4-1.

Table 4-1 Composition of a sub-program

Sub-program	Explanation
O◇◇◇◇;	Sub-program name. The naming requirements are similar to those of the main program name
… … …	Sub-program content
M99;	Sub-program returns instruction. M99 instruction does not have to form a separate program segment

There are several ways to call a sub-program.

(1) When calling a sub-program with a program number of less than 4 digits, the format is:

M98 P ◆◆◆◆◇◇◇◇;

or M98 P◇◇◇◇ L ◆◆◆◆◆◆◆◆◆;

in which, ◆◆◆◆ is the number of calling, and ◇◇◇◇ is the sub-program number.

(2) When calling a sub-program with a program number of more than 5 digits, the format is:

M98 P◇◇◇◇◇◇◇◇◇◇◇ L◆◆◆◆◆◆◆◆◆;

in which, ◆◆◆◆◆◆◆◆◆ is the number of calling, and ◇◇◇◇◇◇◇◇◇ is the sub-program number.

(3) When calling a sub-program by the program name, the format is:

M98 〈○○○○〉 L ◆◆◆◆◆◆◆◆◆;

in which, ◆◆◆◆◆◆◆◆◆ is the number of calling, and ○○○○ is the name of the sub-program.

In our routine programming, we usually call sub-programs with program numbers less than 4 digits, so this course only needs to master the first call method, while the other two only need to be understood.

In the first call method, the number of calling is 1-9999. When the calling number is 1, it can be omitted, and the 0 before the number can also be omitted. The subprogram number should be 4 digits. When the number is less than 4 digits, use 0 to fill in the front. For example, the sub-program "O365" is called five times, and the call instruction is M98 P50365.

When the main program calls a sub-program, if it is considered as a 1 level sub-program call, FANUC 0*i* MF system can call sub-programs up to 10 levels. If combined with macro program instructions, sub-program calls and macro program calls can be nested up to 15 levels. Figure 4-6 is the classification schematic diagram of sub-program call.

2. Coordinate system rotation instructions (G68, G69)

The coordinate system rotation instructions (G68, G69) are used to rotate the

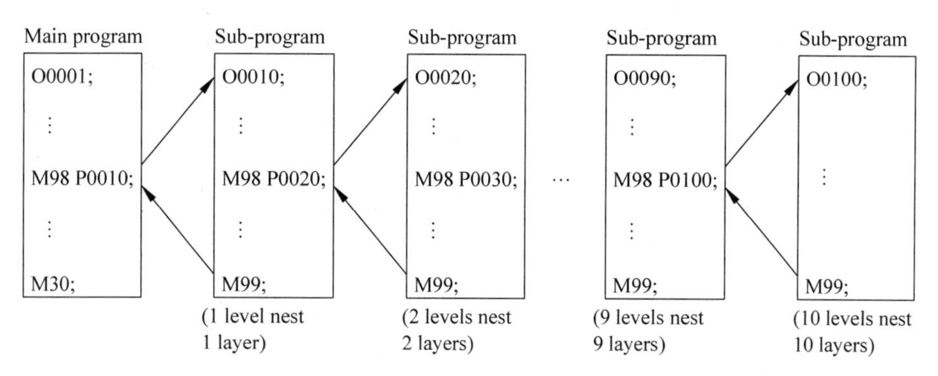

Figure 4-6 classification schematic diagram of sub-program call

programmed shape. By using this function, when the installed workpiece is in the position relative to the rotation of the machine tool, it can be compensated by the rotation instruction, as shown in Figure 4-7. In addition, when there is a figure that makes a shape rotate, the time required for programming and the length of the program can be shortened by writing a shape sub-program and calling the sub-program after making it rotate.

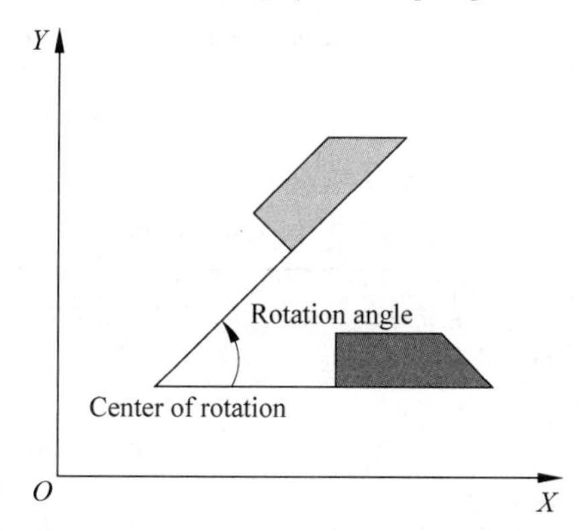

Figure 4-7 Coordinate system rotation

In a FANUC 0*i* MF system, the coordinate rotation programming format is:

G68 X_ Y_ R_; In the *XOY* plane, the shape rotates around the center of rotation (x, y) at angle R.

...

G69; Cancel the coordinate rotation.

In some versions of the system, the G68 instruction can be used with the relative value (incremental value) programming function to make a relative rotation of a certain angle. The unit of angle is (°).

You can set if the coordinate rotation is absolute or incremental by modifying the system parameter RIN (No. 5400 #0).

3. Programming example of coordinate transformation with sub-program

In this section, we take the part shown in Figure 4-8 as an example to practice the basic CNC programming method of using sub-program instructions and coordinate system rotation instructions.

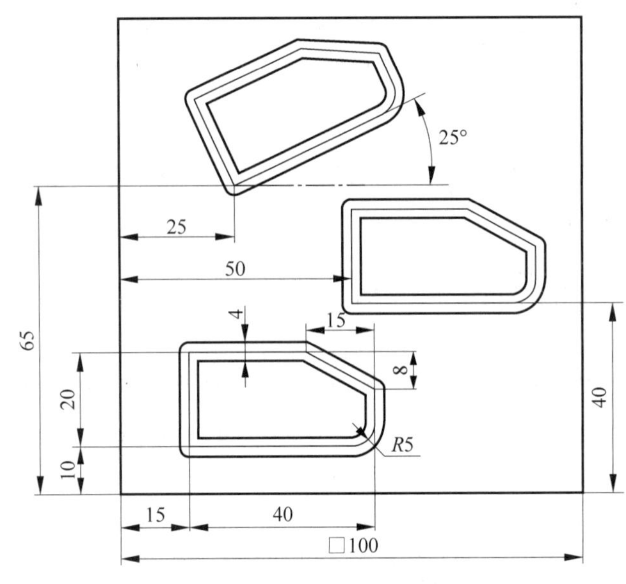

Material: 45 steel　　Track depth: 0.5mm

Figure 4-8　Programming practice of sub-program and coordinate rotation instructions

First, according to the graph in the lower left corner, as shown in Figure 4-8, use the relative coordinate instruction to write the sub-program, as shown in Table 4-2.

Table 4-2　The sub-program

Sub-program	Annotation
O1234;	The sub-program is named "1234"
G1 G90 Z-0.5 F50;	Lower the cutter to the Z-0.5 position at speed of F50
G1 G91 X40 R5 F500;	Cut 40mm along the positive X direction and chamfer the $R5$ at the end point
Y12;	Cut 12mm along the positive Y direction
X-15 Y8;	Simultaneously cut 15mm in the X negative direction and 8mm in the Y positive direction
X-25;	Cut 25mm in the negative X direction
Y-20;	Cut 20mm in the negative Y direction and return to the starting point
G1 G90 Z10;	End of machining. Lift the cutter to the $Z10$ position
M99;	The sub-program ends and returns the calling location

When machining parts, you only need to call the sub-program directly once at the coordinates (15, 10) and (50, 40). At (25, 65) position, first rotate the coordinate system by 25°, and then call the sub-program once to complete the shape processing. The main program is written as shown in Table 4-3.

Table 4-3　The main program

Main program	Annotation
O0001;	The main program is named "0001", and "0" can be omitted
T1 M6;	Call the T1 tool: $\phi 4$ keyway end milling cutter
S3000 M3;	Set the spindle speed to 3000r/min and make the spindle rotate forward

Continued

Main program	Annotation
G0 G90 G54 X15 Y10;	Quickly locate to the absolute coordinate (15, 10) in the G54 coordinate system
G43 Z10 H1;	Call H1 tool length compensation and quickly position to the safety height *Z*10 position
M98 P0011234;	The sub-program "1234" is called once, and the *P* parameter can be simplified to "P1234"
G0 X50 Y40;	Quickly locate to (50, 40)
M98 P0011234;	The sub-program "1234" is called once, and the *P* parameter can be simplified to "P1234"
G0 X25 Y65;	Quickly locate to (25, 65)
G68 X25 Y65 R25;	The coordinate system rotates 25° around the point (25, 65)
M98 P0011234;	The sub-program "1234" is called once, and the *P* parameter can be simplified to "P1234"
G69;	Cancel the coordinate system rotation
G0 Z100;	Lift the cutter to the *Z*100 position
M5;	The spindle stops
M30;	The program ends and returns to the header

G85 X_Y_Z_R_F_K_;
X_Y_: hole position data
Z_: distance from *R*-point to hole bottom
R_: distance from reference plane to *R*-point
F_: feed rate
K_: repetitions (only if necessary)

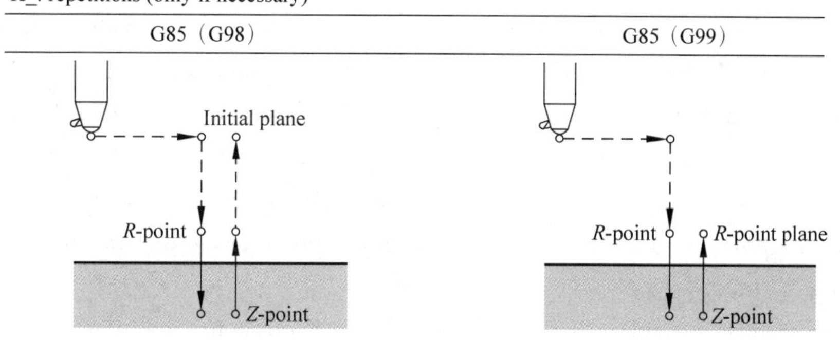

Figure 4-9 Reaming instruction

4. Reaming instruction (G84)

As shown in Figure 4-9, the basic action of reaming processing is to first locate the reaming position along the *X* and *Y* axes, and then feed the tool quickly to the *R*-point plane. After that, ream the hole from the *R*-point plane to the *Z* value setting point. After reaching the specified position of *Z* axis coordinate, the tool returns to the *R*-point plane at the feed rate *F*.

If multiple holes are to be reamed, when specified in G98 instruction, the tool will be raised to the initial plane and then moved to the next hole after machining one hole. When specified in G99 instruction, after machining a hole, the tool is raised to the R-point plane and then moved to the next hole. Therefore, you should pay attention to if there is interference above the processing surface of the workpiece when programming.

4.3 Introduction to programming software

CAM manufacturing engineer software is a software developed by CAXA Technology Co. LTD (CAXA) for NC milling programming. CAXA is an independent industrial software and industrial internet company in China. The company is the contractor of the National Engineering Laboratory of "Intelligent Collaborative Manufacturing Technology and Application", and is also a high-tech enterprise, a high-tech enterprise in Zhongguancun, a credit five-star enterprise in Zhongguancun Demonstration Zone, a demonstration enterprise for the transformation of high-tech achievements in Beijing, a patent demonstration unit in Beijing, and the first batch of intelligent manufacturing system suppliers of the Ministry of Industry and Information Technology.

CAXA has always adhered to technological innovation and independently developed digital design (computer aided design,CAD), product lifecycle management (PLM), and digital manufacturing (manufacturing execution system, MES) software. It is an early software company engaged in this field in China. The R&D team has accumulated many years of professional experience and has the international advanced technology level. It has three R&D centers in Beijing, Nanjing, and Atlanta, the United States. At present, it has more than 330 trademarks, patents, patent application and copyright. The company is a member of the National Intelligent Manufacturing Standardization General Group, a plenipotentiary member of the Big Data Standards Working Group, a member of the China Information Technology Standardization Technical Committee, and a member of the Beijing Standardization Association, leading or participating in the construction of the national intelligent manufacturing standards, industrial cloud, industrial big data, additive manufacturing, etc.

CAXA mainly provides industrial software, intelligent manufacturing solutions, industrial cloud platforms and other products and services for equipment, automobile, electronics, aerospace, education, etc. CAXA's product line is a complete product line, including digital design (CAD), product life cycle management (PLM), and digital manufacturing (MES), which are the core links of enterprise design and manufacturing.

In the modeling direction, CAM manufacturing engineer software adheres to the hybrid modeling method of solid and surface, and the visual design concept. Solid modeling mainly includes extrusion, rotation, guide, lofting, chamfer, fillet, punching, rib plate, draft, parting and other feature modeling methods. You can quickly generate 3D solid models from 2D sketch profiles. It provides a variety of functions to build datum planes. Users can build various datum planes according to known conditions. Surface modeling provides a variety of non-uniform rational B-splines (NURBS) surface modeling methods: complex surfaces can be generated by scanning, lofting, rotating, guiding, isometric, boundary, mesh. It also provides surface line clipping and surface clipping, surface extension, surface stitching according to the average tangent or selected surface tangent, and splicing between multiple surfaces. In addition, it provides powerful surface blending function, which can realize surface blending methods such as two sides, three sides, series of faces, etc., and can also realize equal radius or variable radius blending. The system supports the modeling method of mixing solid and complex surface, which is applied to complex part design or mold design. It provides the functions of surface trimming solid, surface thickening to solid, and closed surface filling to solid. In addition, the system also allows the surface of a solid to be generated into a surface for direct reference by users. The perfect combination of surface and solid modeling methods is a prominent feature of CAM manufacturing engineers software in CAD software. At each operation step, the prompt area of the software has an operation prompt function. No matter beginners or engineers with rich CAD experience, they can quickly grasp the know-how and design their desired part models according to the prompts of the software.

In terms of CNC programming, the CNC milling programming module has convenient code editing functions, which is easy to learn and is very suitable for manual programming. At the same time, it supports automatic import code and hand-written code, including trajectory simulation of macro program code, which can effectively verify the correctness of the code. Support a variety of system code to reverse conversion, to achieve processing programs in different CNC system program sharing, but also has the function of communication transmission, through the RS 232 port can achieve code transfer between the CNC system and programming software. The system provides seven roughing methods: flat area roughing (2D), regional roughing, contour roughing, scanning line, cycloidal line, interpolation milling, guide line (2.5 axis). Provide 14 kinds of finishing methods: plane contour, contour guide, surface profile, surface area, surface parameter line, contour line, projection line, contour line, contour line, guide, scan line, restriction line, shallow plane, three-dimensional offset, deep cavity sidewall and other finishing functions. Provide three kinds of complementary processing: contour line replenishment processing, pen root cleaning, regional remediation and other complementary processing functions. Two kinds of slot machining are available: curved milling groove and scanning milling groove. In terms of multi-axis machining, it is not only possible to carry out four-axis curve and four-axis flat-sectional machining, but also to support five-axis and other parameter lines, five-axis side milling, five-axis curves, five-axis surface areas, and five axes G01 drilling, five-axis orientation, rotary four-axis trajectory and other processing of impellers, blade parts, in addition to the above processing methods, the system also provides special impeller roughing and impeller finishing functions, can achieve the overall processing of impellers and blades.

CAM manufacturing engineer software uses Windows operation system, based on microcomputer platform, using original Windows menus and interactions, multi-language user interface, can be easily and smoothly learned and operated.

In terms of data interface, CAM manufacturing engineers software provides rich data interfaces, including: directly reading the data interface of popular 3D CAD software in the market, such as CATIA, Pro/ENGINEER; DXF and IGES standard graphics interface based on surface, STEP standard data interface based on solid; X-T and x-B format files of Parasolid geometry core; The SAT format of the ACIS geometric core, STL for rapid prototyping equipment and VRML for the Internet and virtual reality, etc. These interfaces ensure two-way data exchange with the world's popular CAD software, enabling enterprises to realize virtual product development and production with partners across platforms and regions.

4.3.1 Introduction of the basic interface

The design environment of CAXA 3D solid design is a window to complete various design tasks, providing various tools and conditions. Figure 4-10 shows the 3D design environment.

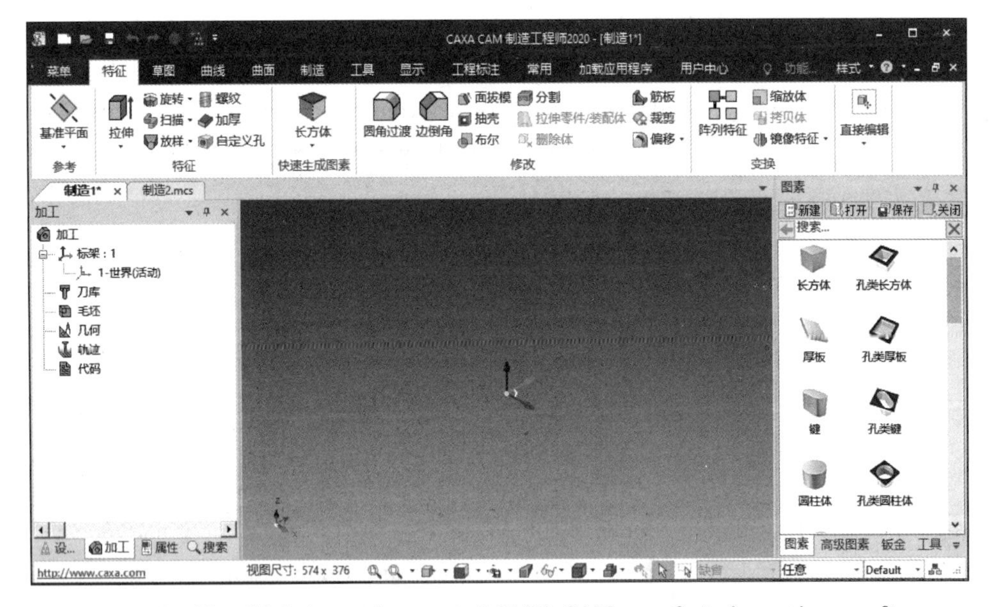

Figure 4-10 3D solid design environment of CAXA-CAM manufacturing engineers software

The top of the CAXA 3D solid design environment is the Quick Launch Bar, software name and current file name. Below is the menu bar divided by function. The middle is the display area of design work. The top of the display area is a multi-document tab, and its left side displays the design tree attributes etc. , and its right side is the design element library that can be automatically hidden. At the bottom is the status bar, which mainly includes the operation prompt, view size, unit, view direction setting, design mode selection, configuration setting, etc.

4.3.2 CAD modeling

3D modeling of the flywheel part is carried out through CAXA 3D. The coordinate system of modeling is the same as that of machining as far as possible, so that the *Z* axis direction is directly selected as the main axis direction during machining. CAXA 3D modeled flywheel part is shown in Figure 4-11.

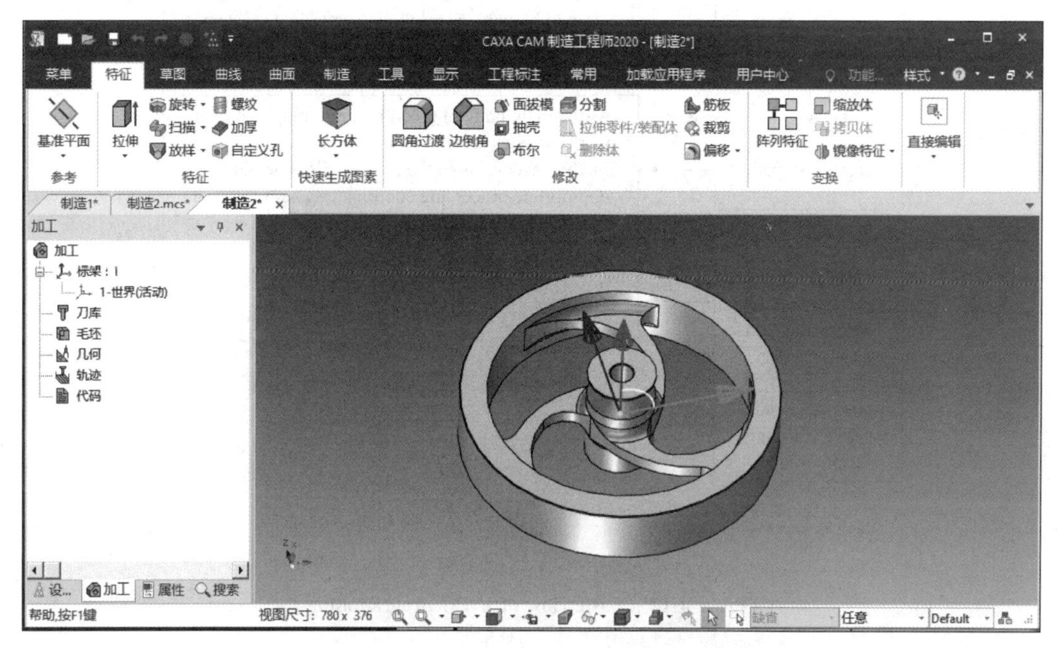

Figure 4-11 CAXA 3D modeled flywheel part

This chapter is mainly about CNC programming technology. The detailed 3D modeling technology will not be described here. Please refer to the relevant material.

4.3.3 CAM Programming

The basic flow of CAM programming is shown in Figure 4-12.

Before programming, appropriate machining coordinate system should be set according to different machining process. It is recommended to overlap the design coordinate system with the machining coordinate system of the main machining process when modeling the parts to ensure the machining accuracy.

The tool definition shall be carried out in strict accordance with the process requirements, and the general tool shall be selected as far as possible. When multiple specifications of tools are required during the processing, the tool size specification shall be as small as possible to reduce the tool change time.

When the part shape is not complex, you can directly select the processing elements on the model without setting the processing geometry.

After the tool path is generated, the cutting process shall be simulated by using the

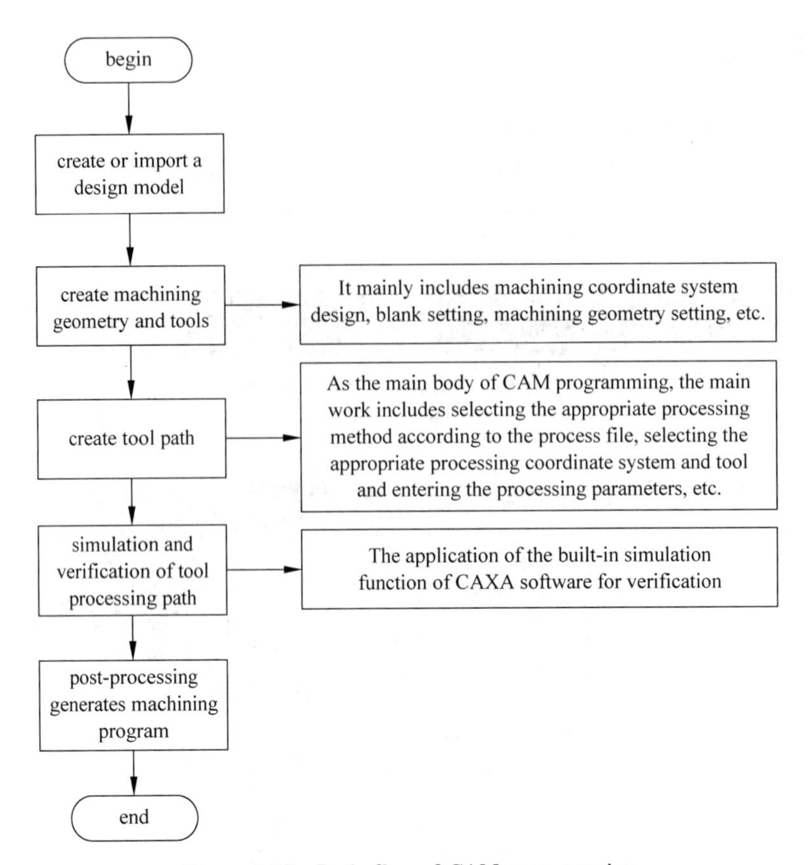

Figure 4-12 Basic flow of CAM programming

tool path simulation function to check whether there is over-cutting, whether there is collision in the rapid positioning process, and whether there is a need to arrange angle cleaning at the position with excessive allowance.

After the simulation is completed, the G code recognized by the machine tool can be generated by post-processing, and the G code program can be input into the CNC machine tool through storage media or distributed numberical control(DNC).

1. Establish the machining coordinate system

Select the icon 加工 in the navigator interface on the left side of the window to enter the machining drop down list. Right-click on the coordinate system 标架:1 and select Create coordinate system. In the pop-up dialog box, set and select parameters to establish the machining coordinate system as shown in Figure 4-13, where the blue arrow represents the *Z* axis direction, the red arrow represents the *X* axis direction, and the green arrow represents the *Y* axis direction. The coordinate system should be defined in accordance with the coordinate direction identified in the process document.

When programming, the coordinate system should be activated according to which machining coordinate system is used.

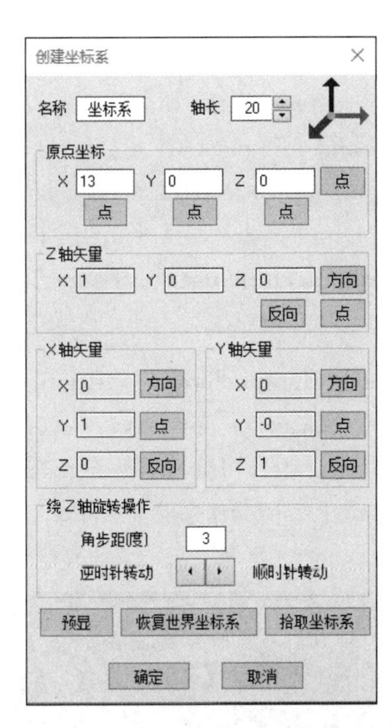

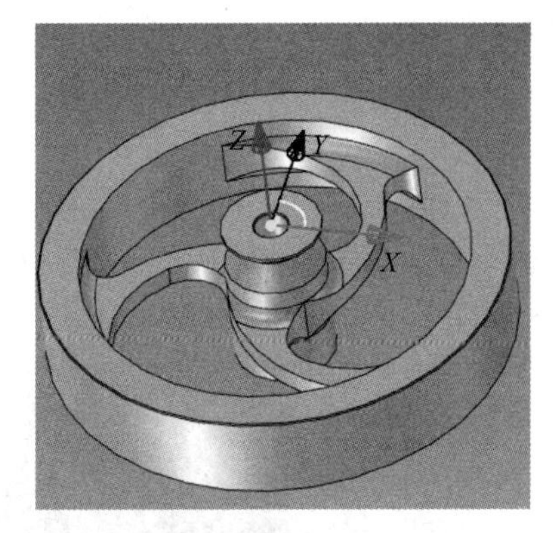

Figure 4-13 Establish the machining coordinate system

2. Create tool

Right-click the icon 刀库 in the machining drop down list and select Create a Tool and pop up Create tool dialog box as shown in Figure 4-14.

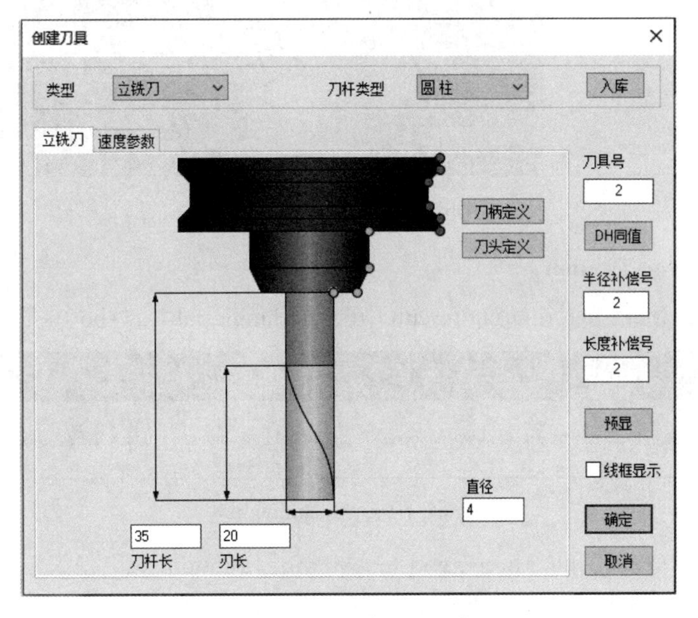

Figure 4-14 Create tool

The selection and creation of tools should be consistent with the process documents, also note that general tools should be selected as far as possible.

3. Create blank

For simulation purposes, a blank should be created for the machining process. In the machining drop down list, select the icon 刀库 and left click Creat blank instruction to create a blank. If the workpiece is relatively simple and there are few processing sites, you can select the automatically generated blank. If the parts are complex and there are many potential collision hazards, the model import method can be used to build the blank.

4. Create geometry

The machining geometry is to precisely specify the machining position. For the part with simple shapes, the boundary line of the part can be directly selected. For more complex parts, the machining boundary can be accurately drawn through the method of sketch drawing. The geometric boundary required for processing a spoke hollow area is shown in Figure 4-15.

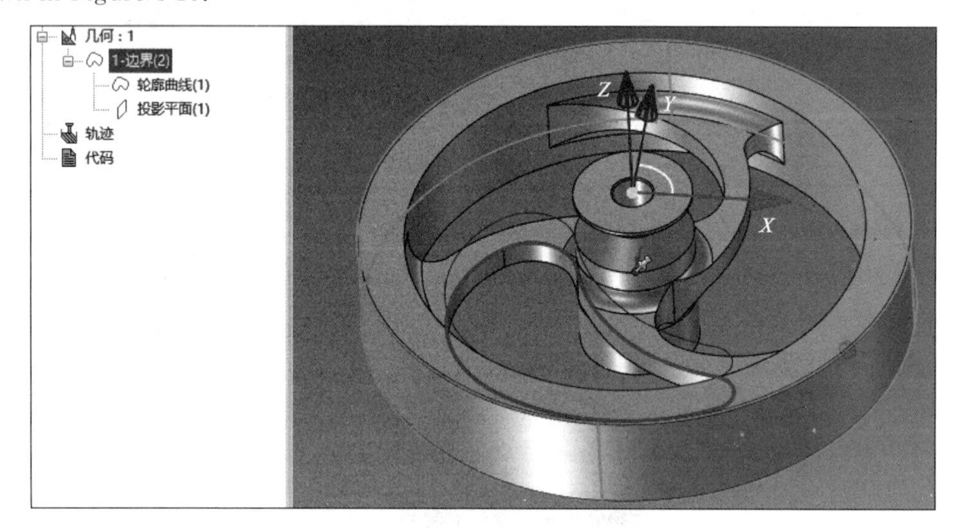

Figure 4-15 Create geometry

5. Trajectory generation

Left click the Manufacturing tag to enter the machining tab, as shown in Figure 4-16.

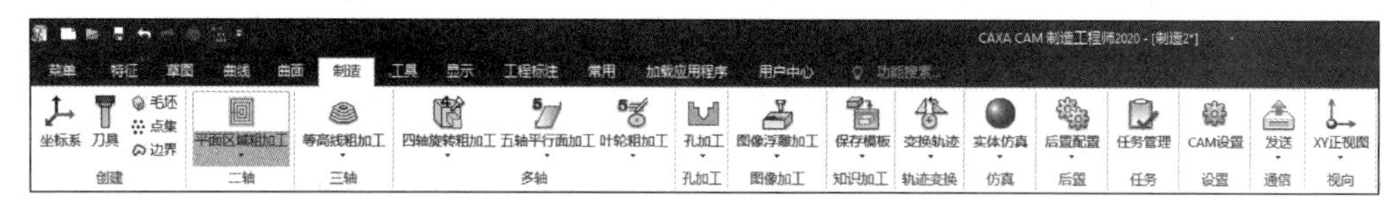

Figure 4-16 Machining tab

This section uses the three-axis machining function of CAXA CAM, and the processing methods are shown in Figure 4-17.

Three-axis machining function includes two rough machining modes and 13 fine machining modes. Contour rough machining is the common rough machining mode, and adaptive rough machining is commonly used in high-speed milling.

Taking contour rough machining as an example, the basic process of CAM

Figure 4-17 Three-axis drop down list

programming is introduced. Select Contour Rough Machining from the Three Axis drop-down menu and pop up the dialog box of Create: contour rough machining, as shown in Figure 4-18.

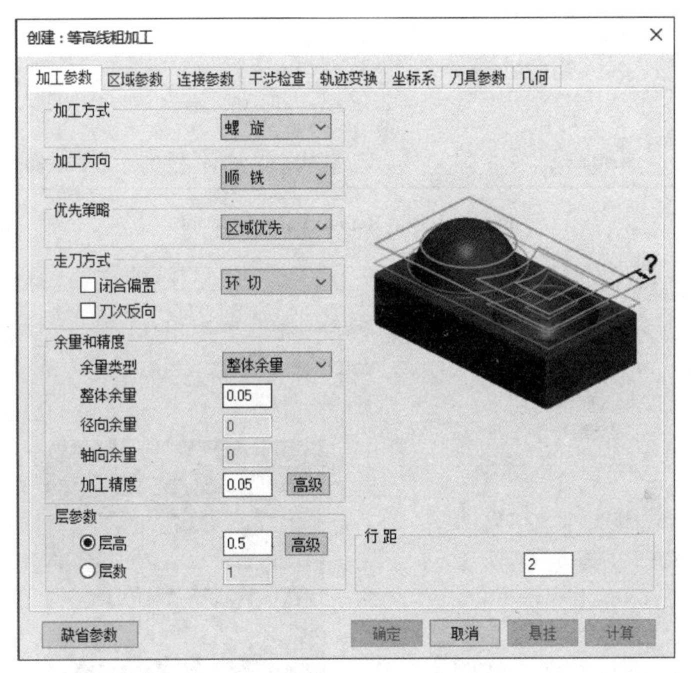

Figure 4-18 Create: contour rough machining dialog box

The machining parameter tab is used to define the machining mode, machining direction, priority strategy, feed mode, machining allowance and precision, cutting layer parameters, cutting line space, etc.

The region parameter tab is used to define the machining height range, start point, processing boundary, workpiece boundary, supplementary processing, etc. As shown in Figure 4-19, the starting height and ending height of machining are defined by picking.

Select the boundary of the machining geometry using the machining boundary tab, as shown in Figure 4-20. When generating tool path, only the interior of the selected machining boundary is processed.

The Connection Parameters tab is used to set the connection mode, tool feeding mode, empty cutting area, smooth and other parameters of each area in the machining

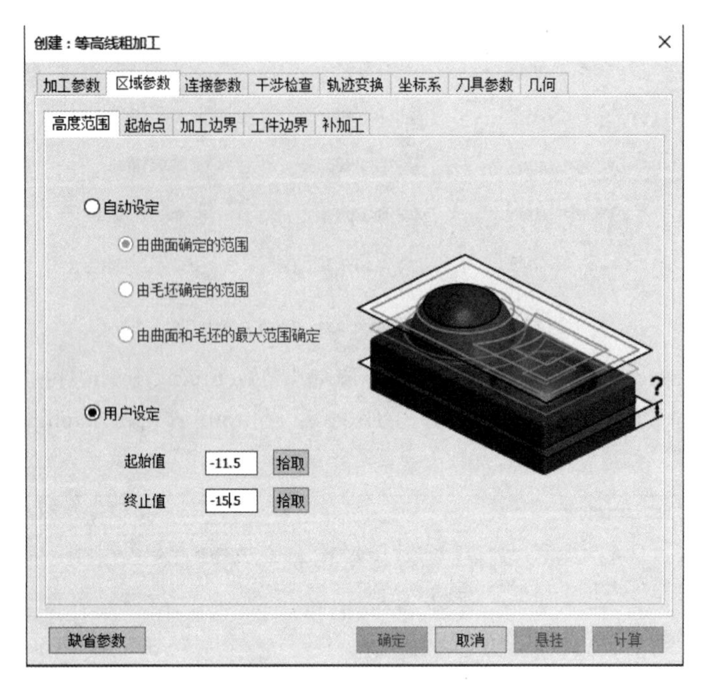

Figure 4-19　Region parameter tab

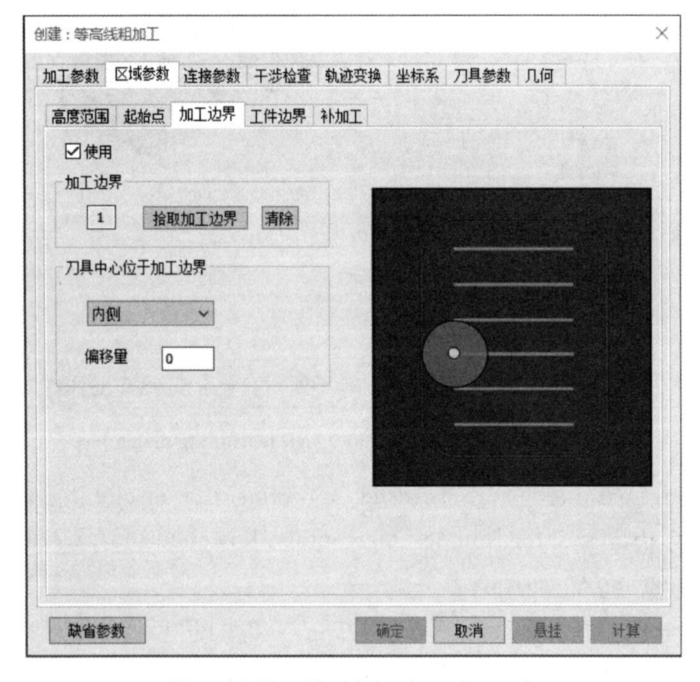

Figure 4-20　Machining boundary tab

process. The parameter setting in the connection mode of inside and outside the area is shown in Figure 4-21.

Interference check tab is used to verify the interference between tool shank, tool shank and cutting area. It is widely used in multi-axis machining, and it cannot be set in machining below three axes. Path transformation tab is used to translate, rotate, mirror

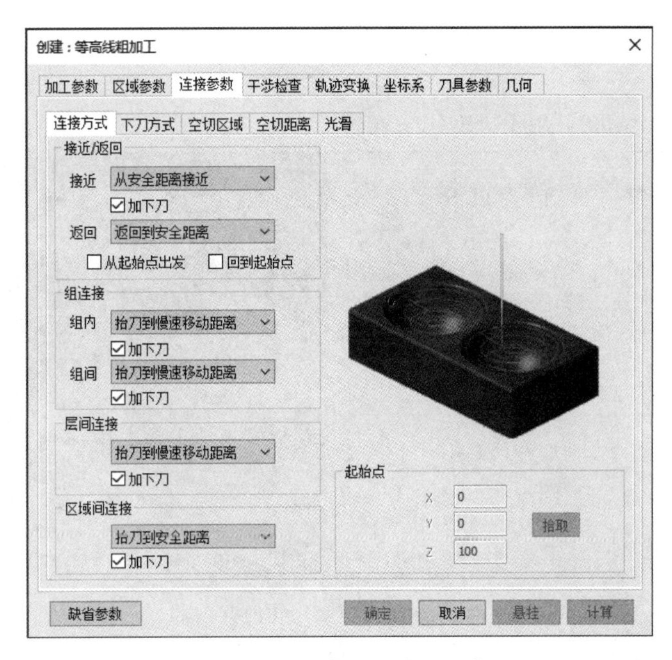

Figure 4-21 Connection mode

the tool path. The coordinate system tab is used when the machining coordinate system is not set before. If the first step is set and activated, it will be directly brought in. Geometry tab is used to select machining parts and blanks, and can directly select part models and blanks.

Tool parameters tab are used to select tools and define cutting parameters, which are generally done when the tool is defined, and can be modified in the speed parameters as needed, as shown in Figure 4-22.

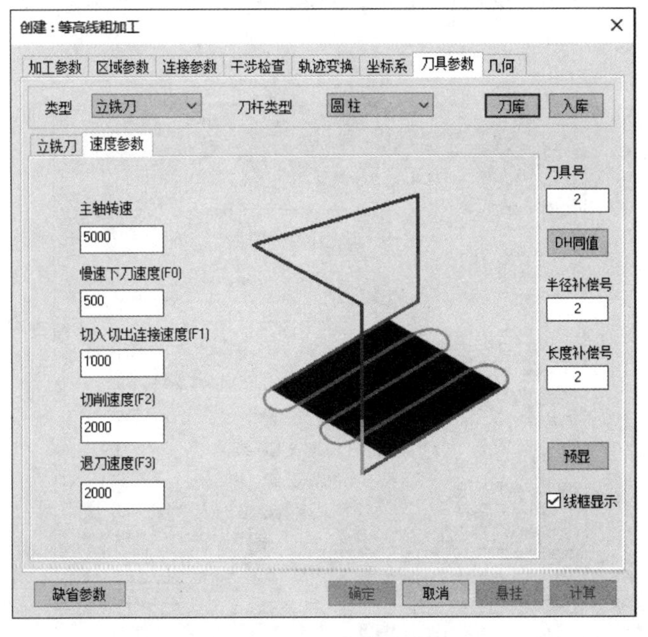

Figure 4-22 Tool parameters tab

After the setting is completed, left click OK button, and the generated tool path is shown in Figure 4-23. If the machining parameters are modified, left click the calculation button to regenerate the tool path.

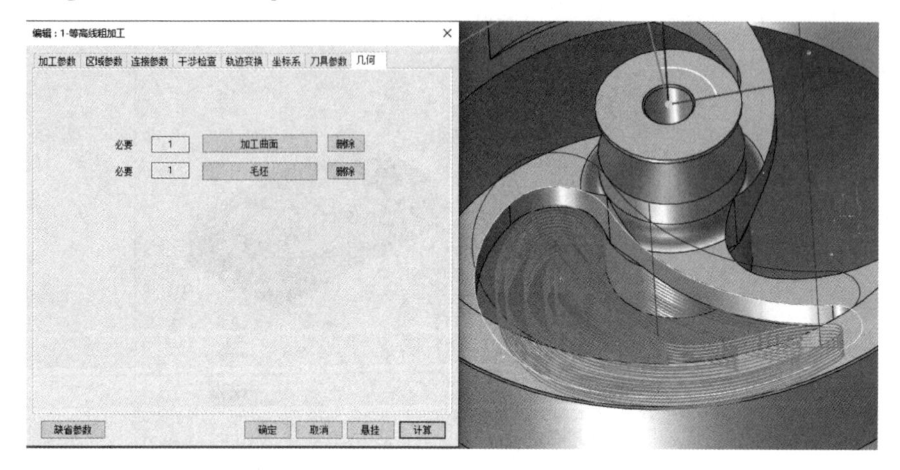

Figure 4-23　Tool path

6. Simulation machining

Right-click on the Tool path, select Solid Simulation, select the tool path name for simulation in the pop-up dialog box, and pick the blank and part to verify the material clearance, and observe whether there is over-cutting.

7. Post processing

Right-click Post Processing in the tool path, select Fanuc in the control system, select Milling center_3X in the machine tool configuration, pick the tool path to post processing, define the starting and ending height during processing, and left click Post, as shown in Figure 4-24.

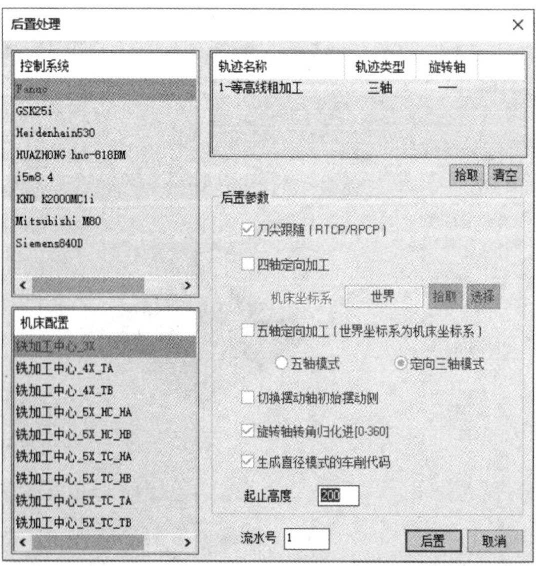

Figure 4-24　Fanuc three axis post processing

8. Edit the code

After post processing, the Edit Code dialog box will pop up automatically. The generated program head, program tail, start position, cutting parameters, etc. can be edited as needed, as shown in Figure 4-25. The program should be stored for processing.

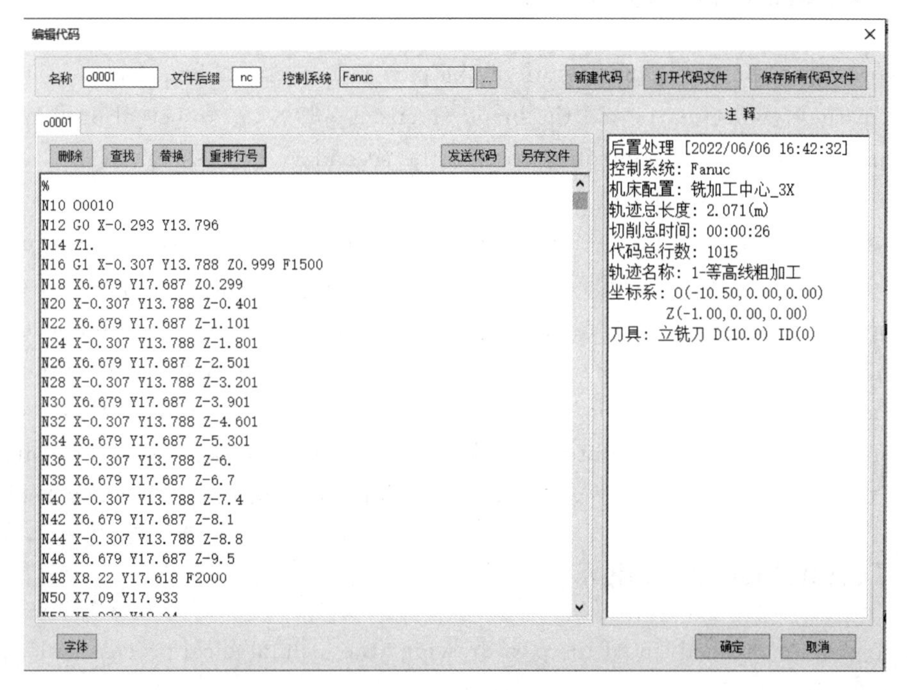

Figure 4-25 Editing the CNC machining program

9. Post configuration

If you are not satisfied with the generated G code, you can perform post configuration. Left click the icon to control program output, machine tool, output variables, etc. by modifying the post processing file parameters, as shown in Figure 4-26.

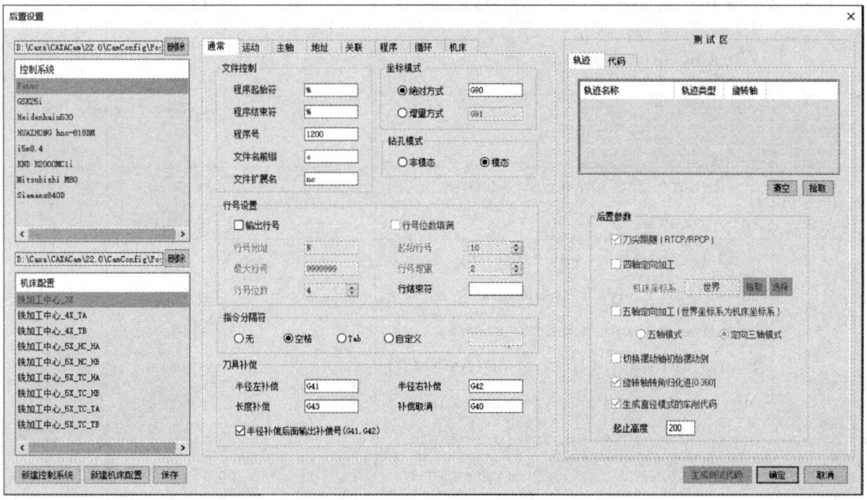

Figure 4-26 The dialog box of post configuration

Task 2 Technological Preparation

4.4 Part drawing analysis

According to the use requirements of the part, 45 steel can be selected as the blank material of the flywheel part, and the blanking size is $\phi72\times25$. In actual production, the cutting size can be flexibly controlled so that the secondary processing can be carried out according to the actual situation.

As shown in Figure 4-1, the three-jaw self-centering chuck is used for clamping, and the effective height of the workpiece above the top surface of the three-jaw self-centering chuck should not be less than 17mm. Use $\phi12$ end milling cutter to rough and finish the top surface and outer circle. Use a $\phi12$ $R2$ fillet end milling cutter to machine the left side of the spoke. After turning over, use three-jaw self-centering chuck to clamp with soft jaws, and process the right part of the workpiece, and then drill and ream the center hole. Finally, benching the M3 set screw hole, and it will not be introduced in this book.

4.5 Technological design

According to the analysis of the part drawing, the technological process is designed as shown in Table 4-4.

Table 4-4 Technological process card

<table>
<tr><td colspan="2" rowspan="2">Machining process card</td><td colspan="2">Product model</td><td colspan="2">STL-01</td><td>Part number</td><td>FL-01</td><td colspan="2">Page 1</td></tr>
<tr><td colspan="2">Product name</td><td colspan="2">Stirling engine</td><td>Part name</td><td>Flywheel</td><td colspan="2">Total 1 page</td></tr>
<tr><td colspan="2">Material grade</td><td>C45</td><td>Blank size</td><td>$\phi72\times25$</td><td>Blank quality</td><td>0.8kg</td><td>Quantity</td><td colspan="2">1</td></tr>
<tr><td colspan="6">Working procedure</td><td rowspan="2">Work section</td><td rowspan="2">Technical equipment</td><td colspan="2">Man-hours/min</td></tr>
<tr><td>No.</td><td>Name</td><td colspan="4">Content</td><td>Preparation & conclusion</td><td>Single piece</td></tr>
<tr><td>5</td><td>Preparation</td><td colspan="4">Prepare the material according to the size of $\phi72\times25$</td><td>Outsourcing</td><td>Sawing machine</td><td></td><td></td></tr>
<tr><td>10</td><td>Milling</td><td colspan="4">Use three-jaw self-centering chuck to clamp the $\phi72$ cylinder, with an effective extension length greater than 17mm. Milling $\phi15$ top surface, $\phi70$ outer circle and side, spoke side by using the forming milling cutter. Ensure the $R2$ size is processed at the same time</td><td>Milling</td><td>Machining center, Vernier caliper</td><td>60</td><td>45</td></tr>
</table>

Continued

Working procedure			Work section	Technical equipment	Man-hours/min	
No.	Name	Content			Preparation & conclusion	Single piece
15	Milling	Clamp the $\phi 70$ outer circle by using three-jaw self-centering chuck with soft three-jaw. Mill the top surface of the part to ensure the size of 23.5mm. Mill the spoke to ensure the size of 3mm and 7mm. Machine the spoke hollow using $\phi 4$ end milling cutter, Machine the 90° groove by using the form milling cutter. Finally drill and ream the $\phi 5$ hole	Milling	Machining center, Vernier caliper	90	60
20	Benching	Drill M3 bottom hole to $\phi 2.5$. Tap M3 fastening threaded hole	Benching	Machining center, Vernier caliper	15	10
25	Cleaning	Clean the workpiece, debur sharp corner	Benching		15	5
30	Inspection	Check the workpiece dimensions	Examination		15	5

The training task is to carry out process design for the 10th milling process and develop the working procedure card for 10th process, as shown in Table 4-5.

Table 4-5 Working procedure card for the 10th process

Machining working procedure card	Product model	STL-01	Part number	FL-01	Page 1
	Product name	Stirling engine	Part name	Flywheel	Total 1 page
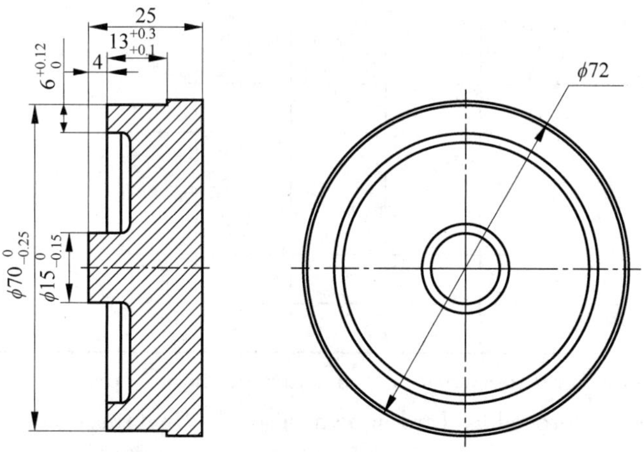				Procedure No.	10
				Procedure name	Milling
				Material	C45
				Equipment	Machining center
				Equipment model	VAL6150e
				Fixture	Three-jaw self-centering chuck
				Measuring tool	Vernier caliper
				Preparation & Conclusion time	60min
				Single-piece time	45min

Continued

Work steps	Content	Cutters	S/ (r/min)	F/ (mm/r)	a_p/ mm	Step hours/min	
						Mechanical	Auxiliary
1	Install the workpiece. Ensure that the effective extension length is greater than 17.3mm						5
2	Milling the $\phi 72$ upper surface, ensure the flatness of the $\phi 15$ cylinder end face	$\phi 12$ end milling cutter	1500	500	0.5	6	5
3	Rough milling $\phi 70$ end face, depth of 4mm, 0.1mm reserved for the bottom and 0.1mm reserved for the side	$\phi 12$ end milling cutter	1500	500	2	6	10
4	Finish milling $\phi 70$ end face, depth of 4mm, 0mm reserved for the bottom and 0.1mm reserved for the side	$\phi 12$ end milling cutter	2500	500	0.1	6	5
5	Rough milling the left side of the spoke, 0.1mm reserved for the bottom and 0.2mm reserved for the side	$\phi 10$ $R2$ end milling cutter	1500	500	0.5	5	10
6	Finish milling the left side of the spoke, ensure the size of 6, $\phi 15$	$\phi 10$ $R2$ end milling cutter					
7	Round off the acute angle						2
8	Disassemble and clean the workpiece						3
9							
10							

Referring to the 10th process, design the working procedure of the 15th process in this training task, and fill in the working procedure card shown in Table 4-6 (the header part has been completed).

Tip: Spoke hollows (three places) are programmed by CAXA CAM manufacturing engineers software.

Table 4-6 Working procedure card for the 15th process

Machining working procedure card	Product model	STL-01	Part number	FL-01	Page 1
	Product name	Stirling engine	Part name	Flywheel	Total 1 page

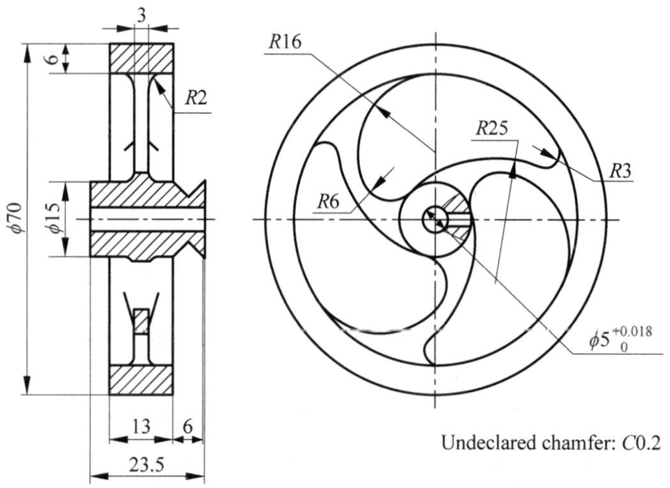

Procedure No.	15
Procedure name	Milling
Material	C45
Equipment	Machining center
Equipment model	VAL6150e
Fixture	three-jaw self-centering chuck
Measuring tool	Vernier caliper
Preparation & Conclusion time	90min
Single-piece time	60min

Undeclared chamfer: $C0.2$

Work steps	Content	Cutters	S/ (r/min)	F/ (mm/r)	a_p/ mm	a_e/ mm	Step hours/min	
							Mechanical	Auxiliary
1								
2								
3								
4								
5								
6								
7								
8								
9								
10								

4.6 Programming for CNC machining

According to working procedure card of the 10th process, the corresponding CNC program for the 10th process milling is prepared as shown in Table 4-7.

Table 4-7 CNC program for the 10th process

No.	Program statement	Annotation
	O0001;	
	T1 M6;	
	S1500 M3;	

Continued

No.	Program statement	Annotation
	G0 G90 G54 X45 Y0;	
	G43 Z10 H1;	
N1	G1 Z0.1 F500;	Feeding for roughing
	G1 X36;	The first cutting
	G2 I-36 J0;	The first cutting for milling the plane, the $\phi72$ circle
	G1 X30;	The second cutting
	G2 I-30 J0;	The second cutting for milling the plane, the $\phi60$ circle
	G1 X24;	The third cutting
	G2 I-24 J0;	The third cutting for milling the plane, the $\phi48$ circle
	G1 X18;	The 4th cutting
	G2 I-18 J0;	The 4th cutting for milling the plane, the $\phi36$ circle
	G1 X12;	The 5th cutting
	G2 I-12 J0;	The 5th cutting for milling the plane, the $\phi24$ circle
	G1 X6;	The 6th cutting
	G2 I-6 J0;	The 6th cutting for milling the plane, the $\phi12$ circle
	G1 X0;	The machining center hole
	G1 X45;	Return to the start point of the milling plane
	G1 Z0 F500;	Feeding for finishing
	G1 X36;	The first cutting
	G2 I-36 J0;	The first cutting for milling the plane, the $\phi72$ circle
	G1 X30;	The second cutting
	G2 I-30 J0;	The second cutting for milling the plane, the $\phi60$ circle
	G1 X24;	The third cutting
	G2 I-24 J0;	The third cutting for milling the plane, the $\phi48$ circle
	G1 X18;	The 4th cutting
	G2 I-18 J0;	The 4th cutting for milling the plane, the $\phi36$ circle
	G1 X12;	The 5th cutting
	G2 I-12 J0;	The 5th cutting for milling the plane, the $\phi24$ circle
	G1 X6;	The 6th cutting
	G2 I-6 J0;	The 6th cutting for milling the plane, the $\phi12$ circle
	G1 X0;	The machining center hole
	G1 Z10;	Finishing is complete, lift the cutter to the $Z10$ position
N2	G0 X45 Y0;	Machining the side 13
	G1 Z0.1;	
	M98 P0080010;	Call the sub-program “0010” four times, roughing the side 13
	G1 Z-3.5;	
	M98 P0010010;	Call the sub-program “0010” four times, finishing the side 13
N3	G1 Z-17;	Processing the $\phi70$ outer circle
	G1 G41 X45.1 Y10 D1;	In the rough machining stage, the tool radius compensation instruction is called
	G3 X35.1 Y0 R10;	Arc feed
	G2 I-35.1 J0;	Rough cutting of the outer circle

Continued

No.	Program statement	Annotation
	G3 X45. 1 Y-10 R10;	Arc retract
	G1 G40 X45 Y0;	Cancel the tool radius compensation
	G1 G41 X44. 95 Y10 D1;	In the finishing stage, the tool radius compensation instruction is called
	G3 X34. 95 Y0 R10;	Arc feed
	G2 I-34. 95 J0;	Finish cutting of the outer circle
	G3 X44. 95 Y-10 R10;	Arc retract
	G1 G40 X45 Y0;	Cancel the tool radius compensation instruction
	Z10;	Lift the cutter
	G0 Z100;	
	M5;	
	T2 M6;	Call the $\phi 10R2$ arc end milling cutter
	S1500 M3;	
	G0 G90 G54 X18 Y0;	Calibrate to the starting point of processing
	G43 Z10 H2;	
N4		Rough and finish machining the left side of the spoke plate
	G1 Z-4 F500;	Move to the starting point of the cutting
	M98 P0100011;	Call the cutting subprogram "0011" ten times for roughing and finishing
	G1 Z10;	
	G0 Z100;	
	M5;	
	M30;	
	%	
	%	
	O0010;	The sub-program for machining the left side of 13
	G1 G91 Z-0. 5 F500;	
	G1 X36;	The first feeding
	G2 I-36 J0;	
	G1 X30;	The 2nd feeding
	G2 I-30 J0;	
	G1 X24;	The 3rd feeding
	G2 I-24 J0;	
	G1 X18;	The 4th feeding
	G2 I-18 J0;	
	G1 X13. 6;	Machine the last circle, leaving a processing allowance of 0. 1mm on one side
	G2 I-13. 6 J0;	
	G1 X45 Y0;	Return to the starting point
	M99;	Sub-program returns
	%	
	O0011;	Machine the left side of the spoke plate

Continued

No.	Program statement	Annotation
	G3 G91 Z-0.5 I-18 J0;	Spiral feeding, 0.5mm per layer
	G3 G90 I-18 J0;	Milling spiral feeding layer
	G1 G41 X18.94 Y-10 D2;	Call the tool radius compensation instruction
	G3 X28.94 Y0 R10;	Arc feeding
	I-28.94 J0;	Processing the outer edge
	X18.94 Y10 R10;	Arc retract
	G1 G40 X18 Y0;	Cancel the tool radius compensation instruction
	G1 G41 X17.5375 Y10;	Call the tool radius compensation instruction
	G3 X7.5375 Y0 R10;	Arc feeding
	G2 I-7.5375 J0;	Cutting ϕ15
	G3 X7.5375 Y-10 R10;	Arc retract
	G1 G40 X18 Y0;	Cancel the tool radius compensation instruction
	M99;	Sub-program returns
	%	

According to working procedure card of the 15th process, the corresponding CNC program for the 15th process milling is prepared as shown in Table 4-8. This program is generated by combining manual programming with automatic programming. It can be changed as needed in actual production.

Table 4-8 CNC program for the 15th process

No.	Program statement	Annotation
	O0002;	
	T1 M6;	It is recommended to set Z0 from the ϕ70 side (locating surface) up to 19mm during tool setting
	S1500 M3;	
	G0 G90 G54 X45 Y0;	
	G43 Z10 H1;	
N1	G1 Z0.1 F500;	Feeding for roughing
	G1 X36;	The 1st cutting
	G2 I-36 J0;	The 1st cutting for milling the plane, the ϕ72 circle
	G1 X30;	The 2nd cutting
	G2 I-30 J0;	The 2nd cutting for milling the plane, the ϕ60 circle
	G1 X24;	The 3rd cutting
	G2 I-24 J0;	The 3rd cutting for milling the plane, the ϕ48 circle
	G1 X18;	The 4th cutting
	G2 I-18 J0;	The 4th cutting for milling the plane, the ϕ36 circle
	G1 X12;	The 5th cutting
	G2 I-12 J0;	The 5th cutting for milling the plane, the ϕ24 circle
	G1 X6;	The 6th cutting
	G2 I-6 J0;	The 6th cutting for milling the plane, the ϕ12 circle

Continued

No.	Program statement	Annotation
	G1 X0;	The machining center
	G1 X45;	Return to the start point of the milling plane
	G1 Z0 F500;	Feeding for finishing
	G1 X36;	The 1st cutting
	G2 I-36 J0;	The 1st cutting for milling the plane, the ϕ72 circle
	G1 X30;	The 2nd cutting
	G2 I-30 J0;	The 2nd cutting for milling the plane, the ϕ60 circle
	G1 X24;	The 3rd cutting
	G2 I-24 J0;	The 3rd cutting for milling the plane, the ϕ48 circle
	G1 X18;	The 4th cutting
	G2 I-18 J0;	The 4th cutting for milling the plane, the ϕ36 circle
	G1 X12;	The 5th cutting
	G2 I-12 J0;	The 5th cutting for milling the plane, the ϕ24 circle
	G1 X6;	The 6th cutting
	G2 I-6 J0;	The 6th cutting for milling the plane, the ϕ12 circle
	G1 X0;	The machining center hole
	G1 Z10;	Finishing is complete, lift the cutter to the Z10 position
N2	G0 X45 Y0;	Machining the side 13
	G1 Z0.1;	
	M98 P0130010;	Call the sub-program "0010" four times, roughing the right side 13
	G1 Z-6;	
	M98 P0010010;	Call the sub-program "0010" four times, finishing the right side 13
	T2 M6;	Call the ϕ10 R2 arc end milling cutter
	S1500 M3;	
	G0 G90 G54 X18 Y0;	Calibrate to the starting point of processing
	G43 Z10 H2;	
N3		Rough and finish machining the left side of the spoke plate
	G1 Z-6 F500;	Move to the starting point of the cutting
	M98 P0100011;	Call the cutting sub-program "0011" ten times for roughing and finishing
	G1 Z10;	
	G0 Z100;	
	M5;	
	T3 M6;	Call the ϕ20 positive 90° milling cutter to machine the sharp groove
	S800 M3;	
	G0 G90 G54 X20 Y0;	
	G43 Z10 H3;	
N4	G1 Z-3.5 F200;	Feed and align the 90° tool tip with the sharp groove
	G1 G41 X17.5 Y0 D3;	Lateral feed
	G2 I-17.5 J0;	Machining the outer circumference
	G1 X17;	The 2nd feeding

Continued

No.	Program statement	Annotation
	G2 I-17 J0；	
	G1 X16.5；	The 3rd feeding
	G2 I-16.5 J0；	
	G1 X16；	The 4th feeding
	G2 I-16 J0；	
	G1 X15.5；	The 5th feeding
	G2 I-15.5 J0；	
	G1 X15；	The 6th feeding
	G2 I-15 J0；	
	G1 X14.5；	The 7th feeding
	G2 I-14.5 J0；	
	G2 I-14.5 J0；	Refine the circumference
	G3 X24.5 Y-10 R10；	Arc cut out
	G1 G40 Y0；	Cancel the tool radius compensation instruction
	Z10；	
	G0 Z100；	
	M5；	
	T4 M6；	Call the ϕ4 end milling cutter to machine the spoke hollow
	S5000 M3；	
	G0 G90 G54 X0 Y0；	
	G43 Z10 H4；	
N5	P0010012；	Call the sub-program generated by CAXA to process the first hollow
	G68 X0 Y0 R120；	The coordinate system is rotated by 120°
	P0010012；	Call the sub-program generated by CAXA to process the second hollow
	G69；	Cancel the coordinate system rotation
	G68 X0 Y0 R240；	The coordinate system is rotated by 120°
	P0010012；	Call the sub-program generated by CAXA to process the third hollow
	G69；	Cancel the coordinate system rotation instruction
	G0 Z100；	
	M5；	
	T5 M6；	Call the 90° center drill
	S2000 M3；	
	G0 G90 G54 X0 Y0；	
	G43 Z10 H5；	
N6	G83 Z-1 R0.5 Q0.5 F50；	Machining the ϕ5 center hole
	G0 G80 Z100；	
	M5；	
	T6 M6；	Call the ϕ4.8 drill
	S2000 M3；	
	G0 G90 G54 X0 Y0；	

Continued

No.	Program statement	Annotation
	G43 Z10 H6;	
N7	G83 Z-27 R0.5 Q1.5 F100;	Process the ϕ5 bottom hole to ϕ4.8
	G0 G80 Z100;	
	M5;	
	T7 M6;	Call ϕ5 reamer
	S800 M3;	
	G0 G90 G54 X0 Y0;	
	G43 Z10 H7;	
N8	G85 Z-27 R0.5 F50;	Machining the ϕ5 hole
	G0 G80 Z100;	
	M5;	
	%	
	O0010;	The sub-program for machining the left side 13
	G1 G91 Z-0.5 F500;	
	G1 X36;	The 1st feeding
	G2 I-36 J0;	
	G1 X30;	The 2nd feeding
	G2 I-30 J0;	
	G1 X24;	The 3rd feeding
	G2 I-24 J0;	
	G1 X18;	The 4th feeding
	G2 I-18 J0;	
	G1 X13.6;	Machine the last circle, leaving a processing allowance of 0.1mm on one side
	G2 I-13.6 J0;	
	G1 X45 Y0;	Return to the starting point
	M99;	Sub-program returns
	%	
	O0011;	Machine the left side of the spoke plate
	G3 G91 Z-0.5 I-18 J0;	Spiral feeding, 0.5mm per layer
	G3 G90 I-18 J0;	Milling spiral feeding layer
	G1 G41 X18.94 Y-10 D2;	Call the tool radius compensation instruction
	G3 X28.94 Y0 R10;	Arc feeding
	I-28.94 J0;	Processing the outer edge
	X18.94 Y10 R10;	Arc retract
	G1 G40 X18 Y0;	Cancel the tool radius compensation instruction
	G1 G41 X17.5375 Y10;	Call the tool radius compensation instruction
	G3 X7.5375 Y0 R10;	Arc feeding
	G2 I-7.5375 J0;	Cutting ϕ15
	G3 X7.5375 Y-10 R10;	Arc retract
	G1 G40 X18 Y0;	Cancel the tool radius compensation instruction
	M99;	Sub-program returns

Continued

No.	Program statement	Annotation
	%	
	O0012;	The sub-program for processing the spoke hollow. The program is automatically generated and edited by CAXA CAM manufacturing engineer software. The program header is positioned to the starting point of the process, and the "M99" instruction is added after the sub-program is processed

Task 3 Hands-on Training

4.7 Equipment and appliances

Equipment: AVL650e vertical machining center.

Cutters: ϕ12 end milling cutter, $\phi10R2$ fillet milling cutter, ϕ20 positive 90° piece milling cutter, 90° center drill, ϕ4.8 twist drill, ϕ5H8 reamer.

Fixture: K11-320 self-centering three-jaw chuck (with soft claws).

Tools: key file.

Measuring tools: 0-150mm vernier caliper, 0.02mm lever arm test indicator (with mounting rod).

Blank: $\phi72\times25$.

Auxiliary appliances: hook wrench, rubber hammer, brush, etc.

4.8 Check before powering on

Check whether there is any abnormality in various parts of the appearance of the machine tool (such as the scatter shield, the footplate, etc.). Check whether the lubricating oil and coolant of the machine tool are sufficient. Check whether there are foreign objects on the tool holder, the fixture, and the baffle plate of the lead rail. Check the state of each knob on the machine tool panel is normal. Check whether there is an alarm after power on the machine tool. Refer to Table 4-9 to check the machine state.

Table 4-9 Preparing card for machine start-up

Check item		Test result	Abnormal description
Mechanical part	Spindle		
	Feed part		
	Tool holder		
	Three-jaws self-centering chuck		
Electrical part	Main power supply		
	Cooling fan		

Continued

Check item		Test result	Abnormal description
CNC system	Electrical components		
	Controlling part		
	Driving section		
Auxiliary part	Cooling system		
	Compressed air system		
	Lubricating system		

4.9 Preparation before machining

(1) After the machine tool is started, each coordinate axis returns to the original point of the machine tool according to the operation instructions.

(2) Install the workpiece. The effective length of the workpiece protruding from the three-jaw self-centering chuck should be greater than 17mm.

(3) Align the center of the workpiece with the centering method and set it as $X0$ and Y0 in the G54 coordinate system.

(4) Install the tool.

(5) Tool setting. The top surface of the fixed jaw of the precision flat-nose pliers can be used as the benchmark for tool setting.

(6) Program editing. Because the program is relatively long, it is recommended to edit the program on the personal computer, and then use the memory card to import the program into the CNC machine tool after checking that there is no problem.

4.10 Part machining

After the graphic verification process has verified that there is no problem, the parts machining can be carried out. Before the parts are processed, you should understand the safety operation requirements of the machine tool in detail, and wear labor protection clothing and utensils. When processing parts, you should be familiar with the functions and positions of the operation buttons of the CNC lathe, and understand the methods of dealing with emergency situations.

Note: if the graphic verification operation is performed, the machine tool must be returned to the mechanical zero point after the operation is completed, and then other related operations can be performed. If the relevant operations are performed without returning to zero, it will cause coordinate offsets, and even abnormal phenomena such as collisions and program execution confusion, resulting in danger.

Select MEM mode on the operation panel, and load the CNC machining program that needs to be processed, and press the cycle start button to perform automatic processing.

The first automatic operation of the CNC machining program should be carried out in the debugging state. First turn down the machine feed rate override knob to 0%, and press the button 单段执行 on the operation panel, and then switch to the check window of running status by left-clicking the Program Check button at the bottom of the display, as shown in Figure 4-27.

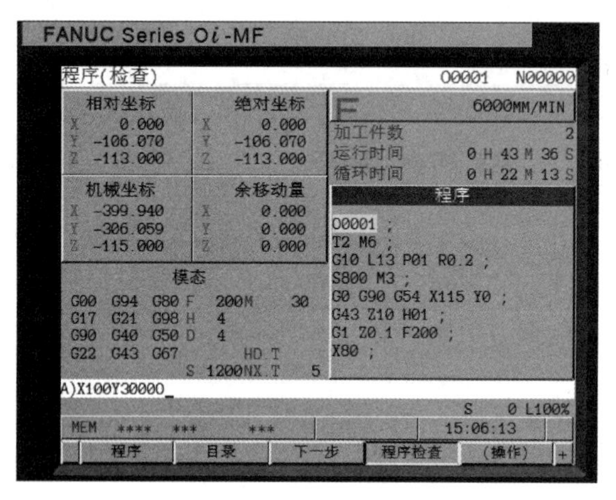

Figure 4-27 Automatic running status inspection of the program

In this state, the program will only automatically execute the line where the cursor is located each time. Before pressing the cycle start button, observe whether the distance between the cutter and the workpiece is safe. After pressing the cycle start button, control the movement speed of the machine tool through the feed override knob, and observe the actual distance between the cutter and the workpiece, at the same time Compare with the remaining movement amount displayed in the "Remaining Movement Amount" column of the display screen. If the difference between the actual distance and the remaining movement amount is too large, the operation should be stopped and checked to avoid a collision accident. In the process of program debugging, you should also pay close attention to the "modal" status on the display screen to ensure that there is no abnormality in the spindle revolution speed, feed speed, workpiece coordinate system number, compensation status and compensation number.

4.11 Part inspection

After the parts are processed, the workpiece should be carefully cleaned, and in accordance with the relevant requirements of quality management, the processed parts should be subject to relevant inspections to ensure the production quality. The "three-level" inspection cards for machined parts is shown in Table 4-10.

Table 4-10 "Three-level" inspection cards for machined parts

<table>
<tr><th>Part drawing number</th><th colspan="2"></th><th>Part name</th><th></th><th>Working step number</th><th></th></tr>
<tr><td>Material</td><td colspan="2"></td><td>Inspection date</td><td></td><td>Working step name</td><td></td></tr>
<tr><td>Inspection items</td><td colspan="2">Self-inspection result</td><td colspan="2">Mutual inspection result</td><td>Professional inspection</td><td>Remark</td></tr>
<tr><td></td><td colspan="2"></td><td colspan="2"></td><td></td><td></td></tr>
<tr><td></td><td colspan="2"></td><td colspan="2"></td><td></td><td></td></tr>
<tr><td></td><td colspan="2"></td><td colspan="2"></td><td></td><td></td></tr>
<tr><td></td><td colspan="2"></td><td colspan="2"></td><td></td><td></td></tr>
<tr><td></td><td colspan="2"></td><td colspan="2"></td><td></td><td></td></tr>
<tr><td></td><td colspan="2"></td><td colspan="2"></td><td></td><td></td></tr>
<tr><td></td><td colspan="2"></td><td colspan="2"></td><td></td><td></td></tr>
<tr><td>Conclusion</td><td colspan="6">□ Qualified □ Unqualified □ Repair □ Concession to receive
Inspection signature:
Date:</td></tr>
<tr><td>Non-conforming item description</td><td colspan="6"></td></tr>
</table>

Project Summary

Through the CNC milling of the flywheel part, master the basic format of CNC milling program and the use of basic cutting instructions. Be able to perform part cutting using the basic cutting instructions, the tool radius compensation instruction, the subprogram and coordinate transformation instruction, etc. Master the application method of G8 * series instructions in high-precision hole processing. Master the use of CNC programming CAXA CAM manufacturing engineer software.

Master the basic operation methods of vertical machining centers, including powering on and off, tool installation, workpiece alignment, cutter setting, program editing, graphic verification, CNC machining program debugging and automatic operation, etc.

Through task training, good professional quality, correct safety operation standards for machining centers, and basic machining quality awareness are developed and cultivated.

Exercises After Class

1. Choice questions

(1) The instruction M98 indicates (   ).

A. returning to the sub-program B. calling the sub-program

C. coolant ON D. coolant OFF

(2) Using general calculation tools and various mathematical methods to manually calculate tool path and program is called (   ).

A. mechanical programming　　B. manual programming

C. CAD programming　　D. CAM programming

(3) The instruction of the subprogram returning to the main program is (　　) instruction.

A. P98　　B. M99　　C. M08　　D. M09

(4) The V-shaped frame is used for positioning the workpiece's outer circle, and the short V-shaped frame limits (　　) degrees of freedom.

A. 10　　B. 12　　C. 2　　D. 9

(5) Machining centers can be classified into composite machining, (　　) centers and drilling machining centers according to their functional characteristics.

A. tool magazine + spindle tool change　　B. horizontal

C. boring and milling　　D. three-axis

(6) The three-jaw self-centering chuck on the lathe and the flat-nose pliers on the milling machine belong to (　　).

A. general fixture　　B. special fixture

C. combined fixture　　D. accompanying fixture

(7) The main function of V-shaped block is to (　　) shaft parts.

A. clamp　　B. measure　　C. position　　D. guide

2. True or false

(1) In the FANUC system, the number of subprogram calls by instructions "M98 P0050012" and "M98 P50012" are the same. (　　)

(2) Generally, the macro programming of the machining center adopts Class A macro instructions, and the macro programming of the CNC milling machine adopts Class B macro instructions. (　　)

(3) K instruction in the fixed cycle function refers to the number of repeated processing, which is generally used in incremental mode. (　　)

(4) The universal three-jaw self-centering chuck is used to clamp the small disc parts. The four-jaw chuck or flower disc is suitable for clamping medium and large disc parts. (　　)

3. Fill in the blanks

(1) If the coordinate values of all coordinate points are measured from a fixed coordinate origin, it is called ________ coordinate programming. If the end point coordinate of the motion track is measured relative to the starting point of the line segment, it is called ________ coordinate programming.

(2) G code can be divided into ________ code and ________ code according to its continuity in the program segment.

(3) The disc parts are mainly composed of the end face, the outer circle, the inner hole, the step surface, the slot four-axis arrangement hole, etc., which belong to the coaxial rotating body, and its main feature is ________.

(4) The methods of modeling parts with CAXA CNC lathe can be divided into three categories: ________, ________ and ________.

(5) The common processing forms of milling are ________, ________, ________, ________, ________, ________, ________ and ________.

4. Short answer questions

(1) Briefly describe the use of CNC tool radius compensation instructions.

(2) What are the clamping methods for circular workpiece on CNC machining center? What are their characteristics?

(3) Briefly describe the difference between the reaming instruction G85 and the drilling instruction G83.

(4) What are the principles for the arrangement of parts processing procedures?

Self-learning test score sheet is shown in Table 4-11.

Table 4-11 Self-learning test score sheet

Task	Task requirements	Score	Scoring rules	Score	Remark
Learn key knowledge points	(1) Master the clamping method of round workpiece (2) Master the use of sub-program instructions (3) Master the use of coordinate system rotation instruction (4) Be able to program by combining coordinate transformation with sub-program (5) Master the use of reaming instructions (6) Understand CAM programming software and be familiar with the basic flow of CAM programming	20	Understand and master		
Technological preparation	(1) Be able to correctly read the shaft part drawings (2) Can be analyzed according to the part drawing, determine the process (3) Be able to write the correct processing program according to the processing process	30	Understand and master		
Hands-on training	(1) The corresponding equipment and utensils will be selected correctly (2) Can correctly operate the CNC lathe and adjust the processing parameters according to the machining situation	50	(1) Understand and master (2) Operation process		

Ideological and Political Classroom

Project 5　Programming and Milling for Cylinder Block Mounting Base

➢ Mind map

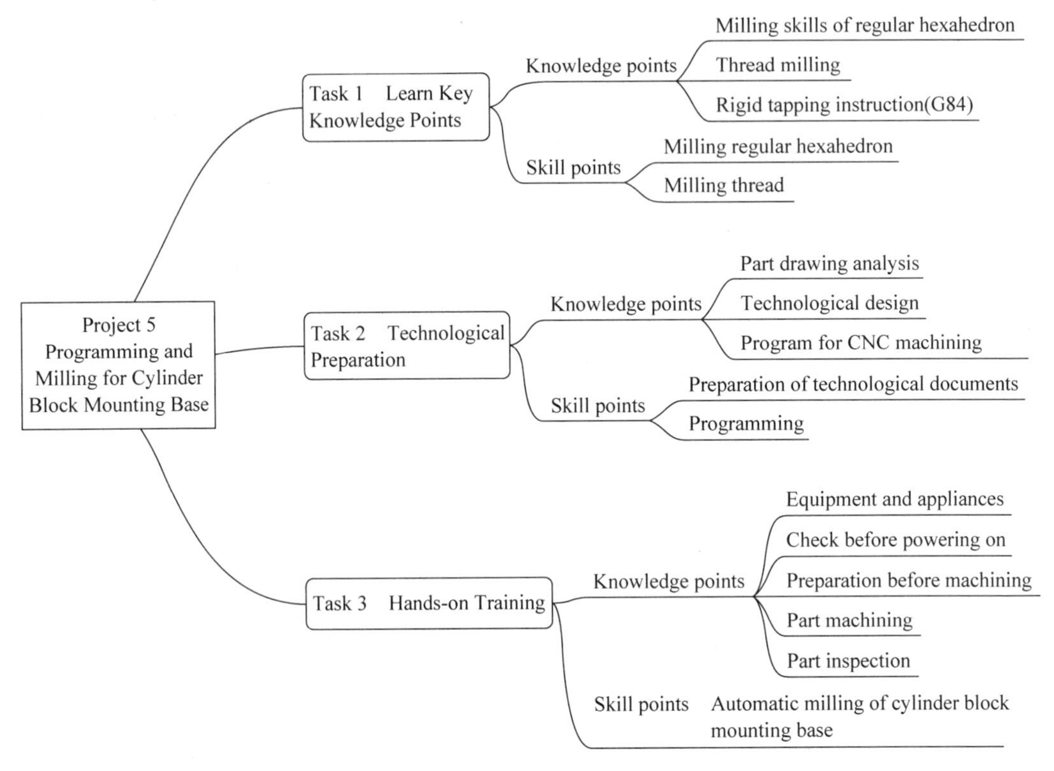

➢ Learning objectives

Knowledge objectives

(1) Understand the working principle of the cylinder body mounting base.

(2) Understand the use and key requirements of the cylinder body mounting base in this mechanism.

Ability objectives

(1) Master the processing skills of the flat plate shaped parts.

(2) Master the selection and use of thread milling tools.

(3) Master the characteristics and programming methods of thread milling.

(4) Be able to independently determine the process routine and fill in the technological documents correctly.

(5) Be able to operate the CNC lathe correctly and adjust the machining parameters according to the machining conditions.

(6) Be able to select measuring tools reasonably according to the accuracy and the structural characteristics of parts, and be able to measure the relevant dimensions correctly and normatively.

Literacy goals

(1) Develop students' scientific spirit and attitude.

(2) Cultivate students' engineering awareness.

(3) Develop students' teamwork skills.

Task 1 Learn Key Knowledge Points

5.1 Milling skills of regular hexahedron

The shape of regular hexahedron looks simple, but it often has high requirement of dimensional accuracy and relative position accuracy. Improper processing methods will cause distortion after assembly or inability to assemble the product, so it is very important to master its processing skills.

5.1.1 Clamp selection

The clamp used for milling hexahedron is usually a flat-nose pliers, as shown in Figure 5-1. A precision flat-nose pliers mainly consists of clamp body, fixed jaw, movable jaw, clamping screw and other main parts. The flat pliers are simple in composition and compact in structure. Turn the screw with a wrench, and drive the movable clamp body to move through the screw nut, thus tightening and loosening the workpiece.

During processing, especially when processing the surfaces connected with each other (such as the cylinder block mounting base of this project), the longitudinal, transverse and horizontal positions of the flat-nose pliers on the workbench should be carefully calibrated before processing. There are three steps to complete the calibration.

1. Correction of longitudinal position

Clamp a parallel sizing block in the flat-nose pliers, install a dial gauge on the tool holder, make the contact of the dial gauge contact with the side of the parallel sizing block, and control the compression of the gauge at about 0.2mm. Then move the ram to see if the pointer of the dial indicator swings. If the pointer does not move, it means that the longitudinal position of the flat pliers is correct.

Figure 5-1 The flat-nose pliers

If the watch needle swings, loosen the nut on the base of the flat-nose pliers, and then turn the flat-nose pliers to adjust until the watch needle does not move.

2. Correction of transverse position

Turn the angle of the flat-nose pliers over 90° so that the dial indicator still contacts the side of the parallel sizing block, and then move the workbench to adjust according to the swing of the needle.

3. Correction of horizontal position

The horizontal position of the flat-nose pliers should also be adjusted in both horizontal and vertical directions. When correcting the longitudinal level, clamp a square in the flat-nose pliers to make the contact of the dial indicator contact with the upper edge of the square, and then move the ram for adjustment. When correcting the horizontal level, place a parallel sizing block on the sliding surface of the caliper body, then contact the dial indicator contact with the upper plane of the parallel sizing block, move the workbench, and adjust it according to the swing of the gauge needle.

In actual production, it is generally necessary to carry out horizontal correction first, and then select one of the *X* axis and *Y* axis directions to carry out longitudinal or transverse correction before processing. Pay attention to the following problems when using the flat-nose pliers.

(1) The machined surface of the workpiece must be higher than the jaw, otherwise the workpiece will be raised with parallel sizing blocks.

(2) In order to be firmly clamped and prevent the workpiece from loosening during processing, the relatively dry plane must be attached to the sizing block and jaw. To make the workpiece close to the sizing block, clamp it while tapping the sub-surface of the workpiece with a hammer. The smooth surface should be struck with a copper bar to prevent damaging the smooth surface.

(3) When using a sizing block to clamp the workpiece, use a wooden hammer or copper hand hammer to lightly hit the upper surface of the workpiece to make the workpiece adhere to the sizing block. After clamping, the sizing block shall be pulled by hand. If it is loose, it means that the workpiece and the sizing block are not well fitted, and the workpiece may move during cutting. The flat-nosed pliers should be loosened and re-clamped.

(4) When clamping a workpiece with poor rigidity, in order to prevent the deformation of the workpiece, the weak part of the workpiece should be supported or solidified first.

(5) If the workpiece is processed by scribing, the workpiece can be corrected with a tosecan or an inside caliper.

The installation of the flat-nose flier in this training program is shown in Figure 5-2. The bottom surface of the flat-nose flier is parallel to the *XOY* plane of the machine tool, and the side of the fixed jaws is parallel to the X axis.

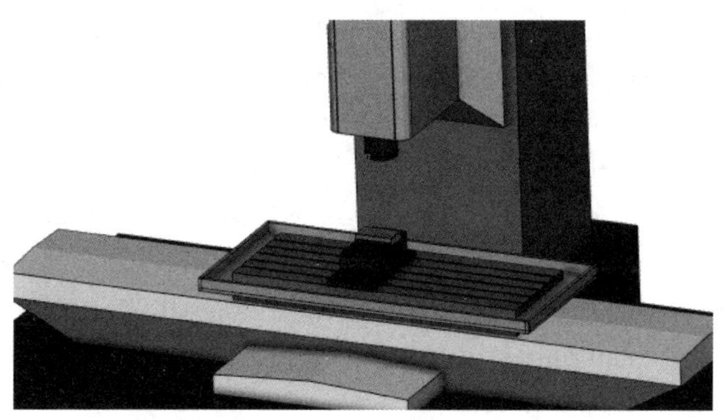

Figure 5-2 Installation of the flat-nose flier

5.1.2 Machining sequence

When processing a hexahedron, generally select a large and flat surface for processing first, such as the top surface① in Figure 5-3.

Then turn the milled top surface① to the fixed jaw side of the flat-nose pliers, and press it tightly to process the front surface②.

After front surface② is processed, it still contacts the fixed jaw side with top surface①, and at the same time, front surface② just processed contacts the bottom surface of the flat-nose pliers, and behind surface③ is processed to ensure the width of the part.

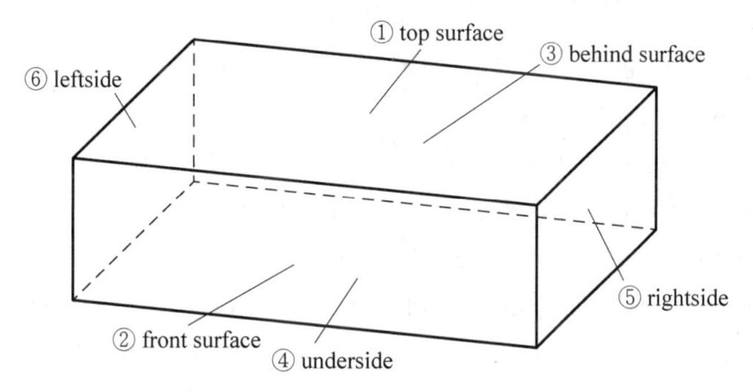

Figure 5-3 Hexahedral milling sequence

Then contact the bottom surface of the flat-nose pliers, the fixed jaw side and the movable jaw side with the top surface①, front surface② and behind surface③ respectively, and process the underside④ to ensure the thickness of the part.

There are two methods to process the leftside⑤. When the workpiece is thick, choose to contact the fixed jaw and movable jaw with the top surface① and underside④ respectively, and use the square or dial indicator to make the top surface① or behind surface③ perpendicular to the bottom of the flat-nose pliers, and mill the rightside⑤. When the workpiece is not too thick, you can choose to make the side of the part extend a short distance from the flat-nose pliers during the clamping process when machining underside④, and use the side edge of the milling cutter to process rightside⑤.

Finally, select top surface① and underside④ to contact the fixed jaw and movable jaw respectively, face rightside⑤ to contact the bottom face of the flat-nose pliers, and cut face leftside⑥ to ensure the length.

Note: When striking the workpiece with a rubber hammer, a wooden hammer or a copper bar (the bottom side of the workpiece is a milled plane), if there is a sizing block at the bottom of the workpiece, it is necessary to tap until the sizing block at the bottom of the vice cannot be moved. If the perpendicularity is measured with a dial indicator, the striking force shall be determined by the movement of the dial indicator to reach the perpendicularity at the fastest speed.

5.2 Thread milling

The research work of thread milling process mainly focuses on the theory and technology. Thread milling generally involves multiple feeds, including rough machining and finish machining. However, for difficult-to-machine materials, the cutting force during CNC milling is larger, which has a great impact on machining accuracy and tool life. Generally, the processing strategy used in practical engineering lacks guidance significance for thread milling of difficult-to-machine materials.

5.2.1 Thread milling tool

The traditional thread processing methods mainly include turning thread with thread turning tool or manual tapping with a taper and a die. With the development of CNC machining technology, especially the emergence of the three-axis linkage CNC machining system, the more advanced thread processing method—CNC milling of thread has been realized. Compared with traditional thread processing, thread milling has great advantages in processing accuracy and efficiency, and is not limited by thread structure and thread direction. For example, a thread milling cutter can process multiple internal and external threads with different directions. The thread is shown in Figure 5-4. For the thread that does not allow transition thread or undercut structure, it is difficult to use traditional turning method or tap and die, but it is very easy to use CNC milling. In addition, the durability of the thread milling cutter is more than ten times or even tens of times that of the tap, and in the process of CNC thread milling, it is very convenient to adjust the thread diameter size, which is difficult to achieve with the tap and die. Due to many advantages of thread milling, the milling process has been widely used in mass thread production in developed countries.

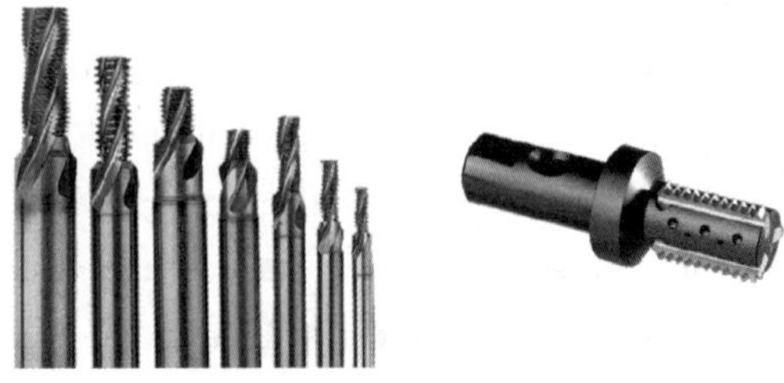

Figure 5-4 Thread milling tools

5.2.2 Thread milling process

The following points should be noted in thread milling process.

(1) Spiral interpolation milling must be used to generate a pitch.

(2) Depending on the thread type and processing method, clockwise or counterclockwise feed direction (right or left), external or internal threads can be used.

(3) It is recommended to use down-milling.

(4) Coolant is recommended, except for thread machining of hardened materials.

(5) The same tool can be used to process internal or external threads, left or right threads.

Milling the internal or external threads is shown in Figure 5-5.

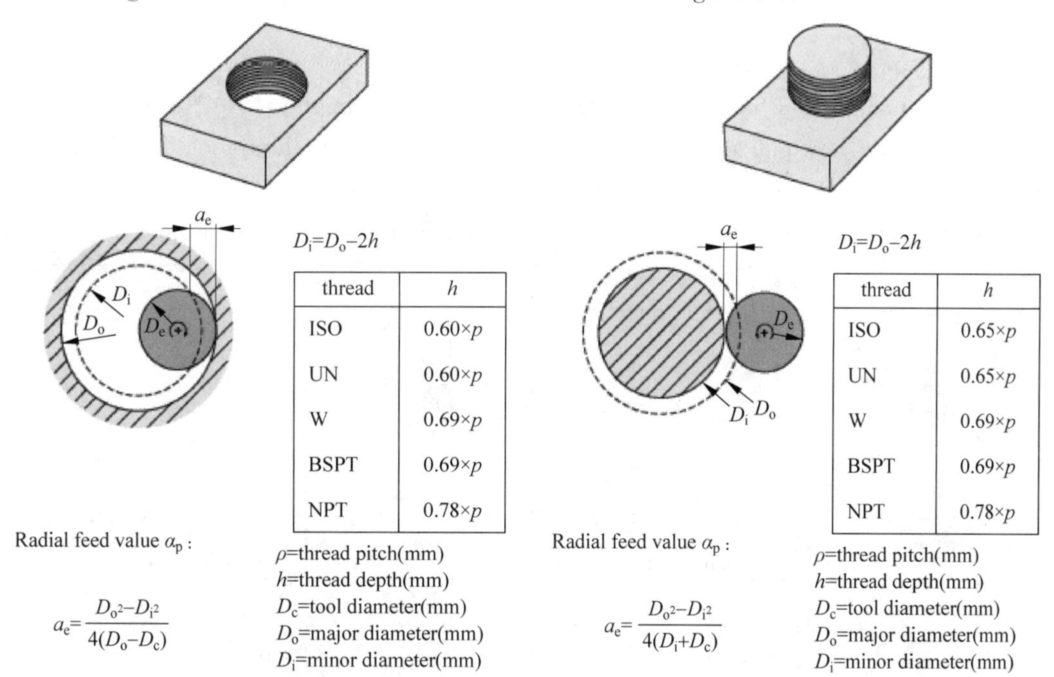

Figure 5-5 Schematic diagram of thread milling process

5.2.3 Programming for thread milling

Spiral interpolation instructions are required for thread milling. The formation of helix is that the tool performs circular interpolation motion and simultaneously performs axial motion. Its command format is:

G2/G3 X_Y_Z_I_J_F_;

The tool path is shown in Figure 5-6.

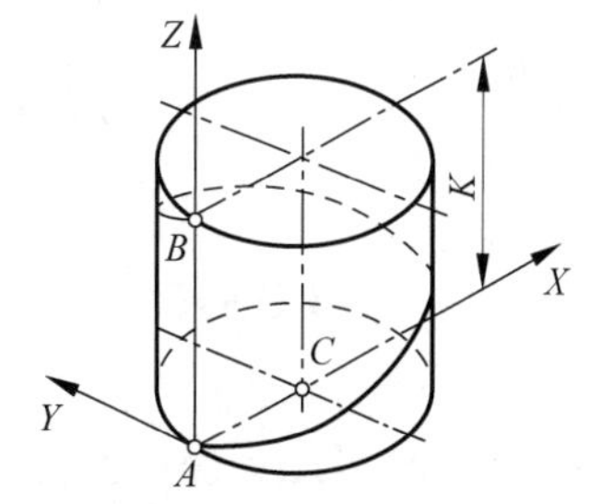

Figure 5-6 Spiral interpolation during thread milling

5.3 Rigid tapping instruction (G84)

Rigid tapping means that the tool handle of tapping is rigid without automatic clearance adjustment, while flexible tapping means that the tool handle of tapping is flexible with clearance adjustment. There are three kinds of instructions for specifying rigid tapping.

(1) The method of instructing "M29 S_" in the same program segment as the tapping instruction. The format is:

G84 X_ Y_ Z_ R_ F_ M29 S_;

(2) The method of instructing "M29 S_" before the tapping cycle. The format is:

M29 S_;

G84 X_ Y_ Z_ R_ F_;

(3) Method without instruction "M29 S_". Set the parameter G84 (No. 5200 # 0) to "1", that is, set the rigid tapping as the default of the main machine tool. The format is:

G84 X_ Y_ Z_ R_ F_;

The movement process of the rigid tapping instruction is shown in Figure 5-7.

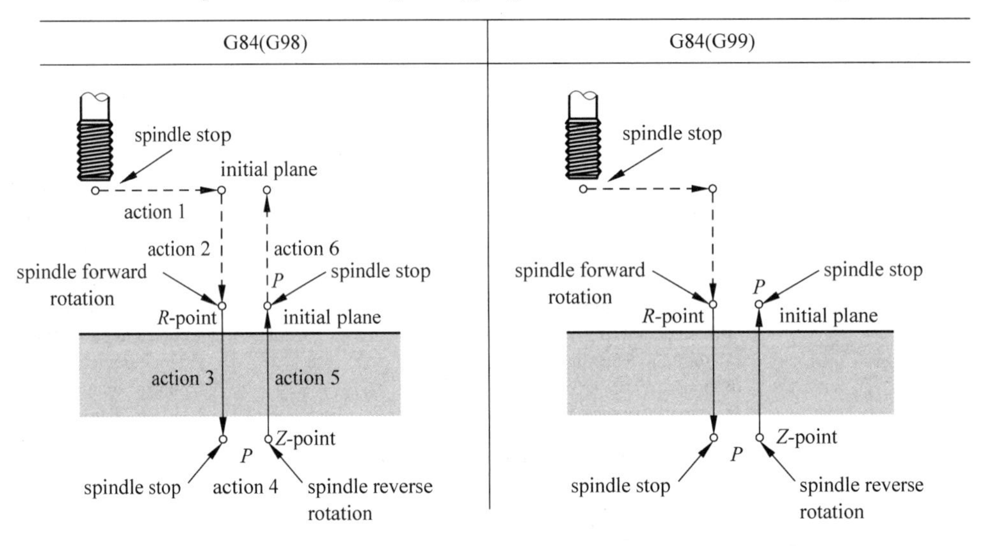

Figure 5-7 Rigid tapping instruction running path

Task 2 Technological Preparation

5.4 Part drawing analysis

Figure 5-8 is the part drawing of the cylinder block mounting base. During the processing, 45 steel is selected as the blank material of cylinder block mounting base part, and the blank size is 75mm×45mm×15mm. The contour is processed first according to the

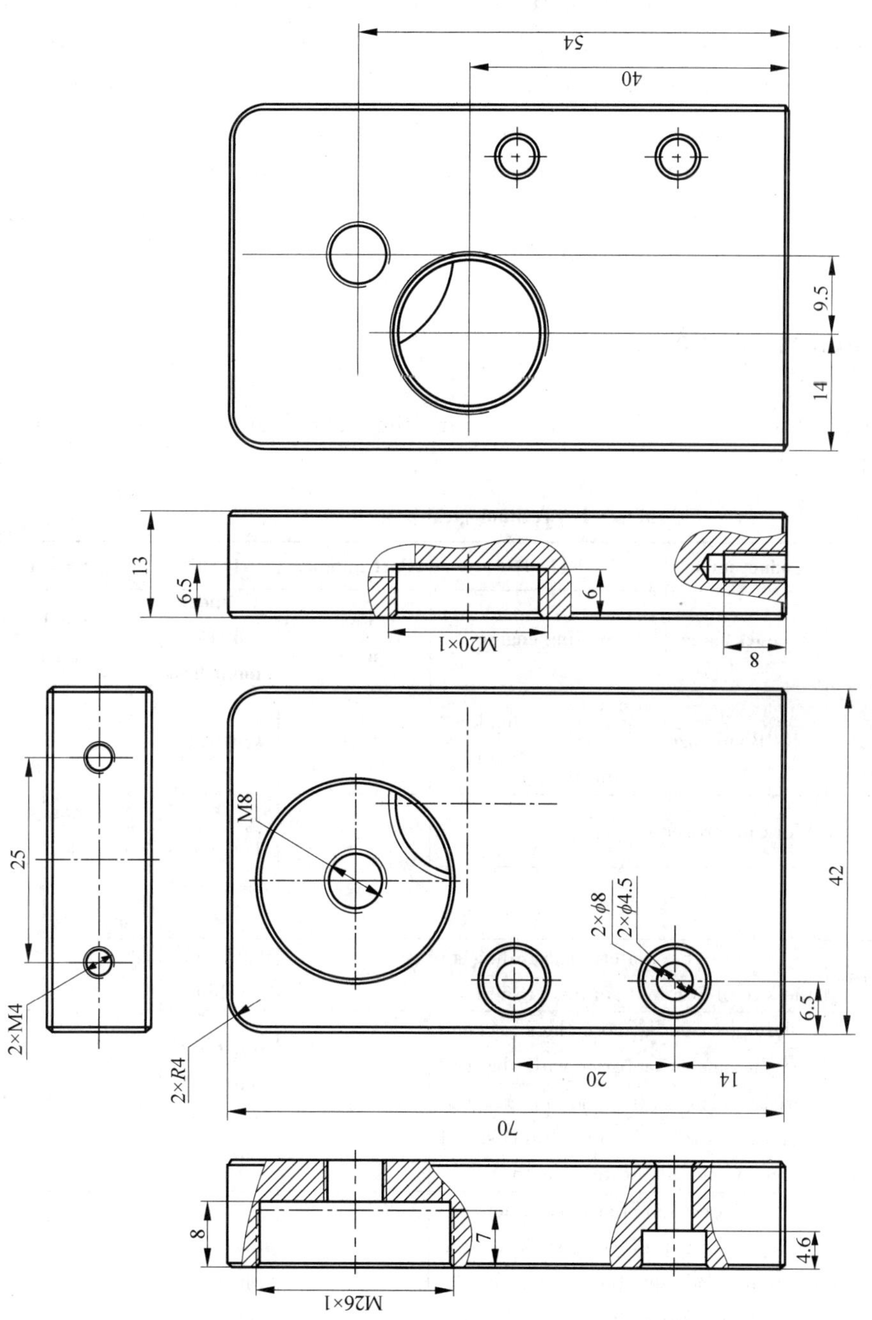

Figure 5-8 Cylinder block mounting base

requirements of hexahedron processing technology to ensure the parallelism and perpendicularity of each surface. When processing one end of the size 70mm, the processing of the $R4$ fillet is completed at the same time. Align the lower left corner of the front, and process two $\phi 4.5$ threaded through-holes, M8 threaded holes, and milling a M26×1 threaded hole. Back processing M20×1 threaded hole. Finally, two M4 screw connection holes shall be machined, and the machining size of the bottom hole is $\phi 3.3$.

Note: When machining $R4$, pay attention to the distance that the part extends out of the flat-nose fliers to prevent the tool from interfering with the jaw. If the chamfer is programmed and machined, pay attention to the position of the milling cutter tip to avoid interference between the bottom and the jaw.

5.5 Technological design

According to the analysis of the part drawing, the technological process is designed as shown in Table 5-1.

Table 5-1 Technological process card

<table>
<tr><td colspan="2" rowspan="2">Machining process card</td><td colspan="2">Product model</td><td colspan="2">STL-01</td><td>Part number</td><td>STL-02</td><td colspan="2">Page 1</td></tr>
<tr><td colspan="2">Product name</td><td colspan="2">Stirling engine</td><td>Part name</td><td>Cylinder block mount base</td><td colspan="2">Total 1 page</td></tr>
<tr><td colspan="2">Material grade</td><td>C45</td><td>Blank size</td><td>75mm×45mm×15mm</td><td>Blank Quality</td><td>0.4kg</td><td>Quantity</td><td colspan="2">1</td></tr>
<tr><td colspan="6">Working procedure</td><td rowspan="2">Work section</td><td rowspan="2">Technical equipment</td><td colspan="2">Man-hours/min</td></tr>
<tr><td>No.</td><td>Name</td><td colspan="4">Content</td><td>Preparation & conclusion</td><td>Single piece</td></tr>
<tr><td>5</td><td>Preparation</td><td colspan="4">Prepare the material according to the size of 75mm×45mm×15mm</td><td>Outsourcing</td><td>Sawing machine</td><td></td><td></td></tr>
<tr><td>10</td><td>Milling</td><td colspan="4">Use precision flat-nose pliers to clamp and machine hexahedron with the $\phi 63$ surface milling cutter and the $\phi 10$ end milling cutter. Ensure dimensional accuracy and geometric tolerance</td><td>Milling</td><td>Machining center, Vernier caliper</td><td>60</td><td>45</td></tr>
<tr><td>15</td><td>Milling</td><td colspan="4">Use precision flat-nose pliers to clamp, align the left corner of the bottom, and set the origin of the machining coordinate system. Drilling $\phi 4.5$ hole and $\phi 6.8$ hole, tapping M8 thread, milling $\phi 8$ counterbore, milling M26 bottom hole to $\phi 24$, milling M26×1 thread</td><td>Milling</td><td>Machining center, Vernier caliper</td><td>90</td><td>60</td></tr>
</table>

Continued

Working procedure			Work section	Technical equipment	Man-hours/min	
No.	Name	Content			Preparation & conclusion	Single piece
20	Milling	Use precision flat-nose pliers to clamp, align the left corner of the bottom, and set the origin of the machining coordinate system. Milling M20 bottom hole to ϕ18, milling M20×1 thread	Milling	Machining center, Vernier caliper	30	10
25	Milling	Use precision flat-nose pliers to clamp, split 45mm×15mm face, set the origin of the machining coordinate system. Drill M4 bottom hole to ϕ3.3. Tapping M4 thread	Milling	Machining center, Vernier caliper	15	10
30	Cleaning	Clean the workpiece and debur sharp corner	Benching		15	5
35	Inspection	Check the workpiece dimensions	Examination		15	5

The 10th process is to process hexahedron, which has been described in the introduction of regular hexahedron processing technology, and it will not be described in detail here. For the milling of the 15th process of this training task, its corresponding working procedure card is formulated as shown in Table 5-2.

Table 5-2 Working procedure card for the 15th process

Machining working procedure card	Product model	STL-01	Part number	STL-02	Page 2
	Product name	Stirling engine	Part name	Cylinder block mount base	Total 4 page
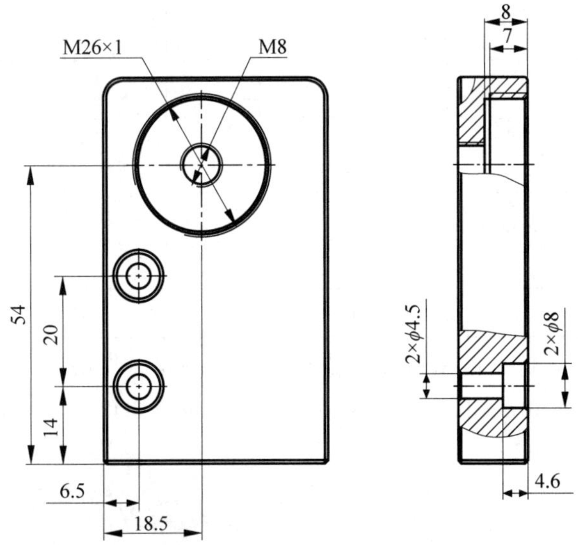				Procedure No.	15
				Procedure name	Milling
				Material	C45
				Equipment	Machining center
				Equipment model	VAL6150e
				Fixture	Precision flat-nose pliers
				Measuring tool	Vernier caliper
				Preparation & Conclusion time	60min
				Single-piece time	90min

Continued

Work steps	Content	Cutters	S/ (r/min)	F/ (mm/r)	a_p/ mm	a_e/ mm	Step hours/min	
							Mechanical	Auxiliary
1	Install the workpiece. The top surface shall extend more than 3mm from the jaw, and the upper left corner shall be aligned							5
2	Drill the center hole, 2×ϕ4.5, M8 position, three in total	ϕ2 center drill	1500	100	0.5	1	5	
3	Drill the through hole 2×ϕ4.5	ϕ4.5 twist drill	1500	200	0.1	2.25	5	
4	Drill M8 threaded bottom hole to ϕ6.8	ϕ6.8 twist drill	1500	200	0.1	3.4	10	
5	Tapping the M8 thread	M8 machine taper	300	100	0.5	1	5	
6	Rough and finish milling ϕ8×4.6 counterbore	ϕ4 end-milling cutter	2500	500	0.5	2	20	
7	Milling M26 threaded bottom hole to ϕ24	ϕ6 end-milling cutter	2500	500	0.5	3	20	
8	Milling M26×1 thread	M10×1 thread milling cutter	2500	500	0.1	0.2	15	
9	Disassemble and clean the workpiece						5	
10								

Referring to the 15th process, design the working procedure of the 20th process in this training task, and fill in the working procedure card shown in Table 5-3 (the header part has been completed).

Table 5-3 Working procedure card for the 20th process

<table>
<tr><td rowspan="2">Machining process cards</td><td>Product model</td><td>STL-01</td><td>Component serial number</td><td>STL-02</td><td>Page 3</td></tr>
<tr><td>The product name</td><td>Stirling engine</td><td>Component name</td><td>Cylinder block mount base</td><td>Total 4 pages</td></tr>
<tr><td colspan="4" rowspan="9">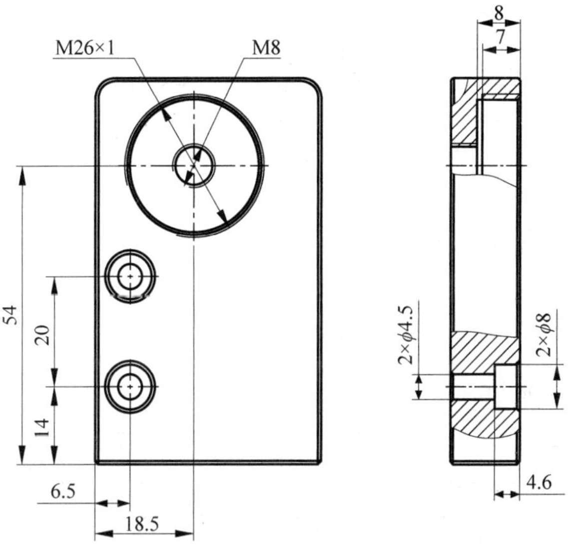
</td><td>The operation number</td><td>20</td></tr>
<tr><td>The name of the operation</td><td>Milling</td></tr>
<tr><td>Material</td><td>C45</td></tr>
<tr><td>Equipment</td><td>Machining center</td></tr>
<tr><td>Device model</td><td>VAL650e</td></tr>
<tr><td>Jig</td><td>Precision flat nose pliers</td></tr>
<tr><td>Measuring</td><td>Vernier calipers</td></tr>
<tr><td>Quasi-closing hours</td><td>30min</td></tr>
<tr><td>Single-piece man-hours</td><td>10min</td></tr>
</table>

Work steps	Content	Cutters	S/(r/min)	F/(mm/r)	a_p/mm	a_e/mm	Step hours/min	
							Mechanical	Auxiliary
1								
2								
3								
4								
5								
6								
7								
8								
9								
10								

The working procedure design for the 25th process of 2 × M4 threaded hole is relatively simple, and students can practice after class.

5.6 Program for CNC machining

According to working procedure card of the 15th process, the corresponding program for milling is prepared as shown in Table 5-4.

Table 5-4　CNC program for the 15th process

No.	Program statement	Annotation
	O0001；	
N1	T1 M6；	Call the $\phi 2$ center drill
	S1500 M3；	
	G0 G90 G54 X6.5 Y14；	
	G43 Z10 H1；	
	G81 Z-1.5 R1 F100；	Drill the center hole
	Y34；	
	X18.5 Y54；	
	G0 G80 Z100；	
	M5；	
N2	T2 M6；	Call the $\phi 4.5$ twist drill
	S1500 M3；	
	G0 G90 G54 X6.5 Y14；	
	G43 Z10 H2；	
	G83 Z-15 R1 Q1 F200；	Drill the $\phi 4.5$ hole
	Y34；	
	G0 G80 Z100；	
	M5；	
N3	T3 M6；	Call the $\phi 6.8$ twist drill
	S1500 M3；	
	G0 G90 G54 X18.5 Y54；	
	G43 Z10 H3；	
	G83 Z-16 R1 Q1 F200；	Drill the M8 threaded bottom hole
	G0 G80 Z100；	
	M5；	
N4	T4 M6；	Call the M8 tap
	S300 M3；	
	G0 G90 G54 X18.5 Y54；	
	G43 Z10 H4；	
	M29 S300；	Call the rigid tapping instruction
	G84 Z-15 R1 F100；	Rigid tapping M8
	G0 G80 Z100；	
	M5；	
N5	T5 M6；	Call the $\phi 4$ end-milling cutter
	S2500 M3；	
	G0 G90 G54 X6.5 Y14；	
	G43 Z10 H5；	
	G1 Z0 F500；	
	M98 P0090010；	Call the sub-program of counterbore roughing to process the second $\phi 8$ counterbore
	G1 G90 Z-4.6；	Finishing the bottom surface
	G1 G91 X1.95；	

Continued

No.	Program statement	Annotation
	G3 I-1. 95 J0;	
	G1 X-1. 95;	Finishing the side
	G1 G41 X1 Y-1 D5;	
	G3 X3 Y3 R3;	
	I-4 J0;	
	X-3 Y3 R3;	
	G1 G40 X-1 Y-1;	
	G1 G90 Z10;	
	G0 X6. 5 Y34;	
	G1 Z0 F500;	
	M98 P0090010;	Call the sub-program of counterbore roughing to process the second $\phi 8$ counterbore
	G1 G90 Z-4. 6;	Finishing the bottom surface
	G1 G91 X1. 95;	
	G3 I-1. 95 J0;	
	G1 X-1. 95;	Finishing the side
	G1 G41 X1 Y-1 D5;	
	G3 X3 Y3 R3;	
	I-4 J0;	
	X-3 Y3 R3;	
	G1 G40 X-1 Y-1;	
	G1 G90 Z10;	
	G0 Z100;	
	M5;	
N6	T6 M6;	Call the $\phi 6$ end milling cutter
	S2500 M3;	
	G0 G90 G54 X18. 5 Y54;	
	G43 Z10 H6;	
	G1 Z0 F500;	
	M98 P0160011;	Call the sub-program for machining the M26 bottom hole, and machine it to $\phi 24$
	G1 G90 Z10;	
	G0 Z100;	
	M5;	
N7	T7 M6;	Call the thread milling cutter M10×1
	S2500 M3;	
	G52 X18. 5 Y54;	
	G0 G90 G54 X0 Y0;	
	G43 Z10 H7;	
	G1 Z-7 F500;	Milling the starting point of the thread
	G1 G41 X2. 5 Y-10 D7;	Feed 0. 5mm for the first cutting
	G3 X12. 5 Y0 R10;	

Continued

No.	Program statement	Annotation
	Z-6 I-12.5 J0;	Down-milling, counterclockwise from bottom to top, machining right-handed threads
	X2.5 Y10 R10;	
	G1 G40 X0 Y0;	
	G1 Z-7 F500;	Milling the starting point of the thread
	G1 G41 X2.7 Y-10 D7;	Feed 0.2mm for the second cutting
	G3 X12.7 Y0 R10;	
	Z-6 I-12.7 J0;	Down-milling, counterclockwise from bottom to top, machining right-handed threads
	X2.7 Y10 R10;	
	G1 G40 X0 Y0;	
	G1 Z-7 F500;	Milling the starting point of the thread
	G1 G41 X2.9 Y-10 D7;	Feed 0.2mm for the third cutting
	G3 X12.9 Y0 R10;	
	Z-6 I-12.9 J0;	Down-milling, counterclockwise from bottom to top, machining right-handed threads
	X2.9 Y10 R10;	
	G1 G40 X0 Y0;	
	G1 Z-7 F500;	Milling the starting point of the thread
	G1 G41 X3 Y-10 D7;	Feed 0.1mm for the fourth cutting
	G3 X13 Y0 R10;	
	Z-6 I-13 J0;	Down-milling, counterclockwise from bottom to top, machining right-handed threads
	X3 Y10 R10;	
	G1 G40 X0 Y0;	
	Z10;	
	G52 X0 Y0;	
	G0 Z100;	
	M5;	
	M30;	
	%	
	O0010;	The sub-program of roughing the $\phi 8\times 4.6$ counterbore
	G1 G91 Z-0.5 F500;	
	X1.95;	
	G3 I-1.95 J0;	
	G1 X-1.95;	
	M99;	
	%	
	O0011;	The sub-program of machining the M26 bottom hole
	G1 G91 Z-0.5 F500;	
	X3;	
	G3 I-3 J0;	

Continued

No.	Program statement	Annotation
	G1 X3;	
	G3 I-6 J0;	
	G1 X3;	
	G3 I-9 J0;	
	G1 X-9;	
	M99;	

For the milling of the 20th process, you only need to mill the M20 bottom hole to ϕ18, and then use the M10×1 thread milling cutter to mill M20 thread according to the program segment No. "N7" method in the 15th process. The corresponding program is relatively simple and will not be repeated here.

For the milling of the 25th process, you only need to drill the M4 bottom hole to ϕ3.3, and then rigid tapping. You can refer to the program segment No. "N1", "N3" and "N4" method in the 15th process for programming.

Task 3 Hands-on Training

5.7 Equipment and appliances

Equipment: AVL650e vertical machining center.

Cutters: ϕ2 center drill, ϕ4.5 twist drill, ϕ6.8 twist drill, M8 machine taper, ϕ4 end-milling cutter, ϕ6 end-milling cutter, M10×1 thread cutter.

Fixture: 125mm precision flat jaw pliers.

Tools: kry file.

Measuring tools: 0-150mm vernier caliper, 0.02mm lever arm test indicator (with mounting rod), edge finder, etc.

Blank: 75mm×45mm×15mm.

Auxiliary appliances: hook wrench, rubber hammer, brush, etc.

5.8 Check before powering on

Check if there is any abnormality in various parts of the appearance of the machine tool (such as the scatter shield, the footplate, etc.). Check if the lubricating oil and coolant of the machine tool are sufficient. Check if there are foreign objects on the tool holder the fixture and the baffle plate of the lead rail. Check if the state of each knob on the machine tool panel is normal. Check if there is an alarm after power on the machine tool. Refer to Table 5-5 to check the machine state.

Table 5-5 Preparing card for machine start-up

Check item		Test result	Abnormal description
Mechanical part	Spindle		
	Feed part		
	Tool holder		
	Three-jaw self-centering chuck		
Electrical part	Main power supply		
	Cooling fan		
CNC system	Electrical components		
	Controlling part		
	Driving section		
Auxiliary part	Cooling system		
	Compressed air system		
	Lubricating system		

5.9 Preparation before machining

(1) After the machine tool is started, each coordinate axis returns to the original point of the machine tool according to the operation instructions.

(2) Install the workpiece. It should extend more than 3mm from the top surface of the flat-nose pliers.

(3) Use the edge finder to align the lower left corner of the workpiece and set it as $X0$ and $Y0$ in the G54 coordinate system.

(4) Install the tool.

(5) Tool setting. The top surface of the fixed jaw of the precision flat-nose pliers can be used as the benchmark for tool setting.

(6) Program editing. Because the program is relatively long, it is recommended to edit the program on the personal computer, and then use the memory card to import the program into the CNC machine tool after checking there is no problem.

5.10 Part machining

After the graphic verification process has verified that there is no problem, the parts machining can be carried out. Before the parts are processed, you should understand the safety operation requirements of the machine tool in detail, and wear labor protection clothing and utensils. When processing parts, you should be familiar with the functions and positions of the operation buttons of the CNC lathe, and understand the methods of dealing with emergency situations.

Note: If the graphic verification operation is performed, the machine tool must be returned to the mechanical zero point after the operation is completed, and then other related operations can be performed. If the relevant operations are performed without returning to zero, it will cause coordinate offsets, and even abnormal phenomena such as collisions and program execution confusion, resulting in danger.

Select "MEM" mode on the operation panel, and load the CNC machining program that needs to be processed, and press the cycle start button to perform automatic processing.

The first automatic operation of the CNC machining program should be carried out in the debugging state. First turn down the machine feed rate override knob to 0%, and press the button [單節執行] on the operation panel, and then switch to the check interface of running status by pressing the "Program Check" key at the bottom of the display, as shown in Figure 5-9.

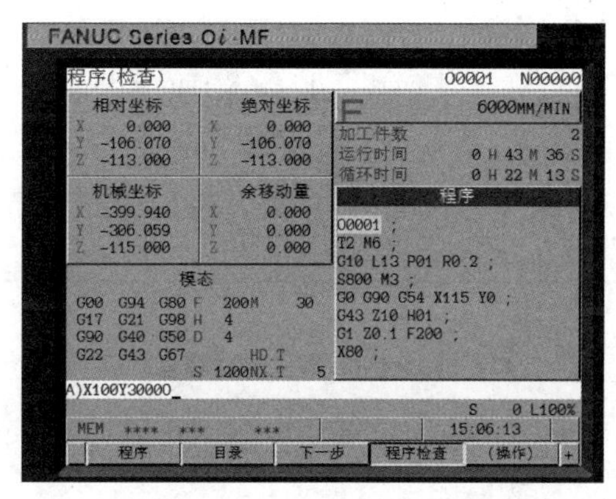

Figure 5-9 Automatic running status inspection of the program

In this state, the program will only automatically execute the line where the cursor is located each time the cycle start button is pressed. Before starting the cycle, observe whether the distance between the cutter and the workpiece is safe. After pressing the cycle start button, control the movement speed of the machine tool through the feed override knob, and observe the actual distance between the cutter and the workpiece, and at the same time compare with the remaining movement amount displayed in the "Remaining Movement Amount" column of the display screen, if the difference between the actual distance and the remaining movement amount is too large, the operation should be stopped and checked to avoid a collision accident. In the process of program debugging, you should also pay close attention to the "modal" status on the display screen to ensure that there is no abnormality in the spindle revolution speed, feed speed, workpiece coordinate system number, compensation status and compensation number.

5.11 Part inspection

After the parts are processed, the workpiece should be carefully cleaned, and in accordance with the relevant requirements of quality management, the processed parts should be subject to relevant inspections to ensure the production quality. The "three-level" inspection cards for machined parts is shown in Table 5-6.

Table 5-6 "Three-level" inspection cards for machined parts

<table>
<tr><th>Part drawing number</th><th></th><th>Part name</th><th></th><th>Working step number</th><th></th></tr>
<tr><td>Material</td><td></td><td>Inspection date</td><td></td><td>Working step name</td><td></td></tr>
<tr><td>Inspection items</td><td>Self-inspection result</td><td colspan="2">Mutual inspection result</td><td>Professional inspection</td><td>Remark</td></tr>
<tr><td></td><td></td><td colspan="2"></td><td></td><td></td></tr>
<tr><td></td><td></td><td colspan="2"></td><td></td><td></td></tr>
<tr><td></td><td></td><td colspan="2"></td><td></td><td></td></tr>
<tr><td></td><td></td><td colspan="2"></td><td></td><td></td></tr>
<tr><td></td><td></td><td colspan="2"></td><td></td><td></td></tr>
<tr><td></td><td></td><td colspan="2"></td><td></td><td></td></tr>
<tr><td></td><td></td><td colspan="2"></td><td></td><td></td></tr>
<tr><td>Conclusion</td><td colspan="5">□ Qualified □ Unqualified □ Repair □ Concession to receive
Inspection signature:
Date:</td></tr>
<tr><td>Non-conforming item description</td><td colspan="5"></td></tr>
</table>

Project Summary

Through the CNC milling of the cylinder block mounting base, master the basic format of CNC milling program and the use of basic cutting instructions. Be able to perform part cutting using the basic cutting instructions, the tool radius compensation instruction, the sub-program and coordinate transformation instruction, etc. Master the application method of the rigid tapping instruction G16.

Master the basic operation methods of vertical machining centers, including powering on and off, tool installation, workpiece alignment, cutter setting, program editing, graphic verification, CNC machining program debugging and automatic operation, etc.

Through project training, good professional quality, correct safety operation standards for machining centers, and basic machining quality awareness are developed and cultivated.

Exercises After Class

1. Choice questions

(1) For FANUC system, if the instruction "G92 X(U) Z(W) F;" is used to process double thread, then "F" in this instruction refers to (　　).

A. thread pitch　　B. thread lead

C. feed per minute　　D. feed per revolution

(2) When milling the vertical planes on both sides of the rectangular workpiece, the workpiece shall be clamped with machine-used flat pliers. If the angle between the milled plane and the reference plane is less than 90°, a sheet of copper paper be placed at the ( ) of the fixed jaw.

A. center B. bottom

C. top D. all the above answers are OK

(3) The machine vice is mainly used for clamping ( ).

A. rectangular workpiece B. shaft parts

C. sleeve parts D. disc workpiece

(4) The automatic tool change device of the machining center is composed of drive mechanism, ( ).

A. tool magazine and manipulator B. tool magazine and control system

C. manipulator and control system D. control system

(5) The machining tool for small diameter internal thread is ( ).

A. thread turning tool B. thread milling tool

C. tapper D. thread rolling tool

2. True or false

(1) Spiral interpolation command is required for thread milling. The formation of the helix is that the tool performs the circular interpolation motion and simultaneously performs the axial motion with it. ( )

(2) When cutting threads in multiple layers, the back feed amount of each layer shall be evenly distributed as far as possible. ( )

(3) When machining multi threads, after machining the first thread, the starting point of the second thread shall be separated by 1 lead from the starting point of the first thread. ( )

(4) The combined fixture is most suitable for processing workpieces with high positional accuracy requirements. ( )

(5) The thread cutting cycle instruction G92 can only be used for turning straight threads, not taper threads. ( )

3. Fill in the blanks

(1) The main types of thread milling cutters are ________, ________ and ________.

(2) Thread milling is completed with the help of the ________ function of machine tool and the spiral interpolation instruction ________.

(3) Compared with traditional thread processing, thread milling has great advantages in processing accuracy and efficiency, and is not limited by ________ and ________.

(4) The flat-nose pliers are mainly composed of ________, ________, ________, ________, etc.

4. Short answer questions

(1) Briefly describe the difference between rigid tapping and flexible tapping.

(2) Briefly describe the use of rigid tapping instruction G84.

(3) Briefly describe the movement mode of thread milling cutter during thread milling programming.

5. Explanatory questions

(1) Explain the meaning of the thread marking of M24×1.5-6g-LH.

(2) Explain the meaning of the thread marking of M20×2-6H-LH.

(3) Explain the meaning of the thread marking of M24×Ph2P1-LH.

(4) Explain the meaning of the thread marking of M30×1.5-LH.

Self-learning test score sheet is shown in Table 5-7.

Table 5-7 Self-learning test score sheet

Task	Task requirements	Score	Scoring rules	Score	Remark
Learn key knowledge points	(1) Master the correct selection of fixture during regular hexahedron milling (2) Be familiar with the milling sequence of regular hexahedron (3) Be familiar with the selection of tools and the processing technology for thread milling (4) Master the programming method of thread milling (5) Master the use of the rigid tapping instruction	20	Understand and master		
Technological preparation	(1) Be able to correctly read the shaft part drawings (2) Can be analyzed according to the part drawing, determine the process (3) Be able to write the correct processing program according to the processing process	30	Understand and master		
Hands-on training	(1) The corresponding equipment and utensils will be selected correctly (2) Can correctly operate the CNC lathe and adjust the processing parameters according to the machining situation	50	(1) Understand and master (2) Operation process		

Ideological and Political Classroom

Project 6 Programming and Machining Training for Eccentric Wheel Milling

➢ Mind map

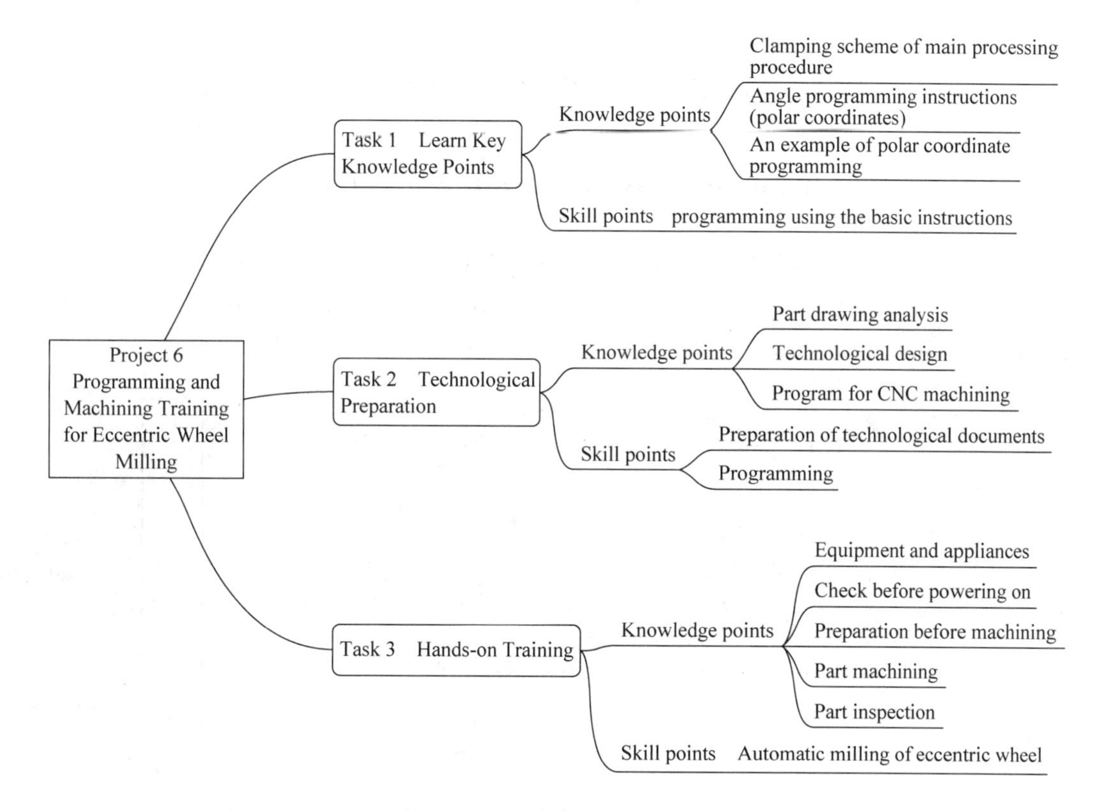

➢ Learning objectives

Knowledge objectives

(1) Understand the working principle of the eccentric wheel.

(2) Understand the use and key requirements of the eccentric wheel in this mechanism.

Ability objectives

(1) Master the processing skills of special-shaped parts, such as flat plate shaped parts.

(2) Master the use of polar coordinate programming instructions.

(3) Be able to independently determine the process routine and fill in the technological

documents correctly.

(4) Be able to operate the CNC lathe correctly and adjust the machining parameters according to the machining conditions.

(5) Be able to select measuring tools reasonably according to the accuracy and the structural characteristics of parts, and be able to measure the relevant dimensions correctly and normatively.

Literacy goals

(1) Develop students' scientific spirit and attitude.

(2) Cultivate students' engineering awareness.

(3) Develop students' teamwork skills.

Task 1 Learn Key Knowledge Points

6.1 Clamping scheme of main processing procedure

As shown in Figure 6-1, the blank of the eccentric wheel is a metal plate with a thickness of 3mm. The part shape is processed through the wire cutting process. The milling part only processes the arc groove and M5 hole.

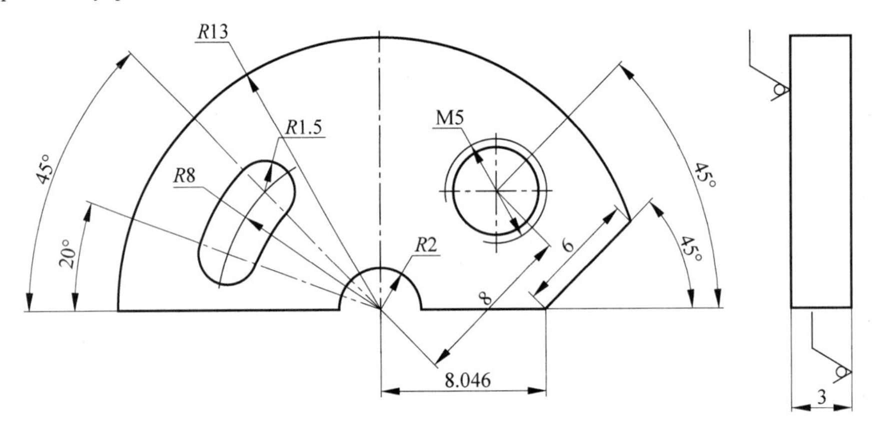

Figure 6-1 Part drawing of the eccentric wheel

Precision flat-nose pliers with a V-groove can be used for milling, as shown in Figure 6-2. The bottom surface of the part is positioned with parallel sizing block, which limits the movement in Z direction and the rotation in X and Y directions, with a total of 3 degrees of freedom. The parallel plane is fitted with the fixed jaw of the flat-nose pliers, which limits the movement in Y direction and rotation in Z direction, with a total of 2 degrees of freedom. The V-block cooperates with $R13$ to limit the movement in the X direction, with 1 degree of freedom. Therefore, the workpiece is fully positioned.

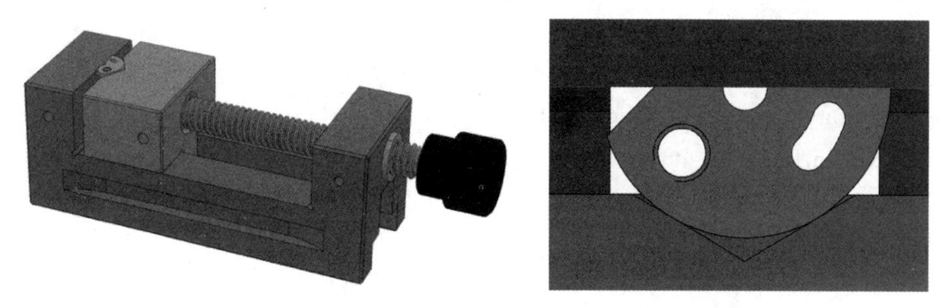

Figure 6-2 The precision flat-nose pliers with a V-groove

6.2 Angle programming instructions (polar coordinates)

Polar coordinates belong to the two-dimensional coordinate system. Its founder is Isaao Newton, and it is mainly used in the field of mathematics. As shown in Figure 6-3, a polar coordinate refers to: take the fixed point O in the plane, which is called the pole, and introduce a ray OX from the point, which is called the polar axis, and then selecting a length unit and the positive direction of the angle (usually the counterclockwise direction). For any point M in the plane, use ρ indicates the length of line segment OM (sometimes expressed in r), θ indicates the angle from OX to OM. ρ is called the polar radius of the point M, and θ is called the polar angle of the point M. The ordered pair (ρ, θ) is called the polar coordinate of the point M. The coordinate system thus established is called the polar coordinate system. Generally, the unit of the polar radius coordinate of the point M is 1 (length unit), and unit of the polar angle coordinate is rad (or°).

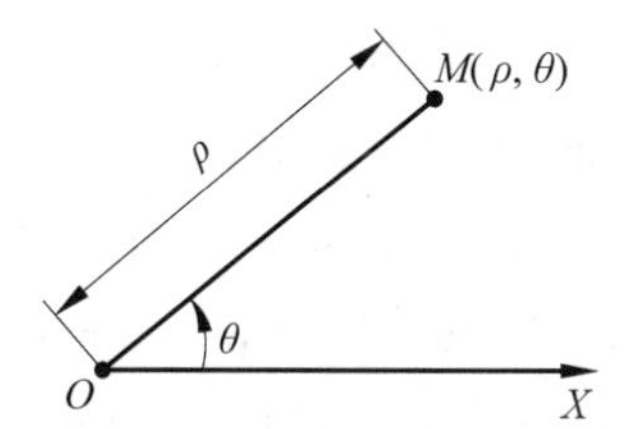

Figure 6-3 Polar coordinate system

In FANUC 0*i* MF system, the basic format of polar coordinate programming is:

G◇◇ G○○ G16;	Start of polar coordinate instruction (polar coordinate mode)
G0 IP_;	
… …	Machining instructions described in polar coordinates
G16;	Cancel the polar coordinate instruction (polar coordinate mode)

Including:

G◇◇: plane selection of polar coordinate command (G17, G18, G19);

G○○: represents the pole (center) selection of the polar coordinate command (G90 represents taking the origin of the workpiece coordinate system as the pole. G91 represents taking the current point position as the pole).

IP_: refers to the polar coordinate command, the first axis represents the polar radius and the second axis represents the polar angle.

As shown in Figure 6-4, there are two common ways to apply the polar coordinates.

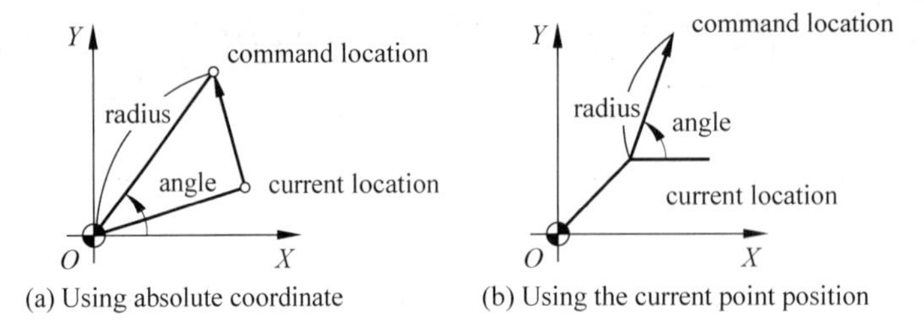

(a) Using absolute coordinate (b) Using the current point position

Figure 6-4 Application methods for the polar coordinates

1. Take the absolute coordinate origin as the pole

When taking the absolute coordinate origin as the pole, the positive X axis direction is 0 degree, and the angle of all lines from the positive X axis to the origin and the command position in the counterclockwise direction is the polar angle θ, and the distance from the origin to the command position is the polar diameter ρ. The program format is:

G16;	To call polar coordinate instruction
G1 Xρ YθF_ ;	To make a linear motion towards the command position
G15;	To cancel polar coordinate instruction

In practical application, G52 instruction can be used to specify the temporary origin of the machining coordinate system, and coordinate with G16 command to carry out polar coordinate machining programming with any position as the pole.

2. Take the position of the current tool as the pole

When taking the position of the current tool as the pole, the current position horizontal to the right is 0 degree, and the angle formed by the straight line from the counterclockwise to the current position and the command position is the polar angle θ, and the distance from the current position to the command position is the polar radius ρ. The program format is:

G91 G16;	To call the polar coordinate instruction
G1 Xρ Yθ F_;	To make a linear motion towards the command position
G90 G15;	To cancel the polar coordinate instruction

6.3 An example of polar coordinate programming

The processing program is written with the polar coordinate command, as shown in Figure 6-5. And the prepared program segment is shown in the Table 6-1.

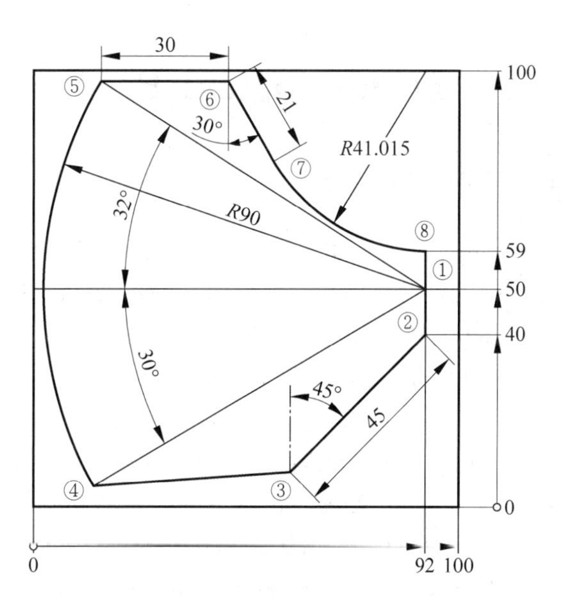

Figure 6-5 Example of polar coordinate programming

Table 6-1 Program segment for polar coordinate programming

No.	Program statement	Annotation
	…	Omit the head and cut in
N1	G1 X92 Y40;	Point ①→point ②
N2	G91 G16;	Defines the current point ② as the pole
	G1 X45 Y225;	Point ②→point ③
	G90 G15;	Cancel the polar coordinate programming
N3	G52 X92 Y50;	Temporarily define the point (92,50) as the machining origin
	G16;	Defines the temporary origin as the pole
	G1 X90 Y210;	Point ③→point ④
	G2 X90 Y148 R90;	*R*90 arc is machined from the point ④ to the point ⑤
	G15;	Cancel the coordinate programming
	G52 X0 Y0;	Restore the machining coordinate system to the point (0,0)
N4	G1 G91 X30;	Relative to the current point ⑤, process 30mm to the right to the point ⑥
N5	G91 G16;	Define the point ⑥ as the pole
	G1 X21 Y300;	Move in a straight line with a pole radius of 21 and a pole angle of 300° to point ⑦
	G90 G15;	Cancel the coordinate programming
N7	G3 X92 Y59 R41.015;	Along the counterclockwise arc, process from the point ⑦ to the point ⑧
N8	G1 X92 Y50;	Move from the point ⑧ to the point ① in a straight line
	…	

Task 2 Technological Preparation

6.4 Part drawing analysis

Figure 6-6 is the part drawing of the eccentric wheel. As shown in the figure, the upper and lower surfaces of the part are non-cutting surfaces, so the blank material can be directly selected from brass plate with thickness of 3mm. In addition, to save material, the strip with thickness of 3mm and width of 15mm can be selected for processing. In order to facilitate clamping on the linear cutting machine, the length of the strip is controlled at 250mm. After the shape is processed by wire cutting, the milling machine processes the arc groove and M5 hole. During the machining process, the flat pliers and V-shaped block are used for clamping, and the polar coordinate command is used for CNC programming.

Note: The clamping force shall be properly controlled during milling to prevent surface damage of parts.

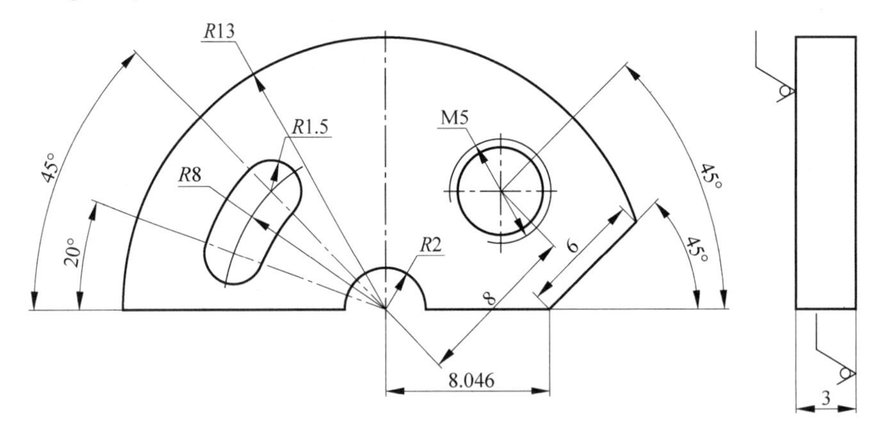

Figure 6-6 The part diagram of the eccentric wheel

6.5 Technological design

According to the analysis of the part drawing, the technological process is designed as shown in Table 6-2.

Table 6-2 The technological process card

<table>
<tr><td rowspan="2">Machining process card</td><td colspan="2">Product model</td><td colspan="2">STL-01</td><td>Part number</td><td>STL-03</td><td>Page 1</td></tr>
<tr><td colspan="2">Product name</td><td colspan="2">Stirling engine</td><td>Part name</td><td>Eccentric</td><td>Total 1 page</td></tr>
<tr><td>Material grade</td><td>H62</td><td>Blank size</td><td>250mm×15mm, $t=3$</td><td>Blank Quality</td><td>0.1kg</td><td>Quantity</td><td>9</td></tr>
</table>

Continued

Working procedure			Work section	Technical equipment	Man-hours/min	
No.	Name	Content			Preparation & conclusion	Single piece
5	Preparation	Prepare the material according to the size of 250mm×15mm. Material thickness $t=3$mm	Outsourcing	Sawing machine		
10	Wire cutting	Clamps are used for clamping, wire cutting is used to process the shape of parts, and the wire connection trace is manually processed after processing. It can be stacked under batch processing conditions	Wire cutting	Wire cutting machine, Vernier caliper	40	30
15	Milling	Use precision flat-nose pliers and V-block clamping, and drill the M5 bottom hole to $\phi 4.2$, and tap the M5 threaded hole, and then milling the circular groove	Milling	Machining center, Vernier caliper	30	20
20	Cleaning	Clean the workpiece, debur sharp corner	Benching		15	5
25	Inspection	Check the workpiece dimensions	Examination		15	5

In this training task, the 15th process, namely milling, is designed detailed, and the corresponding working procedure card is formulated as shown in Table 6-3.

Table 6-3 15 Working procedure card for milling

<table>
<tr><td rowspan="2">Machining working procedure card</td><td>Product model</td><td>STL-01</td><td>Part number</td><td>STL-03</td><td>Page 2</td></tr>
<tr><td>Product name</td><td>Stirling engine</td><td>Part name</td><td>Eccentric</td><td>Total 3 page</td></tr>
<tr><td colspan="4" rowspan="9">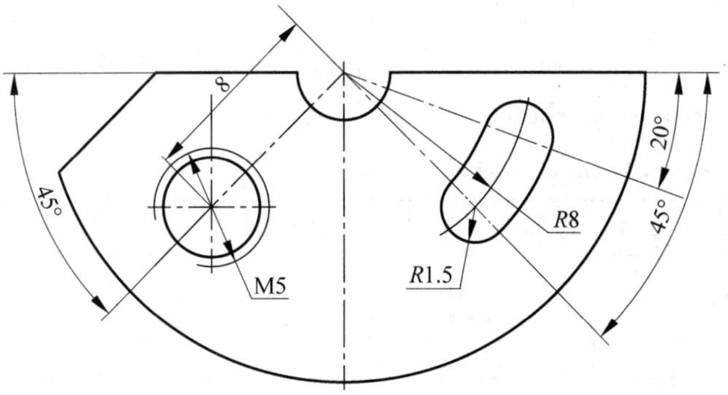
</td><td>Procedure No.</td><td>15</td></tr>
<tr><td>Procedure name</td><td>Milling</td></tr>
<tr><td>Material</td><td>H62</td></tr>
<tr><td>Equipment</td><td>Machining center</td></tr>
<tr><td>Equipment model</td><td>VAL650e</td></tr>
<tr><td>Fixture</td><td>Precision flat-nose pliers</td></tr>
<tr><td>Measuring tool</td><td>Vernier caliper</td></tr>
<tr><td>Preparation & Conclusion time</td><td>40min</td></tr>
<tr><td>Single-piece time</td><td>30min</td></tr>
</table>

Continued

Steps	Content	Cutters	S/ (r/min)	F/ (mm/r)	a_p/ mm	Step hours/min	
						Mechanical	Auxiliary
1	When the workpiece is installed, the bottom surface shall be close to the special parallel sizing block, and attention shall be paid to the avoidance after processing. Fix the jaw tightly with the straight face, V-block clamp $R13$, and pay attention to the clamping force to prevent clamping injury. Take the $R13$ center as the origin of the machining coordinate system						5
2	Drill the center hole	$\phi 2$ Center drill	1500	100	0.5	1	2
3	Drill the M5 threaded bottom hole and the $\phi 4.2$ through hole	$\phi 4.2$ twist drill	1500	200	0.1	2.25	3
4	Tapping the M5 thread	M5 machine taper	300	100	0.5	1	5
5	Rough milling the arc groove	$\phi 2$ end-milling cutter	4000	500	0.3	1.5	10
6	Finish milling the arc groove	$\phi 2$ end-milling cutte3	4000	500	3	0.05	2
7	Dismantling and cleaning the workpiece						3

6.6 Program for CNC machining

According to working procedure card of the 15th milling, working procedure, the corresponding CNC program is written, as shown is Table 6-4.

Table 6-4 Program block for the 15th working procedure

No.	Program statement	Annotation
	O0001;	
N1	T1 M6;	Call the $\phi 2$ center drill
	S1500 M3;	
	G16;	Call polar programming instruction
	G0 G90 G54 X8 Y225;	Positioned to the center of the M5 hole, the pole radius 8mm, the pole angle 225°
	G43 Z10 H1;	

Continued

No.	Program statement	Annotation
	G81 Z-1.5 R1 F100;	Drill the center hole
	G0 G80 Z100;	
	M5;	
N2	T2 M6;	Call the ϕ4.2 twist drill
	S1500 M3;	
	G0 G90 G54 X8 Y225;	Positioned to the center of the M5 hole, the pole radius 8mm, the pole angle 225°
	G43 Z10 H2;	
	G83 Z-5 R1 Q1 F200;	Drill the ϕ4.2 hole
	G0 G80 Z100;	
	M5;	
N3	T3 M6;	Call the M5 machine taper
	S1500 M3;	
	G0 G90 G54 X8 Y225;	Positioned to the center of the M5 hole, the pole radius 8mm, the pole angle 225°
	G43 Z10 H3;	
	M29 S300;	Call the rigid tapping instruction
	G84 Z-5 R1 F100;	Rigid tapping M5
	G0 G80 Z100;	
	M5;	
N4	T4 M6;	Call the ϕ2 end-milling cutter
	S2500 M3;	
	G0 G90 G54 X8.45 Y-45;	Positioned to the beginning of the arc groove, the pole radius 8.45mm, the pole angle −45°
	G43 Z10 H4;	
	G1 Z0.2 F500;	
	M98 P0110010;	Call the sub-program for roughing the arc groove
	G1 X8 Y-20;	Finishing the lateral face
	G1 G41 X6.5 Y-45 D4;	
	G3 X9.5 Y-45 R1.5;	
	X9.5 Y-20 R9.5;	
	X6.5 Y-20 R1.5;	
	G2 X6.5 Y-45 R6.5;	
	G1 G40 X8 Y-45;	
	G1 Z10;	
	G15;	Cancel the polar programming instruction
	G0 Z100;	
	M5;	
	M30;	
	%	
	O0010;	The sub-program for roughing the arc groove
	G3 G91 X0 Y25 Z-0.5 R8.45 F500;	

Continued

No.	Program statement	Annotation
	G90 X7.55 Y-20 R0.45;	
	X7.55 Y-45 R7.55;	
	X8.45 Y-45 R0.45;	
	M99;	
	%	

Task 3 Hands-on Training

6.7 Equipment and appliances

Equipment: AVL650e vertical machining center.

Cutters: ϕ2 center drill, ϕ4.2 twist drill, M5 machine taper, ϕ2 end-milling cutter.

Fixture: 125mm precision flat jaw pliers, movable jaw with V-shaped block.

Tools: kry file.

Measuring tools: 0-150mm vernier caliper, 0.02mm lever arm test indicator (with mounting rod), edge finder, etc.

Blank: semi-finished products after the 10th process.

Auxiliary appliances: hook wrench, rubber hammer, brush, etc.

6.8 Check before powering on

Check if there is any abnormality in various parts of the appearance of the machine tool (such as the scatter shield, the footplate, etc.). Check if the lubricating oil and coolant of the machine tool are sufficient. Check if there are foreign objects on the tool holder the fixture and the baffle plate of the lead rail. Check if the state of each knob on the machine tool panel is normal. Check if there is an alarm after power on the machine tool. Refer to Table 6-5 to check the machine state.

Table 6-5 Preparing card for machine start-up

Check item		Test result	Abnormal description
Mechanical part	Spindle		
	Feed part		
	Tool holder		
	Three-jaw self-centering chuck		
Electrical part	Main power supply		
	Cooling fan		

Continued

Check item		Test result	Abnormal description
CNC system	Electrical components		
	Controlling part		
	Driving section		
Auxiliary part	Cooling system		
	Compressed air system		
	Lubricating system		

6.9 Preparation before machining

(1) After the machine tool is started, each coordinate axis returns to the original point of the machine tool according to the operation instructions.

(2) Install the workpiece.

(3) Use the edge finder to set the left and right sides of the fixed jaw as $X0$, the fitting surface of the alignment fixed jaw as $Y0$, and set as $X0$ and $Y0$ in the G54 coordinate system.

(4) Install the tool.

(5) Tool setting. The top surface of the fixed jaw of the precision flat-nose pliers can be used as the benchmark for tool setting.

(6) Program editing. Because the program is relatively long, it is recommended to edit the program on the personal computer, and then use the memory card to import the program into the CNC machine tool after checking there is no problem.

6.10 Part machining

After the graphic verification process has verified that there is no problem, the parts machining can be carried out. Before the parts are processed, you should understand the safety operation requirements of the machine tool in detail, and wear labor protection clothing and utensils. When processing parts, you should be familiar with the functions and positions of the operation buttons of the CNC lathe, and understand the methods of dealing with emergency situations.

Note: If the graphic verification operation is performed, the machine tool must be returned to the mechanical zero point after the operation is completed, and then other related operations can be performed. If the relevant operations are performed without returning to zero, it will cause coordinate offsets, and even abnormal phenomena such as collisions and program execution confusion, resulting in danger.

Select "MEM" mode on the operation panel, and load the CNC machining program that needs to be processed, and press the cycle start button to perform automatic processing.

The first automatic operation of the CNC machining program should be carried out in the debugging state. First turn down the machine feed rate override knob to 0%, and press the button 单节执行 on the operation panel, and then switch to the check window of running status by pressing the Check key at the bottom of the display, as shown in Figure 6-7.

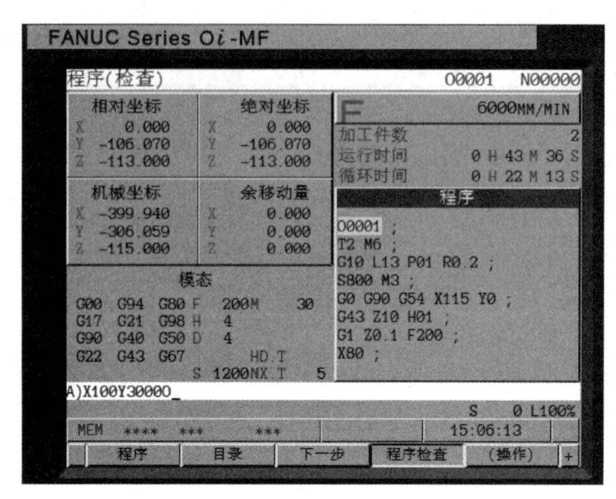

Figure 6-7 Automatic running status inspection of the program

In this state, the program will only automatically execute the line where the cursor is located each time the cycle start button is pressed. Before starting the cycle, observe whether the distance between the cutter and the workpiece is safe. After pressing the cycle start button, control the movement speed of the machine tool through the feed override knob, and observe the actual distance between the cutter and the workpiece, at the same time comparing with the remaining movement amount displayed in the "Remaining Movement Amount" column of the display screen, if the difference between the actual distance and the remaining movement amount is too large, the operation should be stopped and checked to avoid a collision accident. In the process of program debugging, you should also pay close attention to the modal status on the display screen to ensure that there is no abnormality in the spindle revolution speed, feed speed, workpiece coordinate system number, compensation status and compensation number.

6.11 Part inspection

After the parts are processed, the workpiece should be carefully cleaned, and in accordance with the relevant requirements of quality management, the processed parts should be subject to relevant inspections to ensure the production quality. The "three-level" inspection cards for machined parts is shown in Table 6-6.

Table 6-6 "Three-level" inspection cards for machined parts

Part drawing number		**Part name**		**Working step number**	
Material		Inspection date		Working step name	
Inspection items	Self-inspection result	Mutual inspection result	Professional inspection	Remark	
Conclusion	☐ Qualified ☐ Unqualified ☐ Repair ☐ Concession to receive Inspection signature: Date:				
Non-conforming item description					

Project Summary

Through the CNC milling of the eccentric wheel, master the basic format of CNC milling program and the use of basic cutting instructions, and be able to use the basic cutting instructions, the tool radius compensation instruction, the sub-program and coordinate transformation instruction to cut parts, and master the use of polar coordinate programming instruction G16.

Master the basic operation methods of vertical machining centers, including: powering on and off, tool installation, workpiece alignment, cutter setting, program editing, graphic verification, CNC machining program debugging and automatic operation, etc.

Through task training, good professional quality, correct safety operation standards for machining centers, and basic machining quality awareness are developed and cultivated.

Exercises After Class

1. Choice questions

(1) When CNC programming, you should first set (　　).

A. the origin point of the machine tool

B. the reference point of the machine tool

C. the coordinate system of the machine tool

D. the workpiece coordinate system

(2) The core of the machining center to execute the sequence control action and

control the machining process is (　　).

A. the basic component　　B. the spindle component

C. the CNC system　　D. ATC

(3) When using the path method to cut grooved parts, for the surfaces on both sides of the groove, (　　).

A. both sides are up-milling

B. both sides are down-milling

C. one side is down-milling and the other side is up-milling, so the quality of both sides is different

D. one side is down-milling and the other side is up-milling, but the quality of both sides is the same

(4) When measuring the coordinate dimensions of the cam, (　　) should be used.

A. the inner diameter cosine gauge

B. the microscope

C. the inner diameter triangle plate

D. the block gauge

(5) In the program segment "G17 G16 G90 X100.0 Y30.0;" of the FANUC system, the X instruction is (　　).

A. X-axis coordinate position

B. the distance from polar origin to tool center

C. rotation angle

D. time parameters

2. True or false

(1) G16 G17 G90 to establish the polar coordinates G01 X_ Y_ F_ where X refers to the distance from the end point to the origin, that is, the radius. (　　)

(2) The eccentric wheel is the cam. (　　)

(3) Milling can be carried out using the precision flat-nose pliers with a V-groove. The bottom surface of the part is positioned with parallel horns, which only restricts the Z direction to move one degree of freedom. (　　)

(4) The precision flat-nose pliers with a V-groove is a special fixture. (　　)

(5) For angle programming, the angle is the included angle with the negative direction of Z, which is positive clockwise and negative counterclockwise. (　　)

3. Fill in the blanks

(1) If the coordinate values of all coordinate points are measured from the origin of the coordinate system, it is called ________ coordinate. If the end point coordinate of the motion track is measured relative to the starting point of the line segment, it is called ________ coordinate.

(2) The two common ways to apply the polar coordinate are ________ and ________ .

(3) CNC milling machines are mainly used for processing the flat and curved contours of the parts, and it can also process parts of ________, such as cams, plates, molds, spiral

grooves, etc. At the same time, parts can also be drilled, expanded, reamed and bored.

(4) G16 instruction is a polar coordinate programming instruction. after the G16 instruction is used, X represents ________, Y represents ________.

4. Short answer questions

(1) Briefly describe the use method and precautions of the polar coordinate programming instruction.

(2) Consult the data and list the use methods and characteristics of other instructions for polar coordinate programming.

(3) Briefly describe the application of polar coordinates.

(4) Briefly describe the difficulties in machining special-shaped parts and the reasons for the low efficiency of CNC machining of special-shaped parts.

(5) Briefly describe the working principle of eccentric wheel and the difference between eccentric wheel and cam.

Self-learning test score sheet is shown in Table 6-7.

Table 6-7 Self-learning test score sheet

Task	Task requirements	Score	Scoring rules	Score	Remark
Learn key knowledge points	(1) Understand the clamping scheme of the main processing procedure (2) Master the basic format and usage of angle programming instructions (polar coordinates) (3) Be able to use polar coordinate instructions to write corresponding programs	20	Understand and master		
Technological preparation	(1) Be able to correctly read the shaft part drawings (2) Can be analyzed according to the part drawing, determine the process (3) Be able to write the correct processing program according to the processing process	30	Understand and master		
Hands-on training	(1) The corresponding equipment and utensils will be selected correctly (2) Can correctly operate the CNC lathe and adjust the processing parameters according to the machining situation	50	(1) Understand and master (2) Operation process		

Ideological and Political Classroom